REGRESSION ANALYSIS
BY EXAMPLE USING R

REGRESSION ANALYSIS BY EXAMPLE USING R

Sixth Edition

Ali S. Hadi
The American University in Cairo

Samprit Chatterjee
New York University

Copyright © 2024 by John Wiley & Sons, Inc. All rights reserved.

Published by John Wiley & Sons, Inc., Hoboken, New Jersey.
Published simultaneously in Canada.

No part of this publication may be reproduced, stored in a retrieval system, or transmitted in any form or by any means, electronic, mechanical, photocopying, recording, scanning, or otherwise, except as permitted under Section 107 or 108 of the 1976 United States Copyright Act, without either the prior written permission of the Publisher, or authorization through payment of the appropriate per-copy fee to the Copyright Clearance Center, Inc., 222 Rosewood Drive, Danvers, MA 01923, (978) 750-8400, fax (978) 750-4470, or on the web at www.copyright.com. Requests to the Publisher for permission should be addressed to the Permissions Department, John Wiley & Sons, Inc., 111 River Street, Hoboken, NJ 07030, (201) 748-6011, fax (201) 748-6008, or online at http://www.wiley.com/go/permission.

Trademarks: Wiley and the Wiley logo are trademarks or registered trademarks of John Wiley & Sons, Inc. and/or its affiliates in the United States and other countries and may not be used without written permission. All other trademarks are the property of their respective owners. John Wiley & Sons, Inc. is not associated with any product or vendor mentioned in this book.

Limit of Liability/Disclaimer of Warranty: While the publisher and author have used their best efforts in preparing this book, they make no representations or warranties with respect to the accuracy or completeness of the contents of this book and specifically disclaim any implied warranties of merchantability or fitness for a particular purpose. No warranty may be created or extended by sales representatives or written sales materials. The advice and strategies contained herein may not be suitable for your situation. You should consult with a professional where appropriate. Further, readers should be aware that websites listed in this work may have changed or disappeared between when this work was written and when it is read. Neither the publisher nor authors shall be liable for any loss of profit or any other commercial damages, including but not limited to special, incidental, consequential, or other damages.

For general information on our other products and services or for technical support, please contact our Customer Care Department within the United States at (800) 762-2974, outside the United States at (317) 572-3993 or fax (317) 572-4002.

Wiley also publishes its books in a variety of electronic formats. Some content that appears in print may not be available in electronic formats. For more information about Wiley products, visit our web site at www.wiley.com.

Library of Congress Cataloging-in-Publication Data applied for:

ISBN: 9781119830870 (HB); ePDF:9781119830887; epub: 9781119830894

Cover Design: Wiley
Cover Images: © Ali S. Hadi

Set in 11/13.5pt NimbusRomNo9L by Straive, Chennai, India

Dedicated to:

The memory of my parents – A. S. H.

Allegra, Martha, and Rima – S. C.

It's a gift to be simple ...

                        Old Shaker hymn

True knowledge is knowledge of why things are as they are, and not merely what they are.

                        Isaiah Berlin

CONTENTS

Preface xiii

About the Companion Website xvii

1 Introduction 1
   1.1   What Is Regression Analysis? 1
   1.2   Publicly Available Data Sets 2
   1.3   Selected Applications of Regression Analysis 3
        1.3.1   Agricultural Sciences 3
        1.3.2   Industrial and Labor Relations 4
        1.3.3   Government 5
        1.3.4   History 5
        1.3.5   Environmental Sciences 7
        1.3.6   Industrial Production 7
        1.3.7   The Space Shuttle Challenger 8
        1.3.8   Cost of Health Care 8
   1.4   Steps in Regression Analysis 9
        1.4.1   Statement of the Problem 9
        1.4.2   Selection of Potentially Relevant Variables 10
        1.4.3   Data Collection 10
        1.4.4   Model Specification 11
        1.4.5   Method of Fitting 13
        1.4.6   Model Fitting 14
        1.4.7   Model Criticism and Selection 15
        1.4.8   Objectives of Regression Analysis 16

		1.5 Scope and Organization of the Book	16
		Exercises	19
2	**A Brief Introduction to R**		**21**
	2.1	What Is R and RStudio?	21
	2.2	Installing R and RStudio	22
	2.3	Getting Started With R	23
		2.3.1 Command Level Prompt	23
		2.3.2 Calculations Using R	24
		2.3.3 Editing Your R Code	27
		2.3.4 Best Practice: Object Names in R	27
	2.4	Data Values and Objects in R	28
		2.4.1 Types of Data Values in R	28
		2.4.2 Types (Structures) of Objects in R	31
		2.4.3 Object Attributes	37
		2.4.4 Testing (Checking) Object Type	37
		2.4.5 Changing Object Type	38
	2.5	R Packages (Libraries)	39
		2.5.1 Installing R Packages	39
		2.5.2 Name Spaces	40
		2.5.3 Updating R	40
		2.5.4 Datasets in R Packages	40
	2.6	Importing (Reading) Data into R Workspace	41
		2.6.1 Best Practice: Working Directory	41
		2.6.2 Reading ASCII (Text) Files	42
		2.6.3 Reading CSV Files	44
		2.6.4 Reading Excel Files	44
		2.6.5 Reading Files from the Internet	45
	2.7	Writing (Exporting) Data to Files	46
		2.7.1 Diverting Normal R Output to a File	46
		2.7.2 Saving Graphs in Files	46
		2.7.3 Exporting Data to Files	47
	2.8	Some Arithmetic and Other Operators	48
		2.8.1 Vectors	48
		2.8.2 Matrix Computations	50
	2.9	Programming in R	55
		2.9.1 Best Practice: Script Files	55
		2.9.2 Some Useful Commands or Functions	56
		2.9.3 Conditional Execution	56
		2.9.4 Loops	58
		2.9.5 Functions and Functionals	59
		2.9.6 User-Defined Functions	61
	2.10	Bibliographic Notes	66
		Exercises	67

3 Simple Linear Regression — 71
- 3.1 Introduction — 71
- 3.2 Covariance and Correlation Coefficient — 71
- 3.3 Example: Computer Repair Data — 77
- 3.4 The Simple Linear Regression Model — 79
- 3.5 Parameter Estimation — 80
- 3.6 Tests of Hypotheses — 85
- 3.7 Confidence Intervals — 90
- 3.8 Predictions — 91
- 3.9 Measuring the Quality of Fit — 93
- 3.10 Regression Line Through the Origin — 97
- 3.11 Trivial Regression Models — 100
- 3.12 Bibliographic Notes — 101
- Exercises — 101

4 Multiple Linear Regression — 109
- 4.1 Introduction — 109
- 4.2 Description of the Data and Model — 109
- 4.3 Example: Supervisor Performance Data — 110
- 4.4 Parameter Estimation — 112
- 4.5 Interpretations of Regression Coefficients — 114
- 4.6 Centering and Scaling — 117
  - 4.6.1 Centering and Scaling in Intercept Models — 118
  - 4.6.2 Scaling in No-Intercept Models — 119
- 4.7 Properties of the Least Squares Estimators — 121
- 4.8 Multiple Correlation Coefficient — 122
- 4.9 Inference for Individual Regression Coefficients — 123
- 4.10 Tests of Hypotheses in a Linear Model — 126
  - 4.10.1 Testing All Regression Coefficients Equal to Zero — 128
  - 4.10.2 Testing a Subset of Regression Coefficients Equal to Zero — 131
  - 4.10.3 Testing the Equality of Regression Coefficients — 134
  - 4.10.4 Estimating and Testing of Regression Parameters Under Constraints — 136
- 4.11 Predictions — 138
- 4.12 Summary — 139
- Exercises — 139
- Appendix: 4.A Multiple Regression in Matrix Notation — 146

5 Regression Diagnostics: Detection of Model Violations — 151
- 5.1 Introduction — 151
- 5.2 The Standard Regression Assumptions — 152
- 5.3 Various Types of Residuals — 155

5.4		Graphical Methods	157
5.5		Graphs Before Fitting a Model	160
	5.5.1	One-Dimensional Graphs	160
	5.5.2	Two-Dimensional Graphs	161
	5.5.3	Rotating Plots	163
	5.5.4	Dynamic Graphs	164
5.6		Graphs After Fitting a Model	164
5.7		Checking Linearity and Normality Assumptions	165
5.8		Leverage, Influence, and Outliers	166
	5.8.1	Outliers in the Response Variable	168
	5.8.2	Outliers in the Predictors	169
	5.8.3	Masking and Swamping Problems	169
5.9		Measures of Influence	172
	5.9.1	Cook's Distance	173
	5.9.2	Welsch and Kuh Measure	174
	5.9.3	Hadi's Influence Measure	174
5.10		The Potential–Residual Plot	177
5.11		Regression Diagnostics in R	178
5.12		What to Do with the Outliers?	178
5.13		Role of Variables in a Regression Equation	180
	5.13.1	Added-Variable Plot	180
	5.13.2	Residual Plus Component Plot	181
5.14		Effects of an Additional Predictor	184
5.15		Robust Regression	186
		Exercises	186

6 Qualitative Variables as Predictors — 193

6.1	Introduction	193
6.2	Salary Survey Data	194
6.3	Interaction Variables	197
6.4	Systems of Regression Equations: Comparing Two Groups	202
	6.4.1 Models with Different Slopes and Different Intercepts	203
	6.4.2 Models with Same Slope and Different Intercepts	210
	6.4.3 Models with Same Intercept and Different Slopes	212
6.5	Other Applications of Indicator Variables	213
6.6	Seasonality	214
6.7	Stability of Regression Parameters Over Time	215
	Exercises	218

7 Transformation of Variables — 225

7.1	Introduction	225
7.2	Transformations to Achieve Linearity	227
7.3	Bacteria Deaths Due to X-Ray Radiation	230

		7.3.1	Inadequacy of a Linear Model	231
		7.3.2	Logarithmic Transformation for Achieving Linearity	233
	7.4	Transformations to Stabilize Variance		234
	7.5	Detection of Heteroscedastic Errors		239
	7.6	Removal of Heteroscedasticity		241
	7.7	Weighted Least Squares		243
	7.8	Logarithmic Transformation of Data		244
	7.9	Power Transformation		247
	7.10	Summary		249
		Exercises		250

8 Weighted Least Squares — 257

8.1	Introduction		257
8.2	Heteroscedastic Models		258
	8.2.1	Supervisors Data	259
	8.2.2	College Expense Data	260
8.3	Two-Stage Estimation		262
8.4	Education Expenditure Data		264
8.5	Fitting a Dose–Response Relationship Curve		273
	Exercises		275

9 The Problem of Correlated Errors — 277

9.1	Introduction: Autocorrelation	277
9.2	Consumer Expenditure and Money Stock	278
9.3	Durbin–Watson Statistic	281
9.4	Removal of Autocorrelation by Transformation	283
9.5	Iterative Estimation with Autocorrelated Errors	287
9.6	Autocorrelation and Missing Variables	288
9.7	Analysis of Housing Starts	289
9.8	Limitations of the Durbin–Watson Statistic	292
9.9	Indicator Variables to Remove Seasonality	294
9.10	Regressing Two Time Series	296
	Exercises	298

10 Analysis of Collinear Data — 301

10.1	Introduction		301
10.2	Effects of Collinearity on Inference		302
10.3	Effects of Collinearity on Forecasting		308
10.4	Detection of Collinearity		311
	10.4.1	Simple Signs of Collinearity	312
	10.4.2	Variance Inflation Factors	315
	10.4.3	The Condition Indices	318
	Exercises		321

11 Working With Collinear Data — 327
- 11.1 Introduction — 327
- 11.2 Principal Components — 328
- 11.3 Computations Using Principal Components — 332
- 11.4 Imposing Constraints — 335
- 11.5 Searching for Linear Functions of the β's — 338
- 11.6 Biased Estimation of Regression Coefficients — 342
- 11.7 Principal Components Regression — 343
- 11.8 Reduction of Collinearity in the Estimation Data — 345
- 11.9 Constraints on the Regression Coefficients — 347
- 11.10 Principal Components Regression: A Caution — 348
- 11.11 Ridge Regression — 351
- 11.12 Estimation by the Ridge Method — 353
- 11.13 Ridge Regression: Some Remarks — 358
- 11.14 Summary — 359
- 11.15 Bibliographic Notes — 360
- Exercises — 360
- Appendix: 11.A Principal Components — 363
- Appendix: 11.B Ridge Regression — 365
- Appendix: 11.C Surrogate Ridge Regression — 369

12 Variable Selection Procedures — 371
- 12.1 Introduction — 371
- 12.2 Formulation of the Problem — 372
- 12.3 Consequences of Variables Deletion — 373
- 12.4 Uses of Regression Equations — 374
  - 12.4.1 Description and Model Building — 374
  - 12.4.2 Estimation and Prediction — 374
  - 12.4.3 Control — 375
- 12.5 Criteria for Evaluating Equations — 376
  - 12.5.1 Residual Mean Square — 376
  - 12.5.2 Mallows C_p — 376
  - 12.5.3 Information Criteria — 377
- 12.6 Collinearity and Variable Selection — 379
- 12.7 Evaluating All Possible Equations — 379
- 12.8 Variable Selection Procedures — 380
  - 12.8.1 Forward Selection Procedure — 381
  - 12.8.2 Backward Elimination Procedure — 381
  - 12.8.3 Stepwise Method — 382
- 12.9 General Remarks on Variable Selection Methods — 382
- 12.10 A Study of Supervisor Performance — 383
- 12.11 Variable Selection with Collinear Data — 388
- 12.12 The Homicide Data — 388
- 12.13 Variable Selection Using Ridge Regression — 391

	12.14	Selection of Variables in an Air Pollution Study	392
	12.15	A Possible Strategy for Fitting Regression Models	398
	12.16	Bibliographic Notes	400
		Exercises	400
		Appendix: 12.A Effects of Incorrect Model Specifications	404

13 Logistic Regression — 409
	13.1	Introduction	409
	13.2	Modeling Qualitative Data	410
	13.3	The Logit Model	410
	13.4	Example: Estimating Probability of Bankruptcies	413
	13.5	Logistic Regression Diagnostics	415
	13.6	Determination of Variables to Retain	417
	13.7	Judging the Fit of a Logistic Regression	420
	13.8	The Multinomial Logit Model	422
		13.8.1 Multinomial Logistic Regression	422
		13.8.2 Example: Determining Chemical Diabetes	423
		13.8.3 Ordinal Logistic Regression	426
		13.8.4 Example: Determining Chemical Diabetes Revisited	426
	13.9	Classification Problem: Another Approach	428
		Exercises	430

14 Further Topics — 433
	14.1	Introduction	433
	14.2	Generalized Linear Model	434
	14.3	Poisson Regression Model	435
	14.4	Introduction of New Drugs	436
	14.5	Robust Regression	437
	14.6	Fitting a Quadratic Model	438
	14.7	Distribution of PCB in U.S. Bays	440
		Exercises	443

References — 445

Index — 455

PREFACE

I have been feeling a great sense of sadness while I was working alone on this edition of the book after Professor Samprit Chatterjee, my longtime teacher, mentor, friend, and co-author, passed away in April 2021. Our first paper was published in 1986 (Chatterjee and Hadi, 1986). Samprit and I also co-authored our 1988 book (Chatterjee and Hadi, 1988) as well as several other papers. My sincere condolences to his family and friends. May God rest his soul in peace.

Regression analysis has become one of the most widely used statistical tools for analyzing multifactor data. It is appealing because it provides a conceptually simple method for investigating functional relationships among variables. The standard approach in regression analysis is to take data, fit a model, and then evaluate the fit using statistics such as t, F, R^2, and Durbin–Watson test. Our approach is broader. We view regression analysis as a set of data analytic techniques that examine the interrelationships among a given set of variables. The emphasis is not on formal statistical tests and probability calculations. We argue for an informal analysis directed toward uncovering patterns in the data. We have also attempted to write a book for a group of readers with diverse backgrounds. We have also tried to put emphasis on the art of data analysis rather than on the development of statistical theory.

The material presented is intended for anyone who is involved in analyzing data. The book should be helpful to those who have some knowledge of the basic concepts of statistics. In the university, it could be used as a text for a course on regression analysis for students whose specialization is not statistics, but, who nevertheless use regression analysis quite extensively in their work. For students whose major emphasis is statistics, and who take a course on regression analysis from a book at the level of Rao (1973), Seber (1977), or Sen and Srivastava (1990), this book can be used to balance and complement the theoretical aspects of the subject with practical applications. Outside the university, this book can be profitably used by those people whose present approach to analyzing multifactor data consists of looking at standard computer output (t, F, R^2, standard errors, etc.), but who want to go beyond these summaries for a more thorough analysis.

We utilize most standard and some not-so-standard summary statistics on the basis of their intuitive appeal. We rely heavily on graphical representations of the data and employ many variations of plots of regression residuals. We are not overly concerned with precise probability evaluations. Graphical methods for exploring residuals can suggest model deficiencies or point to troublesome observations. Upon further investigation into their origin, the troublesome observations often turn out to be more informative than the well-behaved observations. We notice often that more information is obtained from a quick examination of a plot of residuals than from a formal test of statistical significance of some limited null hypothesis. In short, the presentation in the chapters of this book is guided by the principles and concepts of exploratory data analysis.

As we mentioned in previous editions, the statistical community has been most supportive, and we have benefitted greatly from their suggestions in improving the text. Our presentation of the various concepts and techniques of regression analysis relies on carefully developed examples. In each example, we have isolated one or two techniques and discussed them in some detail. The data were chosen to highlight the techniques being presented. Although when analyzing a given set of data it is usually necessary to employ many techniques, we have tried to choose the various data sets so that it would not be necessary to discuss the same technique more than once. Our hope is that after working through the book, the reader will be ready and able to analyze their data methodically, thoroughly, and confidently.

The emphasis in this book is on the analysis of data rather than on plugging numbers into formulas, tests of hypotheses, or confidence intervals. Therefore no attempt has been made to derive the techniques. Techniques are described, the required assumptions are given and, finally, the success of the technique in the particular example is assessed. Although derivations of the techniques are not included, we have tried to refer the reader in each case to sources in which such discussion is available. Our hope is that some of these sources will be followed up by the reader who wants a more thorough grounding in theory.

Recently there has been a qualitative change in the analysis of linear models, from model fitting to model building, from overall tests to clinical examinations of data, from macroscopic to the microscopic analysis. To do this kind of analysis a computer is essential and, in previous editions, we have assumed its availability, but did not wish to endorse or associate the book with any of the commercially available statistical packages to make it available to a wider community.

We are particularly heartened by the arrival of the language R, which is available on the Internet under the General Public License (GPL). The language has excellent computing and graphical features. It is also free! For these and other reasons, I decided to introduce and use R in this edition of the book to enable the readers to use R on their own datasets and reproduce the various types of graphs and analysis presented in this book. Although a knowledge of R would certainly be helpful, no prior knowledge of R is assumed.

Major changes have been made in streamlining the text, removing ambiguities, and correcting errors pointed out by readers and others detected by the authors. Chapter 2 is new in this edition. It gives a brief but, what we believe to be, sufficient introduction to R that would enable readers to use R to carry out the regression analysis computations as well as the graphical displays presented in this edition of the book. To help the readers out, we provide all the necessary R code in the new Chapter 2 and throughout the rest of the chapters. Section 5.11 about regression diagnostics in R is new. New references have also been added. The index at the end of the book has been enhanced. The addition of the new chapter increased the number of pages. To offset this increase, data tables that are larger than 10 rows are deleted from the book because the reader can obtain them in digital forms from the Book's Website at http://www.aucegypt.edu/faculty/hadi/RABE6. This Website contains, among other things, all the data sets that are included in this book, the R code that are used to produce the graphs and tables in this book, and more. Also, the use of R enabled us to delete the statistical tables in the appendix because the reader can now use R to compute the p-values as well as the critical values of test statistics for any desired significance level, not just the customary ones such as 0.1, 0.05, and 0.01.

We have rewritten some of the exercises and added new ones at the end of the chapters. We feel that the exercises reinforce the understanding of the material in the preceding chapters. Also new to accompany this edition a

Solution Manual and Power Point files are available only for instructors by contacting the authors at ahadi@aucegypt.edu or ali-hadi@cornell.edu.

Previous editions of this book have been translated to Persian, Korean, and Chinese languages. We are grateful to the translators Prof. H. A. Niromand, Prof. Zhongguo Zheng, Prof. Kee Young Kim, Prof. Myoungshic Jhun, Prof. Hyuncheol Kang, and Prof. Seong Keon Lee. We are fortunate to have had assistance and encouragement from several friends, colleagues, and associates. Some of our colleagues and students at New York University, Cornell University, and The American University in Cairo have used portions of the material in their courses and have shared with us their comments and comments of their students. Special thanks are due to our friend and former colleague Jeffrey Simonoff (New York University) for comments, suggestions, and general help. The students in our classes on regression analysis have all contributed by asking penetrating questions and demanding meaningful and understandable answers. Our special thanks go to Nedret Billor (Cukurova University, Turkey) and Sahar El-Sheneity (Cornell University) for their very careful reading of an earlier edition of this book.

We also appreciate the comments provided by Habibollah Esmaily, Hassan Doosti, Fengkai Yang, Mamunur Rashid, Saeed Hajebi, Zheng Zhongguo, Robert W. Hayden, Marie Duggan, Sungho Lee, Hock Lin (Andy) Tai, and Junchang Ju. We also thank Lamia Abdellatif for proofreading parts of this edition, Dimple Philip for preparing the Latex style files and the corresponding PDF version, Dean Gonzalez for helping with the production of some of the figures, and Michael New for helping with the front and back covers.

Cairo, Egypt   ALI S. HADI
September 2023

ABOUT THE COMPANION WEBSITE

This book is accompanied by a companion website.

www.wiley.com/go/hadi/regression_analysis_6e

This website includes:

- Table of contents
- Preface
- Book cover
- Places where you can purchase the book
- Data sets
- Stata, SAS or SPSS users
- R users
- Errata/Comments/Feedback
- Solutions to Exercises

CHAPTER 1

INTRODUCTION

1.1 WHAT IS REGRESSION ANALYSIS?

Regression analysis is a conceptually simple method for investigating functional relationships among variables. A real estate appraiser may wish to relate the sale price of a home from selected physical characteristics of the building and taxes (local, school, county) paid on the building. We may wish to examine whether cigarette consumption is related to various socioeconomic and demographic variables such as age, education, income, and price of cigarettes. The relationship is expressed in the form of an equation or a model connecting the *response* or *dependent* variable and one or more *explanatory* or *predictor* variables. In the cigarette consumption example, the response variable is cigarette consumption (measured by the number of packs of cigarette sold in a given state on a per capita basis during a given year) and the explanatory or predictor variables are the various socioeconomic and demographic variables. In the real estate appraisal example, the response variable is the price of a home and the explanatory or predictor variables are the characteristics of the building and taxes paid on the building.

Regression Analysis By Example Using R, Sixth Edition. Ali S. Hadi and Samprit Chatterjee
© 2024 John Wiley & Sons, Inc. Published 2024 by John Wiley & Sons, Inc.
Companion website: www.wiley.com/go/hadi/regression_analysis_6e

We denote the response variable by Y and the set of predictor variables by $X_1, X_2, \ldots, X_p$, where p denotes the number of predictor variables. The true relationship between Y and $X_1, X_2, \ldots, X_p$ can be approximated by the regression model

$$Y = f(X_1, X_2, \ldots, X_p) + \varepsilon, \tag{1.1}$$

where ε is assumed to be a random error representing the discrepancy in the approximation. It accounts for the failure of the model to fit the data exactly. The function $f(X_1, X_2, \ldots, X_p)$ describes the relationship between Y and $X_1, X_2, \ldots, X_p$. An example is the linear regression model

$$Y = \beta_0 + \beta_1 X_1 + \beta_2 X_2 + \cdots + \beta_p X_p + \varepsilon, \tag{1.2}$$

where $\beta_0, \beta_1, \ldots, \beta_p$, called the *regression parameters* or *coefficients*, are unknown constants to be determined (estimated) from the data. We follow the commonly used notational convention of denoting unknown parameters by Greek letters.

The predictor or explanatory variables are also called by other names such as *independent* variables, *covariates*, *regressors*, *factors*, and *carriers*. The name independent variable, though commonly used, is the least preferred, because in practice the predictor variables are rarely independent of each other.

1.2 PUBLICLY AVAILABLE DATA SETS

Regression analysis has numerous areas of applications. A partial list would include economics, finance, business, law, meteorology, medicine, biology, chemistry, engineering, physics, education, sports, history, sociology, and psychology. A few examples of such applications are given in Section 1.3. Regression analysis is learned most effectively by analyzing data that are of direct interest to the reader. We invite the readers to think about questions (in their own areas of work, research, or interest) that can be addressed using regression analysis. Readers should collect the relevant data and then apply the regression analysis techniques presented in this book to their own data. To help the reader locate real-life data, this section provides some sources and links to a wealth of data sets that are available for public use.

A number of data sets are available in books and on the Internet. The book by Hand et al. (1994) contains data sets from many fields. These data sets are small in size and are suitable for use as exercises. The book by Chatterjee et al. (1995) provides numerous data sets from diverse fields. The data are included in a diskette that comes with the book and can also be found at the Website.[1]

Data sets are also available on the Internet at many other sites. Some of the Websites given below allow the direct copying and pasting into the statistical package of choice, while others require downloading the data file and then importing them into a statistical package. Some of these sites also contain further links to yet other data sets or statistics-related Websites.

The Data and Story Library (DASL, pronounced "dazzle") is one of the most interesting sites that contains a number of data sets accompanied by the "story" or background associated with each data set. DASL is an online library[2] of data files and stories that illustrate the use of basic statistical methods. The data sets cover a wide variety of topics. DASL comes with a powerful search engine to locate the story or data file of interest.

Another Website, which also contains data sets arranged by the method used in the analysis, is the Electronic Dataset Service.[3] The site also contains many links to other data sources on the Internet.

Finally, this book has a Website,[4] which contains, among other things, all the data sets that are included in this book and more. These and other data sets can be found at the Book's Website.

1.3 SELECTED APPLICATIONS OF REGRESSION ANALYSIS

Regression analysis is one of the most widely used statistical tools because it provides simple methods for establishing a functional relationship among variables. It has extensive applications in many subject areas. The cigarette consumption and the real estate appraisal, mentioned above, are but two examples. In this section, we give a few additional examples demonstrating the wide applicability of regression analysis in real-life situations. Some of the data sets described here will be used later in the book to illustrate regression techniques or in the exercises at the end of various chapters.

1.3.1 Agricultural Sciences

The Dairy Herd Improvement Cooperative (DHI) in upstate New York collects and analyzes data on milk production. One question of interest here

---

[1] http://www.stern.nyu.edu/~jsimonof/Casebook
[2] http://lib.stat.cmu.edu/DASL
[3] http://www-unix.oit.umass.edu/~statdata
[4] http://www.aucegypt.edu/faculty/hadi/RABE6

Table 1.1 Variables in Milk Production Data

Variable	Definition
Current	Current month milk production in pounds
Previous	Previous month milk production in pounds
Fat	Percent of fat in milk
Protein	Percent of protein in milk
Days	Number of days since present lactation
Lactation	Number of lactations
I79	Indicator variable (0 if Days $\leq$ 79 and 1 if Days $>$ 79)

is how to develop a suitable model to predict current milk production from a set of measured variables. The response variable (current milk production in pounds) and the predictor variables are given in Table 1.1. Samples are taken once a month during milking. The period that a cow gives milk is called lactation. Number of lactations is the number of times a cow has calved or given milk. The recommended management practice is to have the cow produce milk for about 305 days and then allow a 60-day rest period before beginning the next lactation. The data set, consisting of 199 observations, was compiled from the DHI milk production records. The Milk Production data can be found at the Book's Website.

1.3.2 Industrial and Labor Relations

In 1947, the United States Congress passed the Taft–Hartley Amendments to the Wagner Act. The original Wagner Act had permitted the unions to use a *Closed Shop Contract*[5] unless prohibited by state law. The Taft–Hartley Amendments made the use of Closed Shop Contract illegal and gave individual states the right to prohibit union shops[6] as well. These right-to-work laws have caused a wave of concern throughout the labor movement. A question of interest here is: What are the effects of these laws on the cost of living for a four-person family living on an intermediate budget in the United States? To answer this question a data set consisting of 38 geographic locations

---

[5] Under a Closed Shop Contract provision, all employees must be union members at the time of hire and must remain members as a condition of employment.

[6] Under a Union Shop clause, employees are not required to be union members at the time of hire, but must become a member within two months, thus allowing the employer complete discretion in hiring decisions.

Table 1.2 Variables in Right-To-Work Laws Data

Variable	Definition
COL	Cost of living for a four-person family
PD	Population density (person per square mile)
URate	State unionization rate in 1978
Pop	Population in 1975
Taxes	Property taxes in 1972
Income	Per capita income in 1974
RTWL	Indicator variable (1 if there are right-to-work laws in the state and 0 otherwise)

has been assembled from various sources. The variables used are defined in Table 1.2. The Right-To-Work Laws data can be found at the Book's Website.

1.3.3 Government

Information about domestic immigration (the movement of people from one state or area of a country to another) is important to state and local governments. It is of interest to build a model that predicts domestic immigration or to answer the question of why do people leave one place to go to another? There are many factors that influence domestic immigration, such as weather conditions, crime, taxes, and unemployment rates. A data set for the 48 contiguous states has been created. Alaska and Hawaii are excluded from the analysis because the environments of these states are significantly different from the other 48, and their locations present certain barriers to immigration. The response variable here is net domestic immigration, which represents the net movement of people into and out of a state over the period 1990–1994 divided by the population of the state. Eleven predictor variables thought to influence domestic immigration are defined in Table 1.3 and can be found at the Book's Website.

1.3.4 History

A question of historical interest is how to estimate the age of historical objects based on some age-related characteristics of the objects. For example, the variables in Table 1.4 can be used to estimate the age of Egyptian skulls.

Table 1.3 Variables in Study of Domestic Immigration

Variable	Definition
State	State name
NDIR	Net domestic immigration rate over the period 1990–1994
Unemp	Unemployment rate in the civilian labor force in 1994
Wage	Average hourly earnings of production workers in manufacturing in 1994
Crime	Violent crime rate per 100,000 people in 1993
Income	Median household income in 1994
Metrop	Percentage of state population living in metropolitan areas in 1992
Poor	Percentage of population who fall below the poverty level in 1994
Taxes	Total state and local taxes per capita in 1993
Educ	Percentage of population 25 years or older who have a high school degree or higher in 1990
BusFail	The number of business failures divided by the population of the state in 1993
Temp	Average of the 12 monthly average temperatures (in degrees Fahrenheit) for the state in 1993
Region	Region in which the state is located (northeast, south, midwest, west)

Table 1.4 Variables in Egyptian Skulls Data

Variable	Definition
Year	Approximate year of skull formation (negative = B.C.; positive = A.D.)
MB	Maximum breadth of skull
BH	Basibregmatic height of skull
BL	Basialveolar length of skull
NH	Nasal Height of skull

Here the response variable is Year and the other four variables are possible predictors. There are 150 observations in this data set. The original source of the data is Thomson and Randall-Maciver (1905), but they can be found in Hand et al. (1994, pp. 299–301). An analysis of the data can be found in Manly (1986). The Egyptian Skulls data can be found at the Book's Website.

Table 1.5 Variables in Study of Water Pollution in New York Rivers

Variable	Definition
Y	Mean nitrogen concentration (mg/liter) based on samples taken at regular intervals during the spring, summer, and fall months
X_1	Agriculture: percentage of land area currently in agricultural use
X_2	Forest: percentage of forest land
X_3	Residential: percentage of land area in residential use
X_4	Commercial/Industrial: percentage of land area in either commercial or industrial use

1.3.5 Environmental Sciences

In a 1976 study exploring the relationship between water quality and land use, Haith (1976) obtained the measurements (shown in Table 1.5) on 20 river basins in New York State. A question of interest here is how the land use around a river basin contributes to the water pollution as measured by the mean nitrogen concentration (mg/liter). The dataset can be found at the Book's Website.

1.3.6 Industrial Production

Nambe Mills in Santa Fe, New Mexico, makes a line of tableware made from sand casting a special alloy of metals. After casting, the pieces go through a series of shaping, grinding, buffing, and polishing steps. Data was collected for 59 items produced by the company. The relation between the polishing time and the product diameters and the product types (Bowl, Casserole, Dish, Tray, and Plate) are used to estimate the polishing time for new products which are designed or suggested for design and manufacture. The variables representing product types are coded as binary variables (1 corresponds to the type and 0 otherwise). Diam is the diameter of the item (in inches), polishing time is measured in minutes, and price in dollars. The polishing time is the major item in the cost of the product. The production decision will be based on the estimated time of polishing. The data is obtained from the DASL library, can be found there and also in the file Industrial.Production.csv at the Book's Website.

1.3.7 The Space Shuttle Challenger

The explosion of the space shuttle Challenger in 1986 killing the crew was a shattering tragedy. To look into the case a Presidential Commission was appointed. The O-rings in the booster rockets, which are used in space launching, play a very important part in its safety. The rigidity of the O-rings is thought to be affected by the temperature at launching. There are six O-rings in a booster rocket. The data consists of two variables: the number of rings damaged and the temperature at launchings of the 23 flights. The data set can be found at the Book's Website. The analysis performed before the launch did not include the launches in which no O-ring was damaged and came to the wrong conclusion. A detailed discussion of the problem is found in The *Flight of the Space Shuttle Challenger* in Chatterjee et al. (1995, pp. 33–35). Note here that the response variable is a proportion bounded between 0 and 1.

1.3.8 Cost of Health Care

The cost of delivery of health care has become an important concern. Getting data on this topic is extremely difficult because it is highly proprietary. These data were collected by the Department of Health and Social Services of the State of New Mexico and cover 52 of the 60 licensed facilities in New Mexico in 1988. The variables in these data are the characteristics which describe the facilities size, volume of usage, expenditures, and revenue. The location of the facility is also indicated, whether it is in the rural or nonrural area. Specific definitions of the variables are given in Table 1.6 and the data can be found at the Book's Website. There are several ways of looking at a

Table 1.6 Variables in Cost of Health Care Data

Variable	Definition
RURAL	Rural home (1) and nonrural home (0)
BED	Number of beds in home
MCDAYS	Annual medical in-patient days (hundreds)
TDAYS	Annual total patient days (hundreds)
PCREV	Annual total patient care revenue ($100)
NSAL	Annual nursing salaries ($100)
FEXP	Annual facilities expenditures ($100)
NETREV	PCREV − NSAL − FEXP

body of data and extracting various kinds of information. For example, (a) Are rural facilities different from nonrural facilities? and (b) How do the hospital characteristics affect the total patient care revenue?

1.4 STEPS IN REGRESSION ANALYSIS

Regression analysis includes the following steps:

- Statement of the problem
- Selection of potentially relevant variables
- Data collection
- Model specification
- Choice of fitting method
- Model fitting
- Model validation and criticism
- Using the chosen model(s) for the solution of the posed problem.

These steps are examined below.

1.4.1 Statement of the Problem

Regression analysis usually starts with a formulation of the problem. This includes the determination of the question(s) to be addressed by the analysis. The problem statement is the first and perhaps the most important step in regression analysis. It is important because an ill-defined problem or a misformulated question can lead to wasted effort. It can lead to the selection of irrelevant set of variables or to a wrong choice of the statistical method of analysis. A question that is not carefully formulated can also lead to the wrong choice of a model. Suppose we wish to determine whether or not an employer is discriminating against a given group of employees, say women. Data on salary, qualifications, and gender are available from the company's record to address the issue of discrimination. There are several definitions of employment discrimination in the literature. For example, discrimination occurs when on the average (a) women are paid less than equally qualified men, or (b) women are more qualified than equally paid men. To answer the question: "On the average, are women paid less than equally qualified men?"

we choose salary as a response variable, and qualification and gender as predictor variables. But to answer the question: "On the average, are women more qualified than equally paid men?" we choose qualification as a response variable and salary and gender as predictor variables, that is, the roles of variables have been switched.

1.4.2 Selection of Potentially Relevant Variables

The next step after the statement of the problem is to select a set of variables that are thought by the experts in the area of study to explain or predict the response variable. The response variable is denoted by Y and the explanatory or predictor variables are denoted by $X_1, X_2, \ldots, X_p$, where p denotes the number of predictor variables. An example of a response variable is the price of a single-family house in a given geographical area. A possible relevant set of predictor variables in this case is: area of the lot, area of the house, age of the house, number of bedrooms, number of bathrooms, type of neighborhood, style of the house, amount of real estate taxes, and so forth.

1.4.3 Data Collection

The next step after the selection of potentially relevant variables is to collect the data from the environment under study to be used in the analysis. Sometimes the data are collected in a controlled setting so that factors that are not of primary interest can be held constant. More often the data are collected under nonexperimental conditions where very little can be controlled by the investigator. In either case, the collected data consist of observations on n subjects. Each of these n observations consists of measurements for each of the potentially relevant variables. The data are usually recorded as in Table 1.7. A column in Table 1.7 represents a variable, whereas a row represents an observation, which is a set of $p + 1$ values for a single subject (e.g., a house); one value for the response variable and one value for each of the p predictors. The notation x_{ij} refers to the ith value of the jth variable. The first subscript refers to observation number and the second refers to variable number.

Each of the variables in Table 1.7 can be classified as either *quantitative* or *qualitative*. Examples of quantitative variables are the house price, number of bedrooms, age, and taxes. Examples of qualitative variables are neighborhood type (e.g., good or bad neighborhood) and house style (e.g., ranch, colonial, etc.). In this book we deal mainly with the cases where the response variable is

Table 1.7 Notation for Data Used in Regression Analysis

Observation Number	Response Variable Y	X_1	X_2	$\cdots$	X_p
1	y_1	x_{11}	x_{12}	$\cdots$	x_{1p}
2	y_2	x_{21}	x_{22}	$\cdots$	x_{2p}
3	y_3	x_{31}	x_{32}	$\cdots$	x_{3p}
$\vdots$	$\vdots$	$\vdots$	$\vdots$	$\vdots$	$\vdots$
n	y_n	x_{n1}	x_{n2}	$\cdots$	x_{np}

quantitative. A technique used in cases where the response variable is *binary*[7] is called *logistic regression*. This is introduced in Chapter 13. In regression analysis, the predictor variables can be either quantitative and/or qualitative. For the purpose of computations, however, the qualitative variables, if any, have to be coded into a set of *indicator* or *dummy* variables as discussed in Chapter 6.

If all predictor variables are qualitative, the techniques used in the analysis of the data are called the *analysis of variance* techniques. Although the analysis of variance techniques can be introduced and explained as methods in their own right,[8] it is shown in Chapter 6 that they are special cases of regression analysis. If some of the predictor variables are quantitative while others are qualitative, regression analysis in these cases is called the *analysis of covariance*.

1.4.4 Model Specification

The form of the model that is thought to relate the response variable to the set of predictor variables can be specified initially by the experts in the area of study based on their knowledge or their objective and/or subjective judgments. The hypothesized model can then be either confirmed or refuted by the analysis of the collected data. Note that the model needs to be specified only in form, but it can still depend on unknown parameters. We need to select the form of the function $f(X_1, X_2, \ldots, X_p)$ in (1.1). This function can

---

[7] A variable that can take only one of two possible values such as yes or no, 1 or 0, and success or failure, is called a binary variable.

[8] See, for example, the books by Scheffé (1959), Iversen (1976), Wildt and Ahtola (1978), Krishnaiah (1980), Iversen and Norpoth (1987), Lindman (1992), and Christensen (1996).

be classified into two types: *linear* and *nonlinear*. An example of a linear function is

$$Y = \beta_0 + \beta_1 X_1 + \varepsilon \tag{1.3}$$

while a nonlinear function is

$$Y = \beta_0 + e^{\beta_1 X_1} + \varepsilon. \tag{1.4}$$

Note that the term *linear* (*nonlinear*) here does not describe the relationship between Y and $X_1, X_2, \ldots, X_p$. It is related to the fact that the regression parameters enter the equation linearly (nonlinearly). Each of the following models is linear:

$$Y = \beta_0 + \beta_1 X + \beta_2 X^2 + \varepsilon,$$
$$Y = \beta_0 + \beta_1 \ln X + \varepsilon,$$

because in each case the parameters enter linearly although the relationship between Y and X is nonlinear. This can be seen if the two models are reexpressed, respectively, as follows:

$$Y = \beta_0 + \beta_1 X_1 + \beta_2 X_2 + \varepsilon,$$
$$Y = \beta_0 + \beta_1 X_1 + \varepsilon,$$

where in the first equation we have $X_1 = X$ and $X_2 = X^2$ and in the second equation we have $X_1 = \ln X$. The variables here are *reexpressed* or *transformed*. Transformation is dealt with in Chapter 7. All nonlinear functions that can be transformed into linear functions are called *linearizable* functions. Accordingly, the class of linear models is actually wider than it might appear at first sight because it includes all linearizable functions. Note, however, that not all nonlinear functions are linearizable. For example, it is not possible to linearize the nonlinear function in (1.4). Some authors refer to nonlinear functions that are not linearizable as *intrinsically* nonlinear functions.

A regression equation containing only one predictor variable is called a *simple regression equation*. An equation containing more than one predictor variable is called a *multiple regression equation*. An example of simple regression would be an analysis in which the time to repair a machine is studied in relation to the number of components to be repaired. Here we have one response variable (time to repair the machine) and one predictor variable (number of components to be repaired). An example of a very complex multiple regression situation would be an attempt to explain the age-adjusted mortality rates prevailing in different geographic regions (response variable) by a large number of environmental and socioeconomic factors (predictor variables). Both types of problems are treated in this book. These two particular examples are studied, one in Chapter 3, the other in Chapter 12.

In certain applications the response variable can actually be a set of variables, $Y_1, Y_2, \ldots, Y_q$, say, which are thought to be related to the same set of predictor variables, $X_1, X_2, \ldots, X_p$. For example, Bartlett et al. (1998) present a data set on 148 healthy people. Eleven variables are measured; six variables represent different types of measured sensory thresholds (e.g., vibration, hand and foot temperatures) and five a priori selected baseline covariates (e.g., age, gender, height, and weight) that may have systematic effects on some or all of the six sensory thresholds. Here we have six response variables and five predictor variables. This data set, which we refer to as the QST (*quantitative sensory testing*) data, is not listed here due to its size (148 observations) but it can be found at the Book's Website. For further description of the data and objectives of the study, see Bartlett et al. (1998).

When we deal only with one response variable, regression analysis is called *univariate* regression and in cases where we have two or more response variables, the regression is called *multivariate* regression. Simple and multiple regressions should not be confused with univariate versus multivariate regressions. The distinction between simple and multiple regressions is determined by the number of predictor variables (simple means one predictor variable and multiple means two or more predictor variables), whereas the distinction between univariate and multivariate regressions is determined by the number of response variables (univariate means one response variable and multivariate means two or more response variables). In this book we consider only univariate regression (both simple and multiple, linear and nonlinear). Multivariate regression is treated in books on multivariate analysis such as Rencher (1995), Johnson and Wichern (1992), and Johnson (1998). In this book the term regression will be used to mean univariate regression.

The various classifications of regression analysis we discussed above are shown in Table 1.8.

1.4.5 Method of Fitting

After the model has been defined and the data have been collected, the next task is to estimate the parameters of the model based on the collected data. This is also referred to as *parameter estimation* or *model fitting*. The most commonly used method of estimation is called the *least squares* method. Under certain assumptions (to be discussed in detail in this book), least

Table 1.8 Various Classifications of Regression Analysis

Type of Regression	Conditions
Univariate	Only one quantitative response variable
Multivariate	Two or more quantitative response variables
Simple	Only one predictor variable
Multiple	Two or more predictor variables
Linear	All parameters enter the equation linearly, possibly after transformation of the data
Nonlinear	The relationship between the response and some of the predictors is nonlinear or some of the parameters appear nonlinearly, but no transformation is possible to make the parameters appear linearly
Analysis of variance	All predictors are qualitative variables
Analysis of covariance	Some predictors are quantitative variables and others are qualitative variables
Logistic	The response variable is qualitative

squares method produce estimators with desirable properties. In this book we will deal mainly with the least squares method and its variants (e.g., weighted least squares). In some instances (e.g., when one or more of the assumptions does not hold) other estimation methods may be superior to least squares. The other estimation methods that we consider in this book are the *maximum likelihood* method, the *ridge regression*, and the *principal components* method.

1.4.6 Model Fitting

The next step in the analysis is to estimate the regression parameters or to fit the model to the collected data using the chosen estimation method (e.g., least squares). The estimates of the regression parameters $\beta_0, \beta_1, \ldots, \beta_p$ in (1.1) are denoted by $\hat{\beta}_0, \hat{\beta}_1, \ldots, \hat{\beta}_p$. The estimated regression equation then becomes

$$\hat{Y} = \hat{\beta}_0 + \hat{\beta}_1 X_1 + \hat{\beta}_2 X_2 + \cdots + \hat{\beta}_p X_p. \qquad (1.5)$$

A *hat* on top of a parameter denotes an estimate of the parameter. The value $\hat{Y}$ (pronounced as *Y-hat*) is called the *fitted* value. Using (1.5), we can compute n fitted values, one for each of the n observations in our data. For example, the ith fitted value $\hat{y}_i$ is

$$\hat{y}_i = \hat{\beta}_0 + \hat{\beta}_1 x_{i1} + \hat{\beta}_2 x_{i2} + \cdots + \hat{\beta}_p x_{ip}, \quad i = 1, 2, \ldots, n, \quad (1.6)$$

where $x_{i1}, \ldots, x_{ip}$ are the values of the p predictor variables for the ith observation.

Note that (1.5) can be used to predict the response variable for any values of the predictor variables not observed in our data. In this case, the obtained $\hat{Y}$ is called the *predicted* value. The difference between fitted and predicted values is that the fitted value refers to the case where the values used for the predictor variables correspond to one of the n observations in our data, but the predicted values are obtained for any set of values of the predictor variables. It is generally not recommended to predict the response variable for a set of values of the predictor variables far outside the range of our data. In cases where the values of the predictor variables represent future values of the predictors, the predicted value is referred to as the *forecasted* value.

1.4.7 Model Criticism and Selection

The validity of a statistical method, such as regression analysis, depends on certain assumptions. Assumptions are usually made about the data and the model. The accuracy of the analysis and the conclusions derived from an analysis depend crucially on the validity of these assumptions. Before using (1.5) for any purpose, we first need to determine whether the specified assumptions hold. We need to address the following questions:

1. What are the required assumptions?
2. For each of these assumptions, how do we determine whether or not the assumption is valid?
3. What can be done in cases where one or more of the assumptions does not hold?

The standard regression assumptions will be specified and the above questions will be addressed in great detail in various parts of this book. We emphasize here that validation of the assumptions must be made *before* any conclusions are drawn from the analysis. Regression analysis is viewed here as an *iterative* process, a process in which the outputs are used to diagnose, validate, criticize, and possibly modify the inputs. The process has to be repeated until a satisfactory output has been obtained. A satisfactory output is an estimated model that satisfies the assumptions and fits the data reasonably well. This iterative process is illustrated schematically in Figure 1.1. This dynamic iterative regression process can be implemented with the aid of the flowchart in Figure 1.2. This process will be illustrated by examples in later chapters.

1.4.8 Objectives of Regression Analysis

The explicit determination of the regression equation is the most important product of the analysis. It is a summary of the relationship between Y (the response variable) and the set of predictor variables $X_1, X_2, \ldots, X_p$. The equation may be used for several purposes. It may be used to evaluate the importance of individual predictors, to analyze the effects of policy that involves changing values of the predictor variables, or to forecast values of the response variable for a given set of predictors. Although the regression equation is the final product, there are many important by-products. We view regression analysis as a set of data analytic techniques that are used to help understand the interrelationships among variables in a certain environment. The task of regression analysis is to learn as much as possible about the environment reflected by the data. We emphasize that what is uncovered along the way to the formulation of the equation may often be as valuable and informative as the final equation.

1.5 SCOPE AND ORGANIZATION OF THE BOOK

This book can be used by all who analyze data. A knowledge of matrix algebra is not necessary. We have seen excellent regression analysis done by people who have no knowledge of matrix theory. A knowledge of matrix algebra is certainly very helpful in understanding the theory. We have provided appendices which use matrix algebra for readers who are familiar with that topic. Matrix algebra permits expression of regression results much more compactly and is essential for the mathematical derivation of the results.

Lack of knowledge of matrix algebra should not deter anyone from using this book and doing regression analysis. For readers who are not familiar with matrix algebra but who wish to benefit from the material in the appendices, we recommend reading the relatively short book by Hadi (1996), *Matrix Algebra as a Tool*. We believe that the majority, if not all, of our readers can read it entirely on their own or with minimal assistance.

There are no formal derivations in the text and readers interested in mathematical derivations are referred to a number of books that contain

SCOPE AND ORGANIZATION OF THE BOOK   17

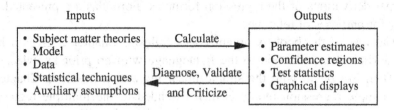

Figure 1.1 A schematic illustration of the iterative nature of the regression process.

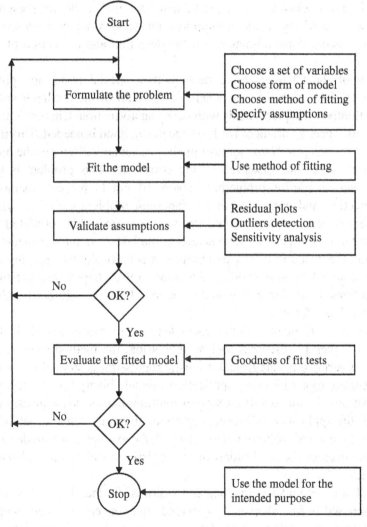

Figure 1.2 A flowchart illustrating the dynamic iterative regression process.

formal derivations of the regression formulas. Formulas are presented, but only for purposes of reference.

The rest of the book is organized as follows: Chapter 2 gives a brief but sufficient introduction to the R language with no prior knowledge of R. Then, we begin with the simple linear regression model in Chapter 3. The simple regression model is then extended to the multiple regression model in Chapter 4. In both chapters, the model is formulated, assumptions are specified, and the key theoretical results are stated and illustrated by examples. For simplicity of presentation and for pedagogical reasons, the analysis and conclusions in Chapters 3 and 4 are made under the presumption that the standard regression assumptions are valid. Chapter 5 addresses the issue of assumptions validation and the detection and correction of model violations.

Each of the Chapters 6–12 deals with a special regression problem. Chapter 6 deals with the case where some or all of the predictor variables are qualitative. Chapter 7 deals with data transformation. Chapter 8 presents situations where a variant of the least squares method is needed. This method is called the *weighted least squares* method. Chapter 9 discusses the problem that arises when the observations are correlated. This problem is known as the *autocorrelation* problem. Chapters 10 and 11 presents methods for the detection and correction of an important problem called *collinearity*. Collinearity occurs when the predictor variables are highly correlated.

Chapter 12 presents variable selection methods – computer methods for selecting the best and most parsimonious model(s). Before applying any of the variable selection methods, we assume in this chapter that questions of assumptions validation and model violations have already been addressed and settled satisfactorily.

Chapters 3–12 dealt with the case where the response variable is quantitative. Chapter 13 discusses logistic regression, the method used when the response variable is categorical. Logistic regression is studied because it is an important tool with many applications. Beside binary logistic regression, we have now included a discussion of multinomial logistic regression. This extends the application of logistic regression to more diverse situations. The categories in some multinomial are ordered, for example, in attitude surveys. We also discuss the application of the logistic model to ordered response variables.

The book concludes with Chapter 14 entitled Further Topics. Two topics are discussed in this chapter. One extends the concept of linear models so that regression and logistic models are all viewed as special cases of the linear model. This extends the range of applications of linear models to more diverse situations. We also discuss Poisson regression, often used to model count data. A brief discussion of robust regression with illustrative examples is also given in this chapter.

We recommend that the chapters be covered in the same sequence as they are presented, although Chapters 6–13 can be covered in any order after Chapter 5, as long as Chapter 10 is covered before Chapter 11, and Chapter 8 is covered before Chapters 13 and 14.

EXERCISES

1.1 Classify each of the following variables as either quantitative or qualitative. If a variable is qualitative, state the possible categories.

(a) Geographical region   (b) Number of children in a family
(c) Price of a house   (d) Race
(e) Temperature   (f) Fuel consumption
(g) Employment rate   (h) Political party preference

1.2 Give two examples in any area of interest to you (other than those already presented in this chapter) where regression analysis can be used as a data analytic tool to answer some questions of interest. For each example:

(a) What is the question of interest?
(b) Identify the response and the predictor variables.
(c) Classify each of the variables as either quantitative or qualitative.
(d) Which type of regression (see Table 1.8) can be used to analyze the data?
(e) Give a possible form of the model and identify its parameters.

1.3 In each of the following sets of variables, identify which of the variables can be regarded as a response variable and which can be used as predictors? (Explain)

(a) Number of cylinders and gasoline consumption of cars
(b) SAT scores, grade point average, and college admission
(c) Supply and demand of certain goods
(d) Company's assets, return on a stock, and net sales
(e) The distance of a race, the time to run the race, and the weather conditions at the time of running

(f) The weight of a person, whether or not the person is a smoker, and whether or not the person has a lung cancer

(g) The height and weight of a child, his/her parents' height and weight, and the gender and age of the child.

1.4 For each of the sets of variables in Exercise 1.3:

(a) Classify each variable as either quantitative or qualitative.

(b) Which type of regression (see Table 1.8) can be used in the analysis of the data?

CHAPTER 2

A BRIEF INTRODUCTION TO R

2.1 WHAT IS R AND RSTUDIO?

R is a universal language specially suited for statistical computing and graphics, which makes it of interest to researchers and practitioners in Mathematics, Statistics, and Data Science fields. R is also a collection of programs (called functions), which are organized in different but specialized packages (or libraries). These packages are written by professional developers who provide comprehensive collection of help pages that explains every element in R.

Example of packages are stats for statistical calculations and random number generation, DescTools for descriptive statistics, olsrr for regression models, and nlme for linear and nonlinear mixed effects models. Examples of functions include lm() for fitting linear models and glm for generalized linear models. R is an open-source and open-ended software, where continual downloads and updates are made available all for free. If you do not find the already-made programs that you need, you can write your own programs in R. Several other advantages of R can be read in the article by Simplilearn at https://www.simplilearn.com/what-is-r-article. Who could ask for anything more!

Regression Analysis By Example Using R, Sixth Edition. Ali S. Hadi and Samprit Chatterjee
© 2024 John Wiley & Sons, Inc. Published 2024 by John Wiley & Sons, Inc.
Companion website: www.wiley.com/go/hadi/regression_analysis_6e

RStudio is an integrated development environment (IDE) for R. It is an interactive desktop-friendly environment to run R. R is compatible with several platforms including Windows, Macintosh, UNIX, and Linux. There are some differences among these platforms, but fortunately they are not a lot. In this book we use the Windows platform simply because we are Windows users.

This chapter requires no prior knowledge of R. It gives a brief but, what we believe to be sufficient introduction to R that enables you to use R to carry out the regression analysis computations as well as the graphical displays presented in this edition of the book. To help you out, we provide you with all the necessary R code.

2.2 INSTALLING R AND RSTUDIO

To be able to work with R or RStudio, you need first to install R then RStudio. To install R please do the following:

1. Open your Browser and visit the CRAN (Comprehensive R Archive Network) Website: https://cran.r-project.org/
2. Select the version for your operating system (e.g., Windows, Mac OS, or Linux)
3. Click on the link for the latest version to start downloading R.
4. Once downloading the installer file is done, you need to find where the file is downloaded.
5. Click on that file to start the installation process and follow through different choices to finish the installation. Selecting default options are recommended. We also recommend selecting a short cut on the Desktop.

To install RStudio please do the following:

1. Go to the Website: https://www.rstudio.com/products/rstudio/download/.
2. Scroll down and click on the "Download RStudio" button and follow the simple instructions that follow.
3. Click on the version for your operating system to start downloading RStudio.

4. Once downloading the installer file is done, you need to find where the file is downloaded.

5. Click on that file to start the installation process and follow through different choices to finish the installation. Selecting default options are recommended. We also recommend selecting a short cut on the Desktop.

Now you are ready to use R or RStudio. We start with R. You can open R like any other program on your computer, usually by clicking its icon or shortcut on your desktop. If you encounter problems or if you have questions about R, please read the frequently asked questions at https://cran.r-project.org/faqs.html. Once you learn R, it would not be difficult for the users to learn RStudio on their own or with some help from online resources.

2.3 GETTING STARTED WITH R

This chapter assumes no previous knowledge of R, but let us start at the outset with a disclaimer. The objective of this chapter is not to make you an R programmer, but rather, to provide you with sufficient information to enable you to apply the methods discussed in this book to your datasets using R. You can, however, take advantage of this boost to sharpen your programming skills on your own by taking one or more of the freely available sources on the Internet for learning R.

2.3.1 Command Level Prompt

Start R by double-clicking on its icon. Whenever you open R, you will see an introductory message displayed in a window (known as the R console) followed by the prompt symbol ">". This symbol means that R is ready for you to issue R commands. The first line in the introductory message is the current R version. The first time you install R, your installed version will be the most recent version. But from time to time a new version of R becomes available. If you wish to update your version to the most recent one, type or copy and paste the following two lines after the prompt symbol ">":

```
install.packages("installr")
installr::updateR(T)
```

Because R is available at many sites around the world, you will be asked to select a CRAN mirror for use in this session. You could choose any site you like (e.g., 0-Cloud [https]) then click the Enter (return) key. If your current version of R is still the most recent version, you will see the message: "No need to update"; otherwise, the new version of R will start to download.

Now you are in an R Workspace that you will use for your data analysis. You need to create objects to store information. The symbols "<–" are used create the object on its left in your R Workspace and assign to it the values on its right. For example, if you type:

```
height <- 1.5
```

an object named "height" will be created and assigned the value 1.5. You will see no output or response. But you will see the prompt symbol ">" again. Now if you wish to display the contents of the object height type print(height) or simple type height and click the Enter key:

```
print(height)
[1] 1.5
height
[1] 1.5
```

R will display the contents of the object height as shown in the second and fourth lines. The height has one value (one line as indicated by [1]). Note that you could use "=" instead of "<–", that is, x = 5 instead of x <– 5, but the use of "=" is not recommended for beginners.

2.3.2 Calculations Using R

R could be used as a calculator, but of course this is not the intended purpose of R. For example if you type:

```
2 + 3
[1] 5
```

you will see the output 5 in the second line above. Similarly, if you type the first three lines below:

```
x <- 9
y <- 16
(x + y)^0.5     # same as sqrt(x+y)
[1] 5
```

then in the above R code, two objects x and y are created and assigned the values 9 and 16, respectively. In the third line, their sum is raised to the power 0.5, which is equivalent to taking the square root of their sum. The result of this operation is 5 as indicated by the output in the last line above. The third line in the above code deserves several comments. First it contains the hashtag character, #. Anything written after this symbol will be ignored by R. It is simply a comment written for the sake of the reader. Second, sqrt() is a command (also known as a function), which is reserved for the computation of the square root of the argument inside the parenthesis. Thus, for example sqrt(9) will return [1] 3. Finally, the symbol "^" will raise the object on its left to the power on its right. For example, 5^2 will return 25.

Note also that the three lines of code above can be written on the same line using the semicolon separator ";" as follows:

```
x <- 9; y <- 16; (x + y)^0.5
[1] 5
```

Thus, a semicolon indicates the end of the line. This is tempting but not recommended because it may make reading the code more difficult.

The assignment operator can also be replaced by the equal sign "=". So, for example, x <- 9 and x = 9 are equivalent. The operators "<-" and "=" are called the left assignment. The right assignment is "->", that is, the above two left assignments can be replaced by 9 -> x, which reads as the number 9 is assigned to the object x. Thus, these three operators are used to assign values to objects.

When carrying out arithmetic calculations, R uses the performs the calculations in the following order: Parentheses, Exponentials, Division, Multiplication, Addition, and Subtraction. This is rule is easy to remember by its abbreviated form, the PEDMAS rule. When operations have the same priority, then they are evaluated from left to right.

Examples:

```
2 + 3 * 4      # returns [1] 14
(2 + 3) * 4    # returns [1] 20
3/4*2          # returns [1] 1.5
3*4/2          # returns [1] 6
```

Each of the above objects x and y is assigned to a single value. We can create objects with more than one value using the function c(), which is short for concatenate. If we wish to assign an object age to three values 7, 4, and 9, we issue the comment:

```
age <- c(7, 4, 9)
```

The values inside the parentheses must be separated by commas (not spaces) and are written between double or single quotes. Thus, issuing the command

```
color <- c("black", "red", "green", "blue")
```

creates an object named color consisting of four character elements. An object with more than one value is called a vector or one-dimensional array.

The object age is a numeric vector and the object color is a character vector. To display the object color, type color or print(color), you will get: [1] "black" "red" "green" "blue". If you wish to print a character object without the quotation marks use: print(noquote(color)) or just noquote(color) and you will get: black red green blue.

Sometimes we need to create sequences of numbers. The command seq(from = start, to = end, by = step) creates a sequence from start to end with increment = step. For example,

```
x <- seq(from =5, to =15, by = 2); x
[1]  5  7  9 11 13 15

y <- seq(from =-3, to =5, by = 1); y
[1] -3 -2 -1  0  1  2  3  4  5
```

A simpler way to obtain this last sequence is to use the Colon (:) operator as follows:

```
y <- -3:5; y
[1] -3 -2 -1  0  1  2  3  4  5
```

This form is used whenever the increment is 1. If you wish to assign an object and display its value, put the expression between two parentheses. For example, use

```
(v <- -3:5)
[1] -3 -2 -1  0  1  2  3  4  5
```

to create v and print its contents as seen above.

Another interesting feature in R is called recycling. For example, if you want to do some arithmetic operations with vectors of unequal length, then R will repeat the shorter vector until the two vectors become equal in length then do the carry out the operation. To illustrate consider the following vectors: x <– 1:3 and y <– 1:8. If you try to add the two vectors (x + y), R will first add 5 more elements to x (the shorter vector) before performing the addition. The completed vector becomes c(1, 2, 3, 1, 2, 3, 1, 2), that is, R recycled the elements of x until it becomes of length 8 (the length of the longer vector). Therefore, x + y, will return

```
[1]  2  4  6  5  7  9  8 10
Warning message:
In x + y: longer object length is not a multiple of shorter
    object length
```

Luckily, R gives you a warning. If this is what you want to do, just ignore the warning and continue; otherwise, revise your vectors. Be careful, however, because the above error suggests that if the length of one vector is a multiple of the length of the other you will not get any warning. For example, if we replace x above by x <- 2:3, for example, which has two elements, in this case x + y will return

```
[1]  3  5  5  7  7  9  9 11
```

with no warnings! This is because the length of the longer vector (8) is 4 times (multiple) the length of the shorter one (2). Another example of recycling occurs when you try to append two vectors using either rbind(x,y) or cbind(x,y). The first command will put y below x (row bind) and the second will put y to the right of x (column binding). Because a and y are not of equal length, R will recycle the shorter one. For example, here is what you get when you use

```
> rbind(x,y)
  [,1] [,2] [,3] [,4] [,5] [,6] [,7] [,8]
x    2    3    2    3    2    3    2    3
y    1    2    3    4    5    6    7    8
```

Again, no warning for the same reason cited above.

2.3.3 Editing Your R Code

You may need to edit a command line that you have previously typed in the console. Instead of retyping it, you can recall and edit it. For this purpose, click on the Up Arrow ↑ key as many times as needed until you find the line you wish to edit. You can also use the Down Arrow ↓ to recall previous lines. Use the Right Arrow → and/or the Left Arrow ← to move right and/or left in the line to edit it. Admittedly, this is an unfriendly and cumbersome way of editing your R code. Section 2.9.1 provides a much easier way not just to edit your code but also to execute and save them in files known as Script files.

When typing a line of code sometimes you click on the Enter key by mistake. Other times the line is too long to fit in one line on the screen. In these cases, R will display a + prompt instead of the usual command prompt >. This mean that R is expecting you to complete the line of code. You can either complete the line or hit Escape to start over again.

2.3.4 Best Practice: Object Names in R

Data values are stored in objects. These objects have names. Object names are chosen by you. Object names have to start with a letter (lower- or upper-case) and contain no spaces, special characters (such as % or $), or arithmetic operators (such as +, –, or *), but they may contain one or more digits from 0 to 9. To avoid possible confusion, names should not be objects that are predefined in R (key words) such as q, c, t, matrix, factor, levels, and nlevels. Note that q, c, and t are single letter commands (function). R actually does not allow using some key words as names. For example, you will get error messages if you try NA <– 6, NULL <– 2, or TRUE <– 5. If you are in doubt about whether a word is a key word, just type it after the prompt symbol and click Enter. If you get an error message saying object not found, then the word is not a key word; otherwise, it is a key word and it should be avoided.

We recommend that you choose short but meaningful and informative names that describe what each object contains. If you wish to use a name with spaces, use underscores, periods, or any combination of them, as a substitute for spaces. For example, instead of using "New York", use "New_York", "New.York", or other alternatives such as "NewYork". It is important to note here that R is case sensitive. Thus, for example, "New" and "new" are two distinct object names! Accordingly, the recommendation here is to use meaningful words that describe what is stored, use only lower case, and use underscores or periods as a substitute for spaces, and avoid using R key words.

2.4 DATA VALUES AND OBJECTS IN R

2.4.1 Types of Data Values in R

The data values assigned to objects have types. The basic data types in R include: character, numeric, logical, NA, NULL, Inf, –Inf, and NaN. These are explained below:

1. Character: Character data are text data. They are included between double or single quotes such as "yes", "$10", "12", and "false". The function nchar(x) returns the number of characters in every element (component) of the object x. For example,

```
color <- c("black", "red", "green", "blue")
nchar(color)
[1] 5 3 5 4
```

The function as.character(x) takes a numeric object x and returns a character object. For example,

```
x <- c(3, 2.1, 4)
y <- as.character(x)
y
[1] "3"   "2.1"   "4"
```

Conversely, the function as.numeric(x) takes a character object x and tries to convert it to a numeric object. For example, as.numeric(y), returns [1] 3.0 2.1 4.0, which is the same as x above.

You can check the type of object using the class() command. For example, class(x) returns numeric but class(y) returns character. Other examples of character objects are letters and LETTERS, which are reserved for the lowercase and uppercase alphabet letters, respectively. Thus, for example, if you type letters[1:3], you will get the first 3 lowercase letters: [1] "a" "b" "c".

2. Numeric: Numbers including the ones written in scientific notation such as 1.234e2 = 123.4 and 1.234e–2 = 0.01234. Also, the values 21, 32.4, and 1.23E3 are numeric but 2,000, $2000 and 25% are not because the first has a comma, the second has a $ sign, and the third has a % sign. Examples of numeric data are Age (in years), Annual Income (in $1000), Height (in centimeters), and Distance (in kilometer).

Special cases of numeric data include double, integer, and complex numbers. Double data are numbers with decimal points and integer data are whole numbers. Age is an example here. R will treat ID as an integer variable. Double is the default for numbers. To explicitly create integers, add an L to each element, for example, c(2L, 4L). You could also use the command as.integer(). For example,

```
x <-c(1.2,3.4, 5); as.integer(x)
[1] 1 3 5
```

which is equivalent to the integer part of a number (that is, round down). In R, the integer part of a number is obtained by floor(x) but rounding up is given by ceiling(x), which gives [1] 2 4 5. Complex data types will not be considered in this book but for information use help(complex).

3. Logical: Takes only one of two values: TRUE or FALSE or, equivalently, T of F, respectively. These values are not included in quotation marks; otherwise, they will be character data. For example, the response of a student to 5 True–False questions could be entered in R Workspace as

```
x <- c(TRUE, FALSE, FALSE, TRUE, FALSE) # or as
y <- c("Yes", "No","No", "Yes", "No")
```

where "Yes" means True and "No" means False. The object x is logical, whereas the object y is character. In a logical object, the value TRUE evaluates to 1 and FALSE to 0. Consequently, if you issue the command sum(x), you will get 2 but if you use the command sum(y), you will get an error indicating that the object y is not numeric.

4. NA: Not available, e.g., a missing value. Missing values occur, for example, when a respondent left the answer to a question in a questionnaire blank. These values are usually replaced by NA. NAs also occur when we use the function as.numeric() to change character values to numbers whenever possible. If it is not possible, R returns NA with a warning message. For example, the following object is character but you can enforce (coerce) it to be numeric using as.numeric():

```
x <- c("2", "3.2", "\$10", "1.2E-1", "10%")
as.numeric(x)
[1] 2.00  3.20    NA  0.12    NA
Warning message:
NAs introduced by coercion
```

The function is.na() gives a logical vector with TRUE for the missing elements and FALSE otherwise. The function anyNA() returns TRUE if the vector contains any missing values and FALSE otherwise.

5. NULL: An empty value z = NULL or z = { } indicates that the object z is empty (an object with no elements).

6. Inf and **–Inf** are reserved for ∞ and $--\infty$, respectively. These values usually appear whenever we divide nonzero numbers by 0. Examples include: 1/0, –1/–0, and sqrt(Inf) all give Inf, whereas –1/0 and 1/–0 give –Inf.

7. NaN: Not a Number, usually appear as a result of a wrong arithmetic operation such as the square root of a negative number, the log of a negative number, 0/0, or Inf/Inf. Note that Inf, –Inf, and NaN are considered to be numeric values, whereas NA is logical.

8. Date and Time: Date and time are frequently encountered in real data and they come in so many different formats, for example, 1949-07-30, 07/30/49, 13:05:20, 30jul1949, and Sat, 10 Jul 1949. What adds insult to injury is that implementation of the functions that deal with dates and times may be platform dependent. Unfortunately, a reasonable, not to mention full, coverage of this topic will extend to several pages. This chapter has already gone much fatter than we anticipated, and I have to make the difficult decision not to include this topic in this chapter. I refer the interested readers to the documentation of the package lubridate as well as the books and online cites coverage of the topic (see Section 2.10). But just to whet your appetite, I leave you with two simple easy-to-read functions, the first computes the elapsed time between two events and the second computes the day (of the week) in which you were born and the number of days that you have lived to date.

```
elapsed.time <- function(n=1000){
  start.time <- Sys.time() # Returns the current time
  x <- rnorm(n) # Generates a sample of size n
               # from the standard Normal distribution
```

```
    median(x)      # Computes the median of the values in x
    end.time <- Sys.time() # Returns the end computations
    # Concatenate and print on the screen
    cat("Elapsed time:",end.time - start.time," seconds.\n")
}
elapsed.time(n=100000)

birthday = function(bd){
# Converts date from character to calendar date.
bd = strptime(bd, "%d-%b-%Y")
wd <- weekdays(bd) # Returns the weekday
cat("You were born on",wd,"and today you are",Sys.time()
    - bd,"days,\n")
cat("which is equivalent to",round((Sys.time() - bd)/365.25,2),
    "years.\n")
}
birthday("30-July-1949")
```

Try them out!

2.4.2 Types (Structures) of Objects in R

The objects to which the above values are assigned also have types (structures). There are many objects structures in R. The ones we use in this book are:

1. Atomic Vector or Array: One-dimensional arrays containing one or more values. All elements of an atomic vector are of the same type. The vectors x and y defined above are numeric and character vectors, respectively. Note that if one or more elements in a vector is character, the entire vector will be a character vector. For example, if we type the following line after the prompt symbol:

```
v <- c(2, 5, "a", 1); v
[1] "2" "5" "a" "1"
```

you will see that v contains four character values indicated by the double quotes around each element. Note that a vector can take any of the above data types as long as all of its elements are of the same type. Individual elements of a vector can be extracted using brackets. For example, the first element of v above is v[1] and the second is v[2]. Also, we can extract subsets of v such as v[1:2], which gives [1] "2" "5", and v[c(1,3)] which gives [1] "2" "a".

2. Matrix: In R, a vector is one-dimensional array, but a matrix is a two- or more dimensional array. Like an atomic vector, all elements of a matrix are of the same type. For example, the following is a two-dimensional matrix that contains 2 rows and 3 columns:

$$A = \begin{bmatrix} 3 & 1 & 0 \\ 2 & 4 & 5 \end{bmatrix}. \tag{2.1}$$

To create the matrix A in R, we first create a vector v, say, then use v to construct the matrix as follows:

```
v <- c(3, 2, 1, 4, 0, 5)
A <- matrix(v, nrow = 2, ncol = 3)  # Create the matrix A
A                                    # Display A
     [,1] [,2] [,3]
[1,]   3    1    0
[2,]   2    4    5
```

The matrix A has two rows and 3 columns as indicated by the nrow and ncol above. The elements of v are used to fill the matrix A by columns. If you wish to fill the matrix by rows, use

```
B <- matrix(v, nrow = 2, ncol = 3, byrow = TRUE)
B
     [,1] [,2] [,3]
[1,]   3    2    1
[2,]   4    0    5
```

The matrices A and B have the same dimensions, but of course they are not equal. The command dim(A) gives the dimension of A (the number of rows and the number of columns). The command t(A) will transpose the matrix A. Thus,

```
t(A)
     [,1] [,2]
[1,]   3    2
[2,]   1    4
[3,]   0    5
```

which has 3 rows and 2 columns. Like vectors, matrices are atomic. Hence, if one or more elements in a matrix is character, the entire matrix will be a character matrix.

We can extract subsets of a matrix using brackets. The *ij*-th element of a matrix A is A[i,j], where the value before the comma is the row number and the value after the comma is a column number. If no values are before the comma, it means we want all the rows. Similarly, if no values are after the comma, it means we want all the columns. Thus, for example, A[2, 3] is 5, the element in the second row and third column, A[,3] is the third column [1] 0 5, and A[2,] is second row [1] 2 4 5. Also, A[1:2, 2:3] gives

```
     [,1] [,2]
[1,]   1    0
[2,]   4    5
```

3. Factor: Categorical variables occur frequently in real-life applications. The variables Gender and Major are examples of categorical variables. Other examples include Grade and Faculty Rank. Categorical variables have unique categories. If the categories in a categorical variable have natural order, they are called ordinal categorical variables. For example, Income Level and Faculty Rank are ordinal categorical variables, whereas Gender and Major are not because their categories are different from each other but neither one is larger than the others. In R, categorical variables are called factors and the categories are called levels. Let us create a categorical variable, which consists of the income level of five randomly chosen people, and check its type:

```
income <- c("medium", "high", "high", "low", "medium")
class(income)
[1] "character"
```

Notice that income is a character object. We therefore need to explicitly tell R that income is a factor object. For this purpose, we use factor() as follows:

```
(income <- factor(income))
[1] medium high   high   low    medium
Levels: high low medium
```

Now class(income) returns "factor". Note that when we put the first statement above between two parentheses, it causes the printing of whatever inside the parentheses. To find out what are the levels of a factor and how many they are, use the commands levels() and nlevels():

```
levels(income)
[1] "high"    "low"     "medium"
nlevels(income)
[1] 3
```

Note that R orders the levels ordered alphabetically by default. For instance, levels of the levels income above were given as [1] "high" "low" "medium" even though the first element in income is medium. If you wish to use different ordering other than alphabetical, use

```
income <- factor(income,levels=c("low","medium",
"high"))
levels(income)
[1] "low"    "medium"    "high"
```

The above command changed the order of the levels within the factor income. But R does not know if the factors in come in ordinal factors, that is, R does not know that low < medium < high. To make income an ordinal factor use:

```
income1 <- factor(income, ordered=TRUE)
> income1
[1] medium high   high    low    medium
Levels: low < medium < high
```

Now object income1 is an ordinal factor. To check if a factor is ordered use

```
is.ordered(income)
[1] FALSE
is.ordered(income1)
[1] TRUE
```

4. Data Frames: A data frame is very much like a matrix, in the sense that rows represent observations and columns represent variables. But it contains some more advanced attributes (features). Data frames group different types of atomic vectors of equal lengths together into a two-dimensional array, where each vector becomes a column in the array. Therefore, each column of a data frame can contain a different type of data; but since columns are atomic vectors, then the elements of every column must be of the same type. Accordingly, we can combine different data types into one data frame object. This is one reason why most data sets in R are stored in objects of the type data fame. The row names of a data frame must be unique. If not given by the argument row.names, the rows are numbered from 1 to the number of rows. The columns of a data frame should be unique and follow the rules for names in R we mentioned in Section 2.3.4.

A data frame is a very important data type in R. It is the preferred data structure for most tabular data. Data frames have additional attributes, but we will mention them as needed. To see the difference (at least in the appearance), let us create a matrix and data frame objects as follows:

```
M <- matrix(c(2:4, F, T, F, "F", "T", "F"),nrow=3,ncol=3)
print(M)
     [,1] [,2]      [,3]
[1,] "2"  "FALSE"   "F"
[2,] "3"  "TRUE"    "T"
[3,] "4"  "FALSE"   "F"
df <- data.frame(x=2:4, y=c(F, T, F), z=c("F", "T", "F"))
print(df)
  x    y    z
1 2    FALSE F
2 3    TRUE  T
3 4    FALSE F
```

Notice the way the rows and columns of M and df are labeled. All elements of M are characters because the third column is character. The (abbreviated) logical values T and F were converted to TRUE and FALSE, but then they were made character values in M. The third column in both M and df is a character object, but the quotation marks are removed from df. In df, x is numeric (integer in this case), y is logical, and z is character. Thus df gives a better representation of the data.

The columns in a data frame can be converted to a matrix using data.matrix(). The use of data.matrix() produces a matrix of numeric values. For example, consider converting the above df to a matrix as seen below:

```
M1 <- data.matrix(df)
print(M1)
     x y z
[1,] 2 0 1
[2,] 3 1 2
[3,] 4 0 1
```

We see that numeric columns in df remain as numeric in M1, logical columns are converted to their equivalent digits (0 for FALSE and 1 for TRUE), factor columns are converted to their integer representation, and character columns are first converted to factor variables, which are in turn converted to their integer representation. Accordingly, the conversion of a data frame to matrices is as straightforward as the conversion of matrices to data frames.

Table 2.1 shows some useful R functions/commands for dealing with data types, where x stands for a vector, a matrix, data frame, or a list.

5. Lists: An R "list" is a list of objects not necessarily of the same type and/or the same dimensions. For example, s <- list(x = c(3,5,7), y = c("A","B"), z = 5) creates a list named s, which contains three objects x, y, and z, where the objects x, y, and z are assigned values of different type and size. Thus, lists group together R objects. A list does not print to the console like a vector. Instead, each element of the list starts on a new line. You can display the contents of s by simply typing s and click the Enter key. You will get every element of the list preceded by the $-accessor as follows:

```
$x
[1] 3 5 7
$y
[1] "A" "B"
$z
[1] 5
```

Table 2.1 Some Useful R Functions for Information About Data

Function	Description
head(x,n)	Displays the first n elements of a vector or n rows of a matrix or a data frame. The default value of n is 6
tail(x,n)	Displays the last n elements of a vector or n rows of a matrix or a data frame. The default value of n is 6
class(x)	Returns type of the object x
length(x)	Returns the number of elements in a vector or a list and the number of columns in a data frame
nrow(x)	Returns the number of rows
ncol(x)	Returns the number of columns
dim(x)	Dimensions: nrow(x) and ncol(x)
str(x)	Useful information about the structure of data frame
names(x)	To get or set the names of an object
rownames(x)	To get or set the row names of a matrix-like object
colnames(x)	To get or set the column names of a matrix-like object
levels(x)	Returns levels of a factor object in order
nlevels(x)	Returns the number of levels of a factor object

This Table will be updated on the Book's Website when the need arises.

If the elements of a list are named (like s above), then you can also display one of more of its elements by using the $ or the [] accessors. For example,

```
s$x     # or equivalently s[1] returns [1] 3 5 7
s$y     # or equivalently s[2] returns [1] "A" "B"
s$z     # or equivalently s[3] returns [1] 5
```

The content of elements of a list can also be accessed by using double square brackets. For example

```
s[[1]] # equivalently s$x or s[1], returns [1] 3 5 7
s[[2]] # equivalently s$x or s[2], returns [1] "A" "B"
s[[3]] # equivalently s$x or s[3], returns [1] 5
```

Thus, double-bracketed indexes [[]] refer to element of the list, whereas single-bracket indexes [] refer to subelement of an element of the list.

So objects in the database can be accessed by simply giving their names.

A third way of getting a direct access to a name of an object (data frame and list) is to use attach(). For example, attach(s) will make the components of the list s available by just typing the name of its element. Accordingly s[[1]], s$x, and x will result in the same output. Attach() is a more convenient way than the other two alternatives to access elements of objects, but it could cause potential confusion with currently used object names in the Workspace. Accordingly, it is recommended using detach() after you no longer need the elements of the object or better not to use attach() at all.

To display the names of the elements of the list s, use

```
names(s)
[1] "x" "y" "z"
```

Unlike atomic vectors, the contents of a list are not restricted to a single type and can consist of any mixture of data types. A list can also be thought of as a special type of non-atomic vector, where each element can be of different type or different size. Lists do not have to be rectangular as they can contain elements of different structures or lengths (e.g., a list can contain data frames or other types of objects). But in order for the list to be changed to a data frame, the length of each element has to be the same.

We should also mention here that there are other useful object structures in R, such as tibble, but we make no use of them in this book.

2.4.3 Object Attributes

An attribute is some information attached to an R object. Examples of attributes are name, length, dimension (dim), and class. Some R functions will check for specific attributes.

You can find out if an object has names using the command names(). You can also use names to set, change, or remove the names of the object. The following code illustrates this using the list s defined above:

```
names(s)
[1] "x" "y" "z"
names(s) <- c("A", "B", "C")
names(s)
[1] "A" "B" "C"
names(s) <- NULL
names(s)
NULL
```

Other examples of attributes are length(), dim(), and class, where you can use to get, set, or change these attributes. Try them out.

2.4.4 Testing (Checking) Object Type

If you wish to know the type of an object, use class() to find out the type. For example,

Table 2.2 Some Useful R Functions for Testing Object Type

Function	Description
all(x)	Returns TRUE if all entries in the logical vector x are TRUE
all.equal(x, y)	Returns TRUE if x and y are nearly equal
any(x)	Returns TRUE if any of the entries in the logical vector x is TRUE
is.character(x)	Returns TRUE if x is a character object
is.data.frame(x)	Returns TRUE if x is a data frame object
is.double(x)	Returns TRUE if x is double
is.factor(x)	Returns TRUE if x is a factor object
is.integer(x)	Returns TRUE if x is a character object
is.list(x)	Returns TRUE if x is a list object
is.logical(x)	Returns TRUE if x is a logical object
is.matrix(x)	Returns TRUE if x is a matrix
is.na(x)	Returns TRUE if x[i] is NA
is.nan(x)	Returns TRUE if x[i] is NaN
is.null(x)	Returns TRUE if x is NULL
is.numeric(x)	Returns TRUE if x is a numeric object
is.ordered(x)	Returns TRUE if the levels of a factor object are ordered
is.unsorted(x)	Returns FALSE if values in x are sorted
is.vector(x)	Returns TRUE if x is an atomic vector
near(x, y)	Returns TRUE if x and y are nearly equal (near() is in dplyr package)

This Table will be updated on the Book's Website when the need arises.

```
class(s)
[1] "list".
```

But if you wish to test whether an object of a specific type, Table 2.2 provides some useful functions for this purpose arranged in alphabetical order. The function returns a logical value (either TRUE or FALSE) depending on whether the object is of the specified type. Note that all functions return one logical value except for is.na() and is.nan(), where they return one logical value for each element of the object.

2.4.5 Changing Object Type

You can change the type of an object to another type. For example, the lines of code

```
x <- 1:4
y <- as.matrix(x)    # creates one-column matrix
```

change the data type of x from a data frame to a matrix and store the result in an object named y. The type of y is a matrix.

```
is.matrix(x)    # gives FALSE   (Why?)
is.matrix(y)    # gives TRUE    (Why?)
```

2.5 R PACKAGES (LIBRARIES)

2.5.1 Installing R Packages

R is based on numerous specialized packages. When you start R, the base R package is automatically downloaded. This package has a lot of what you need because it includes other packages. To see which of these packages are included in the base R, type library() and click the Enter key.

In addition to the base package, there are other R packages (also known as libraries) that you may need beyond the base package. These packages are written by different authors for different purposes. Examples are:

tidyverse: A collection of packages useful for Data Science

robustbase: Robust Statistics

robustX: Regression and Multivariate outlier identification

nlme: Linear and Nonlinear Mixed Effects Models

nnet: Neural Networks

To use a package, you first need to install it on your computer. You need access to the Internet. For example, to install the packages tidyverse and robustbase type:

install.packages("tidyverse", "robustbase")

R will then ask you to select a secure CRAN mirror. Choose any of the listed options, e.g., the first option on the list and click on OK. R will then start to install the packages tidyverse and robustbase. This may take some time depending on the size of the packages. When done, you will see the prompt symbol ">" again. The packages are now installed on your computer. Installing packages is done only once. After that whenever you need a package, you need to download it to the R Workspace. Type: library(tidyverse) or require(tidyverse). Now you are ready to use tidyverse. The downloading is done only once during an R session, but you need to do this step every time you start a new R session and you need a package.

You can do both the installation and downloading by typing:

```
if(!require("tidyverse")) install.packages("tidyverse")
```

This code will check to see if the package tidyverse is already installed on your computer. If it is already installed, it will download it into the R

Workspace. If it was not installed, it will install it and download it into the R Workspace. This is particularly useful if you are going to share your code with others.

2.5.2 Name Spaces

Sometimes different packages contain functions with the same name. Often these functions do completely different things. For example, when we download tidyverse, using library(tidyverse), we get

```
x dplyr::filter()   masks stats::filter()
x dplyr::lag()      masks stats::lag()
```

This means that the functions filter and lag exist in both packages. Now, if you use filter() or lag(), they will refer to the functions in dplyr not stats. If you wish to execute the functions in stats instead, write:

```
stats::filter()   # instead of filter()
stats::lag()      # instead of lag()
```

R follows a certain order when searching for a function in these?namespaces. You can see the order by typing: search().

2.5.3 Updating R

New and updated R packages are introduced on a regular basis. You could upgrade your current R version to the most recent one without having to install R from scratch by typing the following lines:

```
install.packages("installr")
installr::updateR(T)
```

If you already have the most recent version, you will be told so and there will be no need for the updating. If not, R will start to update your current version to the most recent one.

2.5.4 Datasets in R Packages

Many, if not all, R packages include not only R code but also datasets. Indeed, the package datasets contain numerous datasets for the R users. To examine a complete list, type library(help = "datasets"). The package datasets are included in the base R package. Accordingly, all the datasets in this package are available for immediate use. For example, one of the datasets available is the very famous mtcars. To display first 3 rows (observations) of mtcars type head(mtcars, n = 3). Similarly, to display the last 3 rows, type tail(mtcars, n = 3). In both head() and tail(), the default value for n is 6. So, for example, typing just head(mtcars) displays the first 6 rows. To find more information (help) about the mtcars you type ?mtcars or help(mtcars) and you will be directed to a page where the mtcars is fully described. Alternatively, you could use the command View(mtcars) to view the data in a spreadsheet-style data viewer. Note that View() is with upper case V. This command is suitable when your data are reasonably small in size and dimensions.

If the dataset that you wish to use is not in the base R package, you need first to install and download the package which has it. For example, a dataset named starsCYG is found in a package named robustbase. To install and download robustbase, type install.packages("robustbase") and library(robustbase). Now, the starsCYG dataset is available for immediate use. To find more information about starsCYG, type ?starsCYG and you will be directed to a page where the dataset is fully described.

2.6 IMPORTING (READING) DATA INTO R WORKSPACE

The data you want to analyze using R may reside in a file external to the R Workspace (e.g., on your computer, flash disc, or on the Internet). You could use R to read the data stored in several types of files. We describe below how to read text (also known as ASCII) files, CSV (comma separated values) files, Excel files, and online files. But first, let us talk about another best practice in R.

2.6.1 Best Practice: Working Directory

When working with R, you may need to read (import) files from a certain directory (also known as folder) into your Workspace or you may want to write (export) files to the same directory. R has its own default working directory, where all inputs come from this directory and all outputs go to this directory. To know which directory is the current working directory type getwd() after the prompt symbol and click the Enter key. You will see a line like the one below depending on your computer configurations:

    [1] "C:/Users/OneDrive/Documents"

This line gives the full path of the current working directory, which indicates that the current working directory on my computer is the folder named Documents, which is inside another folder named OneDrive, which in turn is inside a third folder named Users, and the Users folder is in the C drive. We recommend, however, that you create your own working directory, give it an appropriate and informative name, and put it in a place where you can easily find (e.g., on your Desktop or Flash disc). You keep in this directory, for example, the files containing the datasets that you want to analyze, the R script (code) that you have for this purpose, etc. The purpose of the working directory is that input to R comes from this folder and the output of R goes to the same folder.

Once you create your own working directory, you need to change the current working directory to the one you created. If you know the full path of your desired directory, then use setwd(path), where path is a character object containing the full path of your chosen folder. Alternatively, type setwd(choose.dir()), then navigate until you find your created Folder and select it. From now on, this will be your working directory until you change it to another working directory.

After you change your working directory, you may wish to clear all user-defined objects in the Workspace. First, the function ls() will list all user-created objects in the Workspace. You can clear (delete) this list by using the following command:

```
rm(list=ls())
```

Now you are in a new and empty Workspace.

To illustrate, reading data from external files, we use the Computer Repair data in the file Computer.Repair.csv that you can download from the Book's Website at http://www.aucegypt.edu/faculty/hadi/RABE6/Data6/Computer.Repair.csv

You can also open the text file Computer.Repair.csv with Excel and save it twice once as a Computer.Repair.xlsx and the other as Computer.Repair.CSV. Please put all these three files inside your current working directory.

2.6.2 Reading ASCII (Text) Files

The ASCII or Text files with extension.txt are common in practice. The file is expected to be in spreadsheet format where the rows represent observations (one observation per row) and the columns represent variables. Before you read the file, it is important to know if the first line of the file contains the variable names. If you are not sure whether the first line contains the names of the variables, you may examine the first n lines of the file before you read it using

```
read_lines("Computer.Repair.txt", n_max = 3)
   Minutes Units
1       23     1
2       29     2
```

We see from the first output line [1], that the variable names are provided in the first line of Computer.Repair.txt file. You can examine any number of lines by setting the parameter n_max to any positive integer.

To read a text file into R use the function read.table(). If the file does not contain the names of the variables in the first line, use

df <- read.table("Computer.Repair.txt")

In this case, the variables will be automatically named as V1, V2, $\cdots$, Vd, where d is the number of variables in the file. If the first line actually contains the variable names and you still use the above line, the first line of the file will be part of the data, which is wrong, of course, and the variables will be named as above.

But, if the first line contains the variable names, which is the recommended practice as is the case here, then the above line is replaced by

df <- read.table("Computer.Repair.txt", header = TRUE)

This is because the default value of the parameter header in read.table() is FALSE. This line of code will read the contents of the data in the file Computer.Repair.txt and store it in the object named df. The object name df is chosen here to signify that the object created by the read.table() is of the type data frame. But, of course, you could specify any other name for your data. The above simple syntax assumes that the file named Computer.Repair.txt is found in the current working directory; otherwise, you will need to replace the name of the file by the full file path. Note that the file name (or path) is given between quotation marks.

There are other parameters in the read.table() command. For example, if the observations have names, such as names of cities, countries, people, or even IDs, the parameter row.names is set to the column number in which the names of the observations can be found. For example, if the names of the observation are in the first column, then the above line is replaced by

df <- read.table("filename.txt", header = TRUE, row.names = 1)

This way, the first column will not be part of the data. It will serve as an identifier (ID) for the observations. If the parameter row.names is not specified, then the observations are numbered from 1 to n, where n is the number of observations in the data file. There are many other parameters in the function read.table(), some of which are set by default and some are not. To learn more about them type ?read.table, which will take you to the page where a more complete information about this command can be found.

After reading your data, it is recommended that you examine at least part of the data to make sure that you have read the data you intended to read. You can examine your data in several ways such as head(), tail(), or View() as described in Section 2.5.4.

2.6.3 Reading CSV Files

The preferable data file format is the CSV (comma separated values). To read a CSV file, with extension.csv, follow the syntax of the read.table() but using the command read.csv(). Thus for example to read a csv file with the first line containing the names of the variables use

```
df <- read.csv("Exams.csv", header = TRUE)
```

The object df here is also of the type data frame. Note that unlike read.table(), the default value for header in read.csv is TRUE. Accordingly, the above line could be simply written as df <- read.csv("Exams.csv"). If the first line in your csv file does not contain the names of the variable, then use

```
df <- read.csv("Exams.csv", header = FALSE)
```

In this case, the variables will be automatically named as V1, V2, $\cdots$, Vd, where d is the number of variables in the file. Again, it is recommended that you examine the data using head(), tail(), or View() as described in Section 2.5.4.

In addition to file name and header, the functions read.table() and read.csv() have many other useful arguments (parameters) such as stringsAsFactors, as.is, blank.lines.skip, and na.strings. Some of these parameters are set by default. For more information about please parameters type ?read.table or ?read.csv for help. You can also inquire about the arguments of any function using args(fun.name), e.g., args(read.csv) will return

```
function (file, header = TRUE, sep = ",", quote = "\"", 
    dec = ".", fill = TRUE, comment.char = "", ...) 
NULL
```

To know what these argument mean, visit the R help page for the function by typing ?read.csv.

2.6.4 Reading Excel Files

The R package readxl provides useful functions for reading Microsoft Excel files in different formats (e.g., files with extension.xls or.xlsx). You also need to examine your Excel file to see if it contains one or more sheets and if it is the latter case, which sheet(s) you wish to read. To use the package readxl, you need to install it using install.packages("readxl"); library(readxl).

The function excel_sheets("filename.xlsx") gives the names of all sheets in the excel file named filename.xlsx. If your Excel file contains more than one sheet, the function read_excel("filename.xlsx") reads the contents of the first Excel Sheet. To read the second Excel Sheet, use read_excel(filename,

Table 2.3 Some Useful R Functions for Reading File Formats

Function	Format	Extension
read_table()	Reads white space separated values	.txt
read_csv()	Reads comma separated values	.csv
read_tsv()	Reads tab separated columns	.tsv
read_csv2()	Also reads tab separated values	.tsv

sheet="Sheetname"). It is very important to note that the sheet name is case sensitive.

Other types of file formats can also be read using one of the functions in the package reader shown in Table 2.3.

2.6.5 Reading Files from the Internet

You can read data files into R Workspace from the Internet. There are three packages that can facilitate reading data from the Internet. Although the packages RCurl and curl provide comprehensive facilities to download from URLs, we will use here the function download.file() in the package utils. To read a data file from the Internet, you need first to know the URL where the file can be found. For example, the data in the text file Computer.Repair.csv at the Book's Website,[1] contains the Computer Repair Data. The following code downloads the dataset in your Working Directory, reads it in your current R Workspace, and store it in an object named df:

url="http://www.aucegypt.edu/faculty/hadi/RABE6/Data6/Computer.Repair.csv"
download.file(url, "Computer.Repair.csv")
df <- read.csv("Computer.Repair.csv",header=TRUE)
head(df,2) # Display the first 2 lines of the dataset

	Minutes	Units
1	23	1
2	29	2

The first line of the above code specifies the URL where the CSV file is found. The second line downloads the data to your current working directory with the name Computer.Repair.csv. The third line reads this file to the R Workspace and names it df. The fourth line displays the first 2 rows of df.

---

[1] http://www.aucegypt.edu/faculty/hadi/RABE6

Please open the file Computer.Repair.csv using Excel, then save it in your working directory as Computer.Repair.xlsx.

2.7 WRITING (EXPORTING) DATA TO FILES

In this section we explain (a) how to direct the normal R output from the console to a text file, (b) how to create and save a graph in a file, and (c) how to export data to a file.

2.7.1 Diverting Normal R Output to a File

To direct (divert) the normal R output from the screen (console) to a file named, say, "outputFileName.txt", type
　sink(file = "outputFileName.txt", append=TRUE, split = TRUE).
From now own, all R output will be stored in the file "outputFileName.txt" in your current working directory. Only prompts and some messages will continue to appear on the console.

To redirect the normal R output back to the console type sink() without any argument. This will close the last diversion file, if more than one such a file is open. The function sink.number() returns how many diversions are in use.

2.7.2 Saving Graphs in Files

The base R comes with its own set of graphs. But a much more powerful package for graphs is the ggplot2 (the gg indicates that ggplot() constructs the graphs based on the Grammar of Graphics). So, you will need to install (if needed) and download ggplot2 into your R Workspace using install.packages("ggplot2") and library(ggplot2). The function ggplot() is the one that constructs several graphs. It is very versatile for creating numerous aesthetically pleasing, powerful,?and high-quality graphs. We use ggplot() here to construct a simple scatter plot of the variables Units and Minutes in the Computer Repair data as follows:

```
g <- ggplot(df, aes(x = Units, y = Minutes))
g + geom_point()
```

The ggplot in the first line of code above has two arguments. The first is the name of the data that we wish to use to construct a graph (graph type is to be specified in the second line of code). The second argument is actually a function aes(), abbreviation of aesthetic. Its purpose here is to specify which variable is to be plotted on the x-axis and which on the y-axis. The code in first line creates a graphics window and assigns this window to the object g, which is of the type "gg" or "ggplot", a new type of object that we did not introduce before. If you plot the first line and press Enter, you will actually see a graphics window with labeled axes, but no graph yet.

The second line of code takes the graphics object g and adds to it (using "+") the scatter of points as indicated by geom_point(), which means scatter plot. You can add other things on top of this graph using "+" as many times as you wish, but note that the "+" must appear either at the end of a line or within lines but not at the beginning of a line. The plot will then appear in a separate window. If you wish to save this graph, right-click on the graph window, then select copy or save as metafile or copy as bitmap. In either case the graph will be saved on the clipboard so that you can paste it in another document such Microsoft Word or Power Point (.pptx). Alternatively, you can select Save as metafile (emf) or Save as postscript (actually, encapsulated postscript, eps). You will then be asked to provide the name of the location of the file that will contain the graph. You can also save the graph in pdf or eps formats using

```
dev.copy2pdf(file = "Fig.pdf")    # for pdf (Portable Document Format) files
dev.copy2eps(file = "Fig.eps")    # for eps (encapsulated post script) files
```

Other types of graphs are presented where needed in this book.

2.7.3 Exporting Data to Files

In Section 2.6 we learned how to read data from external environment (e.g., external files or the Internet) to the R Workspace. Here we will learn how to do the opposite, that is, we wish to save objects that we created in the R Workspace in a multiple-sheet Excel file in the current working directory. For this purpose we need to install and download the package writexl using install.packages("writexl") and library(writexl). You can use write.xlsx() to save a data.frame object to your current directory. If an object is not a data.frame, you can convert it to a data frame using as.data.frame(). Now, suppose you have two data.frame objects named object1 and object2, and you wish to create an Excel file in your directory named myfile.xlsx and save them in it. Simply use the following code:

```
write_xlsx(list(Sheet1 = object1, Sheet2 = object2), "myfile.xlsx")
```

Of course, you can modify the above code to save one or more sheets in one file.

2.8 SOME ARITHMETIC AND OTHER OPERATORS

In this section we discuss some arithmetic, logical, and relational Operators in R with relationship to (a) vectors and (a) matrices and data frames.

2.8.1 Vectors

Table 2.4 shows the arithmetic, logical, and relational operators supported by the R language. These operators act on each element of the vector a with the corresponding element of the vector b. The results of applying these operators on the vectors:

```
a <- c(4, 3, 4) # and
b <- c(0.5, 2, 0)
```

are shown in the last column of Table 2.4.

Please note that the operator "a == b" will return TRUE in the ith position if a[i] is exactly equal to b[i]. For this reason they should not be used to check the equality of floating point numbers. For example, we know mathematically that $(\sqrt{3})^2 = 3$. Therefore, the code (3 ^ (0.5)) ^ 2 == 2 should return TRUE. But when you actually try it out, it will return FALSE. This is because $\sqrt{3}$ cannot be stored exactly in the computer memory. The value is truncated. The absolute difference between the two sides of the equation is

$$|(\sqrt{3})^2 - 3| = 4.440892e{-}16,$$

which as you can see is practically, but not exactly, equal to 0. A way out of this is to use the function all.equal(a, b) in the base package or near(a, b) in the dplyr package. To use near(), you need to install and download dplyr, if you haven't already done so, using install.packages("dplyr") and library(dplyr) You can then use test the equality using near((3 ^ (0.5)) ^ 2, 3). It will return TRUE as expected.

The logical operators in Table 2.4 are applicable only to vectors of type logical, numeric, or complex. Each element of the vector a is compared with the corresponding element of the vector b. The result of comparison is a logical vector.

Two additional logical operators not mentioned in Table 2.4 are && and || because they work differently from the ones mentioned in the table. They consider only the first element of the vectors and return a single element as output. The && is called logical AND operator. It takes the first element of both vectors and returns TRUE only if both are TRUE, and FALSE otherwise.

SOME ARITHMETIC AND OTHER OPERATORS

Table 2.4 Some Arithmetic, Logical, and Relational Operators in R and Their Results When a <- c(4, 3, 4) and b <- c(0.5, 2, 0).

Operator	Description	Result
a ^ b	Raise a to the power of b	[1] 2 9 1
a / b	Divide a by b	[1] 8.0 1.5 Inf
a * b	Multiply a by b	[1] 2 6 0
a + b	Add a and b	[1] 4.5 5.0 4.0
a − b	Subtract b from a	[1] 3.5 1.0 4.0
round(b)	Rounds to digits = 0	[1] 0 2 0
ceiling(b)	Round up	[1] 1 2 0
floor(b)	Round down	[1] 0 2 0
trunc(b)	Truncate (integer part)	[1] 0 2 0
a %% b	Remainder after dividing a by b	[1] 0 1 NaN
a %/% b	Integer division of a by b	[1] 8 1 Inf
a & b	a AND b	[1] TRUE TRUE FALSE
a | b	a OR b	[1] TRUE TRUE TRUE
!b	NOT b	[1] FALSE FALSE TRUE
a > b	TRUE if a[i] > b[i], FALSE otherwise	[1] TRUE TRUE TRUE
a < b	TRUE if a[i] < b[i], FALSE otherwise	[1] FALSE FALSE FALSE
a == b	TRUE if a[i] exactly equals b[i], FALSE otherwise	[1] FALSE FALSE FALSE
a != b	TRUE if a[i] is ≠ b[i], FALSE otherwise	[1] TRUE TRUE TRUE
a >= b	TRUE if a[i] is ≥ to b[i], FALSE otherwise	[1] TRUE TRUE TRUE
a <= b	Returns TRUE if a[i] ≤ b[i], FALSE otherwise	[1] FALSE FALSE FALSE
a %in% b	TRUE if a[i] is in b, FALSE otherwise	[1] FALSE FALSE FALSE
is.element(a, b)	Same as a %in% b	[1] FALSE FALSE FALSE
intersect(a, b)	Set intersection	numeric(0), means NULL
union(a, b)	Set union	[1] 4.0 3.0 0.5 2.0 0.0
unique(c(a, b))	Same as union(a, b)	[1] 4.0 3.0 0.5 2.0 0.0
setequal(a, b)	TRUE if a and b are equal	[1] FALSE
setdiff(a, b)	Set difference a\b	[1] 4 3

This Table will be updated on the Book's Website when the need arises.

The || is called logical OR operator. It takes the first element of both vectors and returns TRUE if one of them is TRUE, else FALSE.

There are subtle differences between the methods of rounding that you will not see them using the vector b in Table 2.4. The function round(x,digits = 0) has the number of significant decimal points, digits = 0 by default. To help you understand the other differences, examine the following R code and its output matrix d:

```
z=c(-1.5, -1.0, -0.5, 1.0, 1.5, 2.0)
d=cbind(z,round(z,1),round(z),trunc(z),floor(z),ceiling(z))
colnames(d)=c("z","round1","round0","trunc","floor","ceiling")
d
        z  round1 round0 trunc floor ceiling
[1,] -1.5   -1.5    -2    -1    -2     -1
[2,] -1.0   -1.0    -1    -1    -1     -1
[3,] -0.5   -0.5     0     0    -1      0
[4,]  1.0    1.0     1     1     1      1
[5,]  1.5    1.5     2     1     1      2
[6,]  2.0    2.0     2     2     2      2
```

2.8.2 Matrix Computations

As we mentioned in Section 2.4.2, a matrix is a two- or more dimensional array. In linear algebra, a vector is defined to be a matrix with one row or one column. As such, they are special cases of matrices. They are matrices with one row or one column. To see the differences, consider the following code, its outputs, and associated comments:

```
(x <- c(3, 2, 4))  # Creates and displays x
[1] 3 2 4
is.vector(x)       # Gives TRUE because x is one-dimensional
[1] TRUE
is.matrix(x)       # Gives FALSE because x is one-dimensional
[1] FALSE
(length(x))        # Gives the number of element in x
[1] 3
dim(x)             # Gives NULL because x is one-dimensional
NULL
(U <- matrix(x, nrow = 1, ncol = 3))  # U is a row vector
     [,1] [,2] [,3]
[1,]   3    2    4
is.vector(U)       # Gives FALSE because U is two-dimensional
[1] FALSE
is.matrix(U)       # Gives TRUE because U is two-dimensional
[1] TRUE
length(U)          # U has 3 elements
[1] 3
dim(U)             # Gives number of rows and number of columns
[1] 1 3
```

```
(V <- matrix(x, nrow = 3, ncol = 1))  # V is a column vector
     [,1]
[1,]    3
[2,]    2
[3,]    4
is.vector(V)     # Gives FALSE because V is two-dimensional
[1] FALSE
is.matrix(V)     # Gives TRUE because V is two-dimensional
[1] TRUE
length(V)        # V has 3 elements
[1] 3
dim(V)           # Gives number of rows and number of columns
[1] 3 1
```

As you can see, the row vector U and the column vector V are matrices because they are two-dimensional arrays. But the vector x is not a matrix because it is a one-dimensional array. Note also that dim() is appropriate only for two- (or more) dimensional arrays, but length() gives the number of elements in the object regardless of its dimension.

The usual arithmetic operations listed in Table refTR.Operators continue to work element-wise matching the *ij*-th element of one object with the *ij*-th element of the other. Therefore, for element-wise calculations to work the objects must have the same dim(); otherwise, you will get an error saying non-conformable arrays as is illustrated in the below code using the matrices U and V defined in the above code.

```
U + V     # will not work
Error in U + V: non-conformable arrays
U + t(V)  # Element-wise addition. t(V) is transpose of V
     [,1] [,2] [,3]
[1,]    6    4    8
U*V       # will not work
Error in U * V: non-conformable arrays
U*t(V)    # Element-wise multiplication
     [,1] [,2] [,3]
[1,]    9    4   16
1/U       # Element-wise inverse (reciprocal of each element).
          [,1] [,2] [,3]
[1,] 0.3333333  0.5 0.25
```

As we learn in linear algebra, there are other types of products and inverses of matrices. Table 2.5 gives some of the R commands that are designed for matrices.

To see the difference between seemingly identical functions try them out. For example, examine the output of rep(x,times) and replicate(times, x) to see the difference or visit their help pages using ? followed by the function name without parenthesis. Also, the functions sort(), order(), and rank()

Table 2.5 Some R Commands Useful for Matrix Calculations and Manipulations

Operator	Description
dim(A)	The number of rows and number of columns
t(A)	The transpose of A
det(A)	Determinant of A
rank(A)	Rank of A
diag(d)	Returns the diagonal matrix with the vector d on the diagonal
diag(A) <- d	Replaces the diagonal elements of A with the values in d
diag(A)	Returns the diagonal elements of the matrix A
A%*%B	Inner product of two matrices
A%o%B	Outer (Kronecker) product of two matrices $A \otimes B$ (same as outer(A, B))
solve(A)	Inverse, A^{-1}, of a square matrix A if it exists
solve(A, b)	Solve a system of linear equations $Ax = b$
inverse(A)	Computes the inverse of A if it exists
eigen(A)	Computes the eigen system for square matrices

in Table 2.6 deserve further explanation. The function sort(x) arranges the values in a vector x into ascending or descending order. Like most functions in R, sort() has some arguments (parameters) that are set by default to some values. Two important arguments are decreasing = FALSE and na.last = NA. So, if you wish to order the values in increasing order, set decreasing = TRUE. The default values for NA, if present in x, is to remove them. But you have two other options for dealing with NAs: If na.last = TRUE, they are in the data last and if na.last = FALSE, they are put first. For example, some different uses of sort() and their corresponding output are:

```
x <- c(2, -1, NaN, 2, NA)
sort(x)
[1] -1  2  2
sort(x,decreasing = TRUE)
[1]  2  2 -1
sort(x,na.last = FALSE)
[1] NaN  NA  -1  2  2
sort(x, decreasing = TRUE, na.last = TRUE)
[1]  2  2 -1 NaN  NA
```

The function order(x) returns the positions of the minimum, second minimum, ..., the maximum. That is, x[order(x)] is the same as sort(x).

Table 2.6 Some Useful R Commands or Functions

Function	Description
unique(x)	Removes duplicate values of a vector x
unique(X)	Removes duplicate rows of a matrix or data frame X
unique(c(x, y))	Returns the union of x and y
min(x)	Returns the minimum of x
max(x)	Returns the maximum of x
rep(x, times)	Replicate x as many times as in "times"
replicate(times, x)	Replicate x as many times as in "times"
which.min(x)	Returns the position of the minimum value in x
which.max(x)	Returns the position of the maximum value in x
sample(x, n)	Returns a random sample from x of size n without replacement
sample(x, n, replace = T)	Returns a random sample from x of size n with replacement
sum(x)	Returns the sum of a vector x
mean(x)	Returns the mean of a vector x
mean(x, trim)	Returns the trimmed mean of a vector x with trimming 100*trim% of the observations from each side
median(x)	Returns the median of a vector x
colMeans(X)	Returns the column means of a matrix or data frame X
colMedians(X)	Returns the median of each column of a matrix or data frame X. This function is in the package robustbase
weighted.mean(x, w)	Returns the weighted mean of x with weights in w
var(x)	Returns the variance of a vector x
sd(x)	Returns the standard deviation of a vector x
var(X)	Returns the covariance matrix of a matrix/data frame x
cor(X)	Returns the correlation matrix of a matrix/data frame x
quantile(x, probs)	Returns sample quantiles of x corresponding to the the probabilities in probs
IQR(x)	Returns the interquartile range of a vector x
summary(x)	Returns the six-number summary of a vector or for each column of a matrix or data frame

(*continued*)

Table 2.6 (*Continued*)

Function	Description
summary(f, maxsum)	Returns the frequency of the first maxsum − 1 most frequent levels of a factor object and other less frequent levels are summarized under "Others"
table(x)	Returns the frequency breakdown or the contingency table of a factor object
ftable(x)	Creates a "flat" contingency table.
cbind(x, y)	Column bind: Puts y next to x
rbind(x, y)	Row bind: Puts y below x
sort(x)	Sorts the vector x into ascending/descending order
order(x)	Returns the positions of the min, 2nd min., ..., the max.
rank(x)	Returns the sample ranks of the values in the vector x

This Table will be updated on the Book's Website when the need arises.

This function also has arguments that are set by default. Two important arguments are decreasing = FALSE and na.last = TRUE. Ties are broken by providing other vectors as argument. For example order(x,y,z), orders x and break the ties using y. If the values are still tied, then z is used to break the ties, and so on. Here is an example illustrating sort() for the same vector x as in the previous code:

```
x <- c(2, -1, NaN, 2, NA)
order(x)
[1] 2 1 4 3 5
order(x,decreasing = TRUE)
[1] 1 4 2 3 5
order(x,na.last = FALSE)
[1] 3 5 2 1 4
order(x, decreasing = TRUE, na.last = TRUE)
[1] 1 4 2 3 5
```

Finally, the function rank() returns the positions of the first value, second value, ..., last value in the ordering of the vector x. The important arguments that are set by default are na.last = TRUE and ties.method = "average". The average method for breaking the ties by giving them a rank equal to the average of their ranks. For example if two equal values share the 2nd, 3rd ranks, then they are both given the rank $(2 + 3)/2 = 2.5$, as can be seen from the code below. Several other options for breaking the ties can be found by typing ?rank.

```
x <- c(2, -1, NaN, 2, NA)
rank(x)
[1] 2.5  1.0  4.0  2.5  5.0
rank(x,na.last = FALSE)
[1] 4.5  3.0  1.0  4.5  2.0
```

The function rank does not have the argument decreasing because the ranking is done in increasing order. If you wish to rank x in decreasing order use rank(-x) to obtain [1] 1.5 3.0 4.0 1.5 5.0.

2.9 PROGRAMMING IN R

Before we explain how to write programs in R, we first discuss a topic that helps us in writing, editing, executing, and saving R programs. This is Script files.

2.9.1 Best Practice: Script Files

In Section 2.3.3 we discussed the rather cumbersome editing of lines typed in the R Console using the arrow keys. Here we discuss an easier way for not only editing but also for executing and saving your R code in files known as Script files. A file that contains R code is called script file. To open a script file go to the File menu on the top line of your console window, and select New script. A new window will open up, where you can type all your lines of code in them. You could easily edit the code just as you do with a word processor. It is recommended that you write and edit all of your R code in a script file before you run it in the console. To run the code in one or more lines, copy the line(s) and paste them in the console window, or just press on the letter R while holding the ctrl (control) key. The lines will be copied to the console and executed. If you find errors during the execution, you may go back to the script file and edit it. When done, name and save the script file in your current working directory as, say, filename.R, where the extension.R indicates that this is a file containing R codes. Even though it has an extension.R, the script file is just a plain text file that can be read by other word processors. At a later time, you could open the file and further edit it or execute all of its contents using source(filename.R). R will then execute all the code in the file. This way, your work is reproducible. This is an important advantage of using script files. Additionally, script file makes it very easy to check and edit your code. You could also share your work with others.

2.9.2 Some Useful Commands or Functions

In this section we list some useful commands or functions that are used for specific purpose computation. These are listed in Table 2.6. Note that the functions that are relevant to numeric data (e.g., sum(), mean(), quantile(), etc.) will return NA if the data contain missing values (NA and/or NaN). For them to work properly, we need to add the additional parameter "na.rm = TRUE", which means to remove the missing values before the computation. For example, we use

sum(x,na.rm = TRUE); mean(x,na.rm = TRUE)

To learn more about any of these and other R functions type a question mark ? followed by the name of the function without the parentheses (e.g., ?unique). We will use these functions in this chapter and throughout the remaining chapters of the book.

2.9.3 Conditional Execution

Conditional execution is one of the basic features of programming. The most common conditional expressions are if(), and ifelse. The basic syntax of if() is

```
if(test){
    First set of R code if test is TRUE
}else{
    Second set of R code if test is FALSE
}
```

The test will return a logical vale TRUE or FALSE. Thus if the test is TRUE, the first set of R code between the two braces { } will be executed; otherwise, the second set of code between the two braces { } will be executed.

Example: Head/Tail:

```
if(p > 0.5){
    print("Head")
}else{
    print("Tail")
}
```

The output of the above code depends on the predefined value of p. If $p > 0.5$, it prints "Head", else it prints "Tail". Note that the above if() statement does not have to have an else. In this case, if $p > 5$, it prints Head; otherwise, it will do nothing! The ifelse() has the syntax

ifelse(test, do something if TRUE, do something else if FALSE)

Thus the above code can be replaced by
   ifelse(p > 0.5, "Head", "Tail")
The ifelse is particularly useful because it works on vectors. Here is an example:

```
x <- c(0, 1, 2, -4, 5)
ifelse(x > 0, 1/x, NA)
[1]   NA 1.0 0.5   NA 0.2
```

Here is another example, where ifelse is used to select between the mean and median as appropriate measure of location. Let us first define a simple criterion for finding outliers in an univariate sample x. Outliers on the lower side (small values) are the observations smaller than the lower limit LL = Q1 - 1.5*(Q3 - Q1) = Q1 - 1.5*IQR. Similarly, outliers on the upper side (large values) are the observations larger than the upper limit LL = Q3 + 1.5*(Q3 - Q1) = Q3 + 1.5*IQR, where Q1 and Q3 are the first and third quartiles of x and IQR = Q3 – Q1 is the interquartile range of x. Using these criteria, the following self-explanatory code will compute the median if the data contains outliers and the mean otherwise.

```
# Example 1:
x <- c(0, 1, 2, 4, 50, 5)           # The sample values
Q <- quantile(x, c(0.25,0.75))      # Compute Q = c(Q1, Q2)
LL <- Q[1] - 1.5 * (Q[2] - Q[1])    # Lower limit
UL <- Q[2] + 1.5 * (Q[2] - Q[1])    # Upper limit
LO <- sum(x < LL)                   # Number of lower outliers
UO <- sum(x > UL)                   # Number of upper outliers
NO <- LO + UO                       # Total number of outliers
ifelse(NO > 0, median(x), mean(x))  # Compute median if N) > 0
[1] 3                               # or the mean otherwise
```

The value 3 is the median because the data contains NO = LO + UO = 0 + 1 = 1 outlier, which is clearly the fifth value of 50 (by eye balling). It is a recommended practice to use the # to explain the computation of each line as seen in the above code. Note that the code in Example 1 above produced a correct result. But this is because we were lucky that the object x contains no missing values. If you use this code with the data which contains NA and/or NaN, it will produce an NA. As we mentioned earlier this is because the functions quantile(), sum(), mean(), and median all assume that the data are free from missing values. To modify the code, we need to add the argument na.rm = TRUE, which means to remove the missing values before the computation. For example, we use

```
Q <- quantile(x,c(0.25,0.75), na.rm = TRUE)
sum(x < LL, na.rm = TRUE)
sum(x > UL, na.rm = TRUE)
ifelse(NO > 0, median(x, na.rm = TRUE), mean(x, na.rm = TRUE))
```

58   A BRIEF INTRODUCTION TO R

The if-else can also be used to replace all the missing values in a vector with 99 (or any other value that does not exist in the data) using the following code:

```
# Example 2:
x <- c(NaN, -2, 5, NA, 1, 3) # Sample values
is.na(x)                     # TRUE means Missing, FALSE nonmissing
[1]  TRUE FALSE FALSE  TRUE FALSE FALSE
> y <- ifelse(is.na(x),99,x) # If x[i] is TRUE, replace by 99
> y                          # Display y
[1] 99 -2  5 99  1  3
```

Note here that NaN is also considered missing. We should caution here that replacing missing values with numeric values can produce wrong and misleading results. The use of na.rm = TRUE deals with missing values by removing them, not by replacing them by other numeric values.

2.9.4 Loops

When we need to perform exactly the same task over and over again, we need to use loops. Here we discuss two commonly used loops: The for loop and the while loop.

The for Loop   The syntax of the for loop is:

```
for(index in range){
    R Code
}
```

This means that the index takes each value in range and executes code for each value. This mean that we will repeat the R code length(range(times. For example, suppose we wish to sum the first n positive integers, for each $n \leq m$, where m is provided by the user. This can be done using the for loop as follows:

```
m <- 8
s <- {}              # Am empty object s. Same as s <- NULL
range <- 1:m         # Creates the range
for(i in range){     # The index is i and the range is 1, ..., m
    s[i] <- sum(1:i) # Replace s[i] by sum of 1, ..., i
}                    # End of the for loop
rbind(range,s)       # Puts the range 1:m on top of s
      [,1] [,2] [,3] [,4] [,5] [,6] [,7] [,8]
range    1    2    3    4    5    6    7    8
s        1    3    6   10   15   21   28   36
```

But since we know from mathematics that the sum of the first n positive integers is $n(n+1)/2$, then we can replace the loop in the above code simply by

```
for(i in range){s[i] <- i*(i + 1)/2}
```

The while Loop We use the for loop when we know the length of the sequence, that is, when we know in advance how many times we are going to repeat the code. But sometimes we do not know in advance how long the sequence is. For example, suppose you take a certain exam as many times as you need until you pass. We do not know in advance how many times you will take the exam. We cannot use the for loop, and here is where the while loop comes in handy. The syntax of the while loop is simple:

```
while(condition){
  R Code
}
```

Suppose that there is a 50% chance of passing the exam. The code below is an example of how we can use the while loop to find out how many trials it takes to pass this exam.

```
pass <- 0         # Initializing the object pass
trials <- 0       # Initializing the object rials
while(pass < 1) { # Invoke the loop only whenever success < 0
    trials <- trials + 1 # Increase the number of trials by 1
    try <- sample( c("Pass","Fail"),
           size = 1, prob = c(0.5,0.5)) # A sample of size 1
    if(try == "Pass"){
        cat("Passed after", trials,"\n");
        pass = 1 # Signaling exiting from while()
    } # End of if()
} # End while()
```

2.9.5 Functions and Functionals

Functionals are functions that apply the same function to each entry in a vector, matrix, data frame, or list. Examples of functionals are apply(), sweep(), lapply(), sapply(), and vapply(). The functional apply() takes the form apply(X, MARGIN,FUN), where X is a matrix or data frame, MARGIN is 1, for rows and 2, for columns, and FUN is an R function such as sum, mean, and sd. It returns an object obtained by applying the function FUN to the MARGIN of an array X. Suppose, for example, we wish to compute the mean and the standard deviation of each column of the following data matrix X, which is created in the R Workspace:

$$X = \begin{bmatrix} 3.15 & 3.44 \\ 2.76 & 3.46 \\ 3.21 & 3.57 \\ 3.69 & 3.19 \\ 3.92 & 3.15 \end{bmatrix}. \tag{2.2}$$

We have the following code and its outputs:

```
xbar <- apply(X, 2, mean)   # Same as colMeans(X)
cat("The mean of the variables are:",xbar,"\n")
The mean of the variables are: 3.346 3.362
s <- apply(X, 2, sd)   # St. dev.s of X, 2 refers to columns
cat("The st. dev. of the variables are:",round(s, 3),"\n");
The st. dev. of the variables are: 0.46 0.183
```

Note that the sd. dev. are rounded to the nearest 3 decimal places. Other examples of apply() are:

apply(X, 1,mean) # Mean of the rows of X, Same as rowMeans(X)
apply(X, 2,mean) # Same as colMeans(X)
apply(X, 2,var) # Variance of the columns of X. Same as diag(var(X))

The functional sweep() returns an object obtained from an input array X by sweeping out a summary statistic. The syntax for sweep() is

sweep(X, MARGIN, STATS, FUN = "−"),

where STATS is the summary statistic which is to be swept out, and the default value for FUN is subtract "−". The subtract operator can be replaced by other appropriate arithmetic operator. As an example, suppose we wish to center the columns of the data matrix X in (2.2). Centering means we subtract a column mean from each value in the column. We can do this using

```
Y <- sweep(X, MARGIN = 2, STATS = colMeans(X), FUN = "-")
apply(Y, 2, sum)    # Check that the columns sum to zero.
[1] -8.881784e-16 -8.881784e-16
```

Now suppose we wish to standardize the columns of the above matrix, that is for each column, we compute the so-called Z-score,

$$z_{ij} = \frac{x_{ij} - \bar{x}_j}{s_j}, \qquad (2.3)$$

where x_{ij} is the ij-th element of X, $\bar{x}_j$ and s_j are the mean and the standard deviation of the j-th column, respectively. These are given by

$$\bar{x}_j = \frac{1}{n}\sum_{i=1}^{n} x_{ij} \quad \text{and} \quad s_j = \frac{1}{n-1}\sum_{i=1}^{n}(x_{ij}-\bar{x}_j)^2, \quad j=1,\cdots,d, \qquad (2.4)$$

where n is the number of rows and d is the number of columns in X. Every column in Z has a mean zero and standard deviation of 1. A code required to standardize the data matrix X is simply given by the R code: Z <- scale(X). But just to illustrate the functions apply and sweep, we use the following code to standardize the data:

```
Y <- sweep(X, MARGIN = 2, STATS = colMeans(X), FUN = "-")
apply(Y, 2, sum) # Check that the columns sum to zero.
[1] -8.881784e-16 -8.881784e-16
s=apply(X,2,sd)   # Compute column standard deviation
Z <- sweep(Y, MARGIN = 2, STATS = s, FUN = "/")
apply(Z, 2, sum) # Check that the columns sum to zero.
[1] -1.887379e-15 -4.996004e-15
apply(Z, 2, sd) # Check that the columns sd. dev. are 1s.
[1] 1 1
```

Other functionals in R include lapply(), sapply(), and vapply(). Interested readers can find information about them by using ?lapply, for example.

2.9.6 User-Defined Functions

As we have seen above, R includes several predefined functions and these functions are found in different packages. These functions do not appear in the Workspace when we use ls()because we did not define them, but they are available for immediate use. R commands that perform related calculations can be grouped together in one program called a function. The main syntax of a function is:

```
fun.name <- function(arguments){
    R code
}
```

Thus the components of the function are: (a) The name of the function. You select the name of the function making sure to follow the same rules of names that we mentioned in Section 2.3.4. (b) The word function, which indicates that the object that we are about to create is a function. (c) The arguments of the function, if any. If you have more than one argument, they are separated by commas. Some of the arguments may be set to default values. (d) The body of the function which is a set of R code that will do a certain task. This body is inserted between two braces { }. Once it is defined, you can execute it like and other R function. If the function has no arguments or if all the arguments are set to default values, to execute it just type fun.name(); otherwise, you need to specify values for the arguments without default values. We illustrate the user-defined functions by two examples.

Multiplication Table: Let us write a function that computes the multiplication table of size $n \times m$, where n and m are integers provided by the user. A suitable name of this function would be mult.table. We should not call it table because table is a defined R function. To make sure that mult.table is not a defined function type exists("mult.table") at the prompt symbol and click on Enter. If the output is FALSE, then the name is not taken. The arguments here are n and m. An obvious default value for these arguments are $n = n12$ and $m = 12$, which will produce the standard 12×12 multiplication table. To construct the table, we can use either outer(n, m) or the equivalent symbol $n\%o\%m$. Then this function may be created as follows:

```
mult.table <- function(n = 12, m = 12){
    outer(1:n, 1:m) # or equivalently (1:n)%o%(1:m)
}
```

Now that the function has been created, if we type mult.table(), the standard multiplication table will be printed on the screen. On the other hand, if we type:

```
mult.table(n = 2, m = 6) # mult.table(2, 6)
     [,1] [,2] [,3] [,4] [,5] [,6]
[1,]   1    2    3    4    5    6
[2,]   2    4    6    8   10   12
```

Note that in the simpler form, we did not write the names of the arguments explicitly. This is because R will match the given values with each name in the same order they appear in the definition of the function. A second equivalent way to run the function is mult.table(m = 6, n = 4). Here the arguments are given in different order. This is okay as long as they are defined explicitly. You could try mult.table() with different values of n and m including negative integers and see for yourself.

What will happen if the user typed: mult.table(4, 6.8)? That is, the value given for the arguments m is not an integer. Luckily, the function will continue to work because the ":" operator truncates the float to integer, that is, it replaces the float argument by argument = floor(argument). But in general, a good programmer should check the user input to make sure that it is the desired one using the commands listed in Table 2.2.

The above function does not create any objects inside the function, but it returns the computations done in the last line of the code (the above function contains only one line of code). Thus, if you type nm.table <- mult.table(2, 6), an object named nm.table will be created in the Workspace. If we wish to make the function return any other value calculated within the function, we add a return() statement as the last line in the function. Thus, the above function can be modified as follows:

```
mult.table <- function(n = 12, m = 12){
    temp <- outer(1:n, 1:m)
    return(temp)
}
```

```
mult.table(2, 6)
     [,1] [,2] [,3] [,4] [,5] [,6]
[1,]   1    2    3    4    5    6
[2,]   2    4    6    8   10   12
```

Note that an object named temp was created only to be returned by return() when the execution is completed. Any object that is created within a function is called a local object. It will be deleted once the function exits. Thus, the name temp (short for temporary) is an appropriate name here. If you type ls(), you will not see locally defined objects in the list. Objects that are created in the Workspace, i.e., outside the user-defined functions, are called global variables. Global variables are available for use from within user-defined functions. It is important to note that the return() function has to be the last line in the function. Actually, R will exit the function right after the return() and, thus, text after the return(), if any, will be ignored. If you wish to save the output of this function, type:

nm.table <- mult.table(2, 6)?

In this case, the function will not print anything, but its output from the return() is stored in the object nm.table. If you now type ls(), you will see the name nm.table listed among all other created (global) objects in the Workspace. The objects in the Workspace environment are called global objects. If you wish, you could create a global object from within a user-defined function. Here is an example:

```
mult.table <- function(n = 12, m = 12) {
    nm.table <<- outer(1:n, 1:m)
    return(nm.table)
}
mult.table(2, 6)
     [,1] [,2] [,3] [,4] [,5] [,6]
[1,]   1    2    3    4    5    6
[2,]   2    4    6    8   10   12
```

We use "<<-" to create global variables from within a user-defined function. Note that a function call like x <- mult.table(2,6) will create two identical global objects: nm.table and x. Accordingly, we should have used just mult.table(2,6), that is, without an assignment. Actually, there is no need for the return(nm.table) because the function created the global object nm.table.

You might be wondering why bother to create a function that has only one line? The above task can be simply accomplished by typing one of the following lines:

```
nm.table <- outer(1:2, 1:6)     # or equivalently
nm.table <- (1:2)%o%(1:6)
```

The objective here is to illustrate how to create a function by a very simple example that does not involve complicated codes. But function can be most useful if you need to type many lines of code. For example, you can create a function that will include the code for the above example of finding the number of trials until success. This can be done as follows:

```
exam <- function(){
   pass <- 0     # Initializing the object pass
   trials <- 0   # Initializing the object rials
   while(pass < 1){# Invoke the loop only whenever success < 1
   trials <- trials + 1 # Increase the number of trials by 1
   try <- sample( c("Pass","Fail"),
      size = 1, prob = c(0.5,0.5)) # A sample of size 1
   if(try == "Pass"){
     cat("Passed after", trials,"\n");
         pass = 1 # Signaling exiting from while()
     } # End of if()
   } # End while()
}
```

Now all that you need to run the code is to type trial(). You could also make exam() more general by providing as an argument the probability of success. You need to only to make two simple changes. (a) Change the definition as follows: exam <- function($p = 0.5$) and (b) Replace the argument of sample() from prob = c(0.5, 0.5) to prob = c(p, 1 - p). This may be suitable when the exam is harder/easier to pass. Here are some examples of using the modified exam() with different values of the argument p:

```
trials(0.75)
Passed after 1
trials()    # Using the default for p. Same as trials(0.5)
Passed after 1
trials(0.25)
Passed after 4
trials(0.1)
Passed after 53
```

As you can see, it takes more trials as the difficulty of the exam increases (as the probability of success p decreases). Note here we have made the assumption that the probability of success p is the same for all trials. This may be appropriate when you flip the same coin several times until you obtain the first head, but it is certainly not an appropriate assumption for exam trial. Presumably, this is because one expects the probability of passing an exam to increase with the number of trials (at least because of the experience of taking the exam).

Simulation: Let us write a function that simulates rolling a balanced m-sided die n times and returns the result. Simulation would not be a good name because it is not informative (simulation could be done for so many different experiments). Perhaps sim.die is a better name, but of course you could select any name of your choice. The arguments here are also m (the number of sides of the die) and n (the number of rolls of the die). Here is an example of such a user-defined function in this case:

```
sim.die <- function(m = 6, n = 8){
    sides <- 1:m    # Creating the sides of the die
    # Take a random sample with replacement:
    x <- sample(sides, n, replace = TRUE)
    return(x)
}
```

The default arguments are $n = 6$ (the six-sided die) and $m = 8$ (for no good reason but to shorten the output). Here are five calls of this function with the corresponding outputs:

```
sim.die()
[1] 6 2 1 4 3 3 5 1
sim.die(2,8) # Resembles a coin
[1] 1 1 2 1 1 2 2 2
sim.die(3,8) # Have you seen a 3-sided die?
[1] 2 3 3 3 1 1 3 2
sim.die(5,8) # Pyramids of Giza, Egypt, huge butb5-sided!
[1] 2 4 3 3 3 4 3 4
sim.die(100,8)
[1] 28 27 58  5 12 64 73 19
```

The first is the standard 6-sided die, which is the default die. Each of the remaining calls is for an m-sided die with $m = 2, 3, 5,$ and 100. A 2-sided die can be thought of as a coin and, hence, this function can simulate flipping a fair coin n times with $m = 2$, with the understanding that 1 may represent a "Head", say, and 2 a "Tail". You can actually transform them to "Head" and "Tail" as follows:

```
(x <- sim.die(2, 8))    # Simulate and display x
[1] 2 2 2 1 2 1 2 1
ifelse(x == 1,"Head","Tail")
[1] "Tail" "Tail" "Tail" "Head" "Tail" "Head" "Tail" "Head"
```

Recall that x == 1 will return a logical object with the ith value TRUE whenever x[i] is 1 and FALSE otherwise. Therefore, if x[i] == 1 is TRUE, then x[i] will be replaced by "Head" and if it is FALSE, x[i] will be replaced by "Tail". Another point worth mentioning here is that we called sim.die(2, 8) twice above and they did not produce the same output. We should not expect them to produce identical results because of randomness. If for any reason you wish to generate the same sequence of random numbers, you need to set the seed of the random number generator using the function set.seed(), and use the same seed before you make the call. This is illustrated as follows:

```
set.seed(1949); sim.die(2, 8)
[1] 2 2 1 2 1 1 2 1
set.seed(1949); sim.die(2, 8)
[1] 2 2 1 2 1 1 2 1
set.seed(1949); sim.die(2, 8)
[1] 2 2 1 2 1 1 2 1
```

All outputs are the same as expected. The number 1949 used as an argument for the set.seed() is chosen arbitrarily. Any other number would cause all the calls to generate identical sequences. A practical example for the need to set the seed is when you want to conduct a simulation experiment to compare different methods, you should apply each method to the same data; otherwise, the comparisons will not be right because the difference between the performance of the methods may be due to using different data, not due to their actual difference. Another example for the need to set the seed is when you send your code to another person and you want to make sure that you both produce the same results. There are other needs, of course.

You can also use R to generate simulated data from various probability distributions. For example, to simulate n observations from the Normal distribution with mean m and standard deviation s, use rnorm(n, mean = m, sd = s). Examples of other distributions are runif(n, min = 0, max = 1), rexp(n, rate = 1), rpois(n, lambda), and rgeom(n, prob). As you can see, they follow the same pattern, the letter r (for random) followed by an abbreviated name of the distribution with the arguments between the two parentheses. Look up the form for other distributions by typing ?? followed by the name of the distribution (e.g., ??binomial and ??multivariate).

2.10 BIBLIOGRAPHIC NOTES

Of course, it is not our intention of this chapter to teach you how to become a professional programmer in the R language. There are numerous standard books available for studying the R programming language, see, for example, Maindonald and Braun (2003), Dalgaard (2008), Matloff (2011), Crawley (2012), Cotton (2013), Grolemund (2014), Lander (2014), Wickham (2014), Kabacoff (2015), Gillespie and Lovelace (2017), Wickham and Gorlemund (2017), and Grosser et al. (2022), to mention only a few. Readers will also benefit from the very rich documentation available in R Core Team (2022) and the many tutorials and free courses available online for free. Many other books written about the application of R in specific areas. See, for example, Racine and Hyndman (2002), Maindonald and Braun (2003), Verzani (2004), Faraway (2006), Dalgaard (2008), Robert and Casella (2010), Wickham and Bryon (2015), Boehmke (2016), Davies (2016), Wickham (2016), Zumel and Mount (2019), and Lilja and Linse (2022).

EXERCISES

2.1 Execute the following commands and explain the output:
   (a) x <- c(1,2,3)
   (b) y <- c(10, 20, 30, 40, 50, 60, 70)
   (c) x+y

2.2 The following is a line of R commands and the corresponding output for an object named "x" in an R Workspace:
class(x); dim(x); colMeans(x); length(x); mean(x);
[1] "data.frame"
[1] 15 2
height weight
65.0000 136.7333
[1] 2
[1] NA
Warning message:
In mean.default(x): argument is not numeric or logical: returning NA

Using the above information, answer the following questions:
   (a) How many variables are there in the object x? What are they?
   (b) How many observations are there in the object x?
   (c) What is the mean of x[,3]?
   (d) What is the median of the third column of x?
   (e) Compute the sum of each column of x.
   (f) What is the variance of x$height?
   (g) What is the variance of x[,2]?
   (h) What is the standard deviation of x[,2]?
   (i) What *could* you conclude about the relationship between the columns in x?

(j) What *would* you conclude about the relationship between the columns in x?

2.3 What is the best way to represent each of the following in R?
   (a) An object representing levels of exercise undertaken by 5 subjects
   (b) A student took an exam consisting of n True–False questions

2.4 A student took an exam consisting of 50 True–False questions. The results are recorded as 1 if the answer is correct and 0 otherwise. The data were created in R and stored in an object named x. The R command sum(x) resulted in 41. Write an R command that is needed to compute the following quantities and give the corresponding output:
   (a) The length of the object x.
   (b) The mean of the data in x.
   (c) The median of the data in x.
   (d) The number of mistakes the student has made.
   (e) The proportion of the questions answered incorrectly.
   (f) The length of the object x.
   (g) Convert the object x to a logical object.

2.5 Eight students were asked: How many dollars do they have in their wallets? The responses (in dollars) were:

   2,   5,   None,   94,   Refuse to answer,   0, –4,   3

   (a) Which of these values do not make sense to you? Why?
   (b) Which is an appropriate measure of center here, the mean or the median? Why?
   (c) A sample of n students' records of IDs, Age, Weight, Grade in Course 1, Grade in Course 2, and GPA.
      Write an appropriate R command to:
   (d) Create and store these values in an object named x.
   (e) Display the first five values.
   (f) Display the last four values.
   (g) Compute the mean.
   (h) Compute the median.
   (i) The position of the minimum value.
   (j) The position of the second minimum value.
   (k) Sort the data in an increasing order of magnitude putting the NAs first.

2.6 Sequences:
  (a) Write the R command that creates the sequence 4,5, $\cdots$,555.
  (b) Write the R command that creates the sequence 0,2, 4,6, $\cdots$, 1000.

2.7 Consider the following matrices:

$$a = \begin{bmatrix} 2 & 1 \end{bmatrix} \quad b = \begin{bmatrix} 2 \\ 0 \\ 1 \end{bmatrix}. \quad f = \begin{bmatrix} 1 & 2 \\ 3 & -1 \\ 2 & 5 \end{bmatrix} \quad \text{and} \quad g = \begin{bmatrix} 0.123e3 \\ NA \\ \text{``NA''} \\ \text{``B''} \end{bmatrix}.$$

  (a) Write the R commands that are needed to create these matrices as objects with the same name in R Workspace.
  (b) Write the R command that transforms g to a numeric matrix.
  (c) Write the R command that gives the transpose of the matrix f.
  (d) Write the R command that gives the mean of all elements in f.
  (e) Write the R command that gives the median of each column in f.
  (f) Write the R command that displays the first two rows of f.
  (g) Write the R command that displays rows 1 and 3 of f.

2.8 Suppose we have a population that is known to be normally distributed with mean 50 and variance 25. Use R to:
  (a) Simulate drawing a random sample of size 100 from this population and store the results in an object named x.
  (b) Compute the five-number summary of the data in x.
  (c) Compute the number of observations in x that are < 42.
  (d) Compute the number of observations that are < 58.
  (e) Find the value of x such that 15% of values are above it.
  (f) Identify outliers, if any, in the data.

2.9 One version of the Titanic data is found in the R package vcdExtra. Install and download this package if you have not already done so. Then write the R commands to:
  (a) Read this dataset into your R Workspace and rename it df.
  (b) Find the number of rows and columns in df.
  (c) Display the first 3 lines of df.
  (d) Draw the boxplot for the variable age in the dataset df.
  (e) Save the above boxplot in either a pdf or eps files in your current Working directory.

(f) Count the number of outliers in the variable age in the dataset df.
(g) Identify outliers, if any, by number in each of the numeric variables in df.
(h) Compute the contingency table of the two variables pclass and survived and save it in an object named ct.
(i) Compute the flat contingency table of all factor variables in df and save it in an object named ft.
(j) The dimensions of ct and ft.
(k) Save the two objects ct and ft in a file named Titanicp.xlsx in your working Directory.

2.10 The Animals2 dataset is found in the R package robustbase. Read the data into an R workspace and rename it as df. Then, write the R commands needed to:
  (a) Compute the covariance matrix of the variables in df and store it in an object named cov.x.
  (b) Check to see if cov.x is symmetric.
  (c) Compute the correlation matrix of the variables in df and store it in an object named cor.x.
  (d) Standardize all the variables in df and store the result in an object named Z.
  (e) Compute the covariance matrix of the variables in Z and store it in an object named cov.z.
  (f) Check to see if cor.x is equal to cov.z.
  (g) Compute the eigen values and eigen vectors of the correlation matrix cor.z.
  (h) Verify that the sum of the eigen values is equal to the trace of cor.z. Is this always the case?
  (i) Verify that the product of the eigen values is equal to the determinant of cor.z. Is this always the case?
  (j) Create a matrix $\mathbf{V}$ containing the eigen vectors of cor.z.
  (k) Verify that the inverse of $\mathbf{V}$ is simply equal to its transpose, $\mathbf{V}^T$. This is because $\mathbf{V}$ is an orthonormal matrix.
  (l) Create a diagonal matrix $\mathbf{\Lambda}$ containing the eigen values of cor.z on the diagonal.
  (m) Verify that the product of $\mathbf{V\Lambda V}^T$ is equal to cor.z. This is the so-called Spectral Decomposition Theorem.

CHAPTER 3

SIMPLE LINEAR REGRESSION

3.1 INTRODUCTION

We start with the simple case of studying the relationship between a response variable Y and a predictor variable X_1. Since we have only one predictor variable, we shall drop the subscript in X_1 and use X for simplicity. We discuss covariance and correlation coefficient as measures of the direction and strength of the linear relationship between the two variables. A simple linear regression model is then formulated and the key theoretical results are given without mathematical derivations, but illustrated by numerical examples. Readers interested in mathematical derivations are referred to the bibliographic notes at the end of the chapter, where books that contain a formal development of regression analysis are listed.

3.2 COVARIANCE AND CORRELATION COEFFICIENT

Suppose we have observations on n subjects consisting of a dependent or response variable Y and an explanatory variable X. The observations are usually recorded as in Table 3.1. We wish to measure both the *direction*

Regression Analysis By Example Using R, Sixth Edition. Ali S. Hadi and Samprit Chatterjee
© 2024 John Wiley & Sons, Inc. Published 2024 by John Wiley & Sons, Inc.
Companion website: www.wiley.com/go/hadi/regression_analysis_6e

Table 3.1 Notation for the Data Used in Simple Regression and Correlation

Observation Number	Response Variable Y	Predictor X
1	y_1	x_1
2	y_2	x_2
⋮	⋮	⋮
n	y_n	x_n

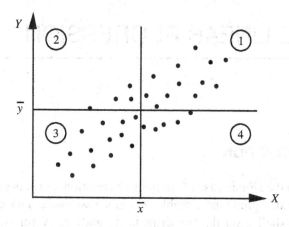

Figure 3.1 Graphical illustration of the correlation coefficient.

and the *strength* of the relationship between Y and X. Two related measures, known as the *covariance* and the *correlation coefficient*, are developed below.

On the scatter plot of Y versus X, let us draw a vertical line at $\bar{x}$ and a horizontal line at $\bar{y}$, as shown in Figure 3.1, where

$$\bar{y} = \frac{\sum_{i=1}^{n} y_i}{n} \quad \text{and} \quad \bar{x} = \frac{\sum_{i=1}^{n} x_i}{n}, \tag{3.1}$$

are the sample mean of Y and X, respectively. The two lines divide the graph into four quadrants. For each point i in the graph, compute the following quantities:

- $y_i - \bar{y}$, the deviation of each observation y_i from the mean of the response variable,

- $x_i - \bar{x}$, the deviation of each observation x_i from the mean of the predictor variable, and
- the product of the above two quantities, $(y_i - \bar{y})(x_i - \bar{x})$.

It is clear from the graph that the quantity $y_i - \bar{y}$ is positive for every point in the first and second quadrants and is negative for every point in the third and fourth quadrants. Similarly, the quantity $x_i - \bar{x}$ is positive for every point in the first and fourth quadrants and is negative for every point in the second and third quadrants. These facts are summarized in Table 3.2.

If the linear relationship between Y and X is positive (as X increases Y also increases), then there are more points in the first and third quadrants than in the second and fourth quadrants. In this case, the sum of the last column in Table 3.2 is likely to be positive because there are more positive than negative quantities. Conversely, if the relationship between Y and X is negative (as X increases Y decreases), then there are more points in the second and fourth quadrants than in the first and third quadrants. Hence the sum of the last column in Table 3.2 is likely to be negative. Therefore, the sign of the quantity

$$\text{cov}(Y, X) = \frac{\sum_{i=1}^{n}(y_i - \bar{y})(x_i - \bar{x})}{n - 1}, \qquad (3.2)$$

which is known as the *covariance* between Y and X, indicates the direction of the linear relationship between Y and X. If $\text{cov}(Y, X) > 0$, then there is a positive relationship between Y and X, but if $\text{cov}(Y, X) < 0$, then the relationship is negative. Unfortunately, $\text{cov}(Y, X)$ does not tell us much about the strength of such a relationship because it is affected by changes in the units of measurement. For example, we would get two different values for the $\text{cov}(Y, X)$ if we report Y and/or X in terms of thousands of dollars instead of dollars. To avoid this disadvantage of the covariance, we *standardize* the data

Table 3.2 Algebraic Signs of the Quantities $(y_i - \bar{y})$ and $(x_i - \bar{x})$

Quadrant	$y_i - \bar{y}$	$x_i - \bar{x}$	$(y_i - \bar{y})(x_i - \bar{x})$
1	+	+	+
2	+	−	−
3	−	−	+
4	−	+	−

before computing the covariance. To standardize the Y data, we first subtract the mean from each observation then divide by the standard deviation, that is, we compute

$$z_i = \frac{y_i - \bar{y}}{s_y}, \tag{3.3}$$

where

$$s_y = \sqrt{\frac{\sum_{i=1}^{n}(y_i - \bar{y})^2}{n-1}} \tag{3.4}$$

is the sample *standard deviation* of Y. It can be shown that the standardized variable Z in (3.3) has mean zero and standard deviation one. We standardize X in a similar way by subtracting the mean $\bar{x}$ from each observation x_i and then divide by the standard deviation s_x. The covariance between the standardized X and Y data is known as the *correlation coefficient* between Y and X and is given by

$$\text{cor}(Y, X) = \frac{1}{n-1} \sum_{i=1}^{n} \left(\frac{y_i - \bar{y}}{s_y}\right)\left(\frac{x_i - \bar{x}}{s_x}\right). \tag{3.5}$$

Equivalent formulas for the correlation coefficient are

$$\text{cor}(Y, X) = \frac{\text{cov}(Y, X)}{s_y s_x} \tag{3.6}$$

$$= \frac{\sum (y_i - \bar{y})(x_i - \bar{x})}{\sqrt{\sum (y_i - \bar{y})^2 \sum (x_i - \bar{x})^2}}. \tag{3.7}$$

Thus, $\text{cor}(Y, X)$ can be interpreted either as the covariance between the standardized variables or the ratio of the covariance to the standard deviations of the two variables. From (3.5), it can be seen that the correlation coefficient is symmetric, that is, $\text{cor}(Y, X) = \text{cor}(X, Y)$.

Unlike $\text{cov}(Y, X)$, $\text{cor}(Y, X)$ is scale invariant, that is, it does not change if we change the units of measurements. Furthermore, $\text{cor}(Y, X)$ satisfies

$$-1 \leq \text{or}(Y, X) \leq 1. \tag{3.8}$$

These properties make the $\text{cor}(Y, X)$ a useful quantity for measuring both the direction and the strength of the relationship between Y and X. The magnitude of $\text{cor}(Y, X)$ measures the strength of the linear relationship between Y and X. The closer $\text{cor}(Y, X)$ is to 1 or -1, the stronger is the relationship between

Table 3.3 Data Set with a Perfect Nonlinear Relationship Between Y and X, Yet $cor(X, Y) = 0$

Y	X	Y	X	Y	X
1	−7	46	−2	41	3
14	−6	49	−1	34	4
25	−5	50	0	25	5
34	−4	49	1	14	6
41	−3	46	2	1	7

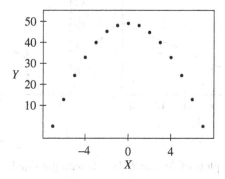

Figure 3.2 Scatter plot of Y versus X in Table 3.3.

Y and X. The sign of $cor(Y, X)$ indicates the direction of the relationship between Y and X. That is, $or(Y, X) > 0$ implies that Y and X are positively related. Conversely, $cor(Y, X) < 0$ implies that Y and X are negatively related.

Note, however, that $cor(Y, X) = 0$ does not necessarily mean that Y and X are not related. It only implies that they are not linearly related because the correlation coefficient measures only *linear* relationships. In other words, the $cor(Y, X)$ can still be zero when Y and X are nonlinearly related. For example, Y and X in Table 3.3 have the perfect nonlinear relationship $Y = 50 - X^2$ (graphed in Figure 3.2)[1], yet $cor(Y, X) = 0$.

Furthermore, like many other summary statistics, the $cor(Y, X)$ can be substantially influenced by one or a few outliers in the data. To emphasize this point, Anscombe (1973) has constructed four data sets, known as Anscombe quartet, each with a distinct pattern, but each having the same set of summary statistics (e.g., the same value of the correlation coefficient).

---

[1] The R code for producing this and all other graphs in this book can be found at the Book's Website at http://www.aucegypt.edu/faculty/hadi/RABE6.

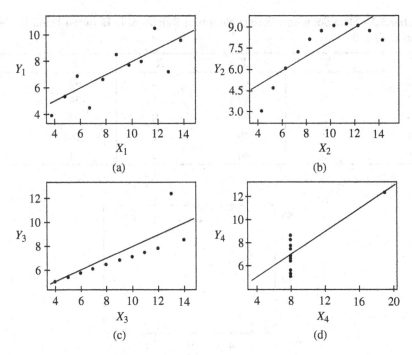

Figure 3.3 Scatter plots of Anscombe's data with the fitted regression lines.

The pairwise scatter plots with the fitted regression lines are given in Figure 3.3. The data can be found at the Book's Website.[2] It can also be obtained from the R package datasets by typing data(anscombe). An analysis based exclusively on an examination of summary statistics, such as the correlation coefficient, would have been unable to detect the differences in patterns.

An examination of Figure 3.3 shows that only the first set, whose plot is given in (a), can be described by a linear model. The plot in (b) shows the second data set is distinctly nonlinear and would be better fitted by a quadratic function. The plot in (c) shows that the third data set has one point that distorts the slope and the intercept of the fitted line. The plot in (d) shows that the fourth data set is unsuitable for linear fitting, the fitted line being determined essentially by one extreme observation. Therefore, it is important to examine the scatter plot of Y versus X before interpreting the numerical value of $\text{cor}(Y, X)$.

---

[2] http://www.aucegypt.edu/faculty/hadi/RABE6

3.3 EXAMPLE: COMPUTER REPAIR DATA

As an illustrative example, consider a case of a company that markets and repairs small computers. To study the relationship between the length of a service call and the number of electronic components in the computer that must be repaired or replaced, a sample of records on service calls was taken. The data consist of the length of service calls in minutes (the response variable) and the number of components repaired (the predictor variable). The data are presented in Table 3.4. The Computer Repair data can also be found at the Book's Website. We use this data set throughout this chapter as an illustrative example. The quantities needed to compute $\bar{y}, \bar{x}$, cov(Y,X), and cor(Y,X) are shown in Table 3.5. We have

$$\bar{y} = \frac{\sum_{i=1}^{n} y_i}{n} = \frac{1361}{14} = 97.21 \quad \text{and} \quad \bar{x} = \frac{\sum_{i=1}^{n} x_i}{n} = \frac{84}{14} = 6,$$

$$\text{cov}(Y,X) = \frac{\sum_{i=1}^{n}(y_i - \bar{y})(x_i - \bar{x})}{n-1} = \frac{1768}{13} = 136,$$

and

$$\text{cor}(Y,X) = \frac{\sum(y_i - \bar{y})(x_i - \bar{x})}{\sqrt{\sum(y_i - \bar{y})^2 \sum(x_i - \bar{x})^2}} = \frac{1768}{\sqrt{27,768.36 \times 114}} = 0.994.$$

The above quantities can be obtained using R as follows: First download the file Computer.Repair.csv from the Book's Website in your Working Directory (see Section 2.6.1) in an object named df (see Section 2.6.5). Then use the following code:

```
attach(df) # To make the variable names in df available
colMeans(df)
cov(Minutes,Units)
cor(Minutes,Units)
detach(df) # To de-attach the object df
```

Before drawing conclusions from this value of cor(Y,X), we should examine the corresponding scatter plot of Y versus X. This plot is given in Figure 3.4. The high value of cor$(Y,X) = 0.996$ is consistent with the strong linear relationship between Y and X exhibited in Figure 3.4. We therefore

Table 3.4 Length of Service Calls (in Minutes) and Number of Units Repaired

Row	Minutes	Units	Row	Minutes	Units
1	23	1	8	97	6
2	29	2	9	109	7
3	49	3	10	119	8
4	64	4	11	149	9
5	74	4	12	145	9
6	87	5	13	154	10
7	96	6	14	166	10

Table 3.5 Quantities Needed for Computation of Correlation Coefficient Between Length of Service Calls, Y, and Number of Units Repaired, X

i	y_i	x_i	$y_i - \bar{y}$	$x_i - \bar{x}$	$(y_i - \bar{y})^2$	$(x_i - \bar{x})^2$	$(y_i - \bar{y})(x_i - \bar{x})$
1	23	1	−74.21	−5	5507.76	25	371.07
2	29	2	−68.21	−4	4653.19	16	272.86
3	49	3	−48.21	−3	2324.62	9	144.64
4	64	4	−33.21	−2	1103.19	4	66.43
5	74	4	−23.21	−2	538.90	4	46.43
6	87	5	−10.21	−1	104.33	1	10.21
7	96	6	−1.21	0	1.47	0	0.00
8	97	6	−0.21	0	0.05	0	0.00
9	109	7	11.79	1	138.90	1	11.79
10	119	8	21.79	2	474.62	4	43.57
11	149	9	51.79	3	2681.76	9	155.36
12	145	9	47.79	3	2283.47	9	143.36
13	154	10	56.79	4	3224.62	16	227.14
14	166	10	68.79	4	4731.47	16	275.14
Total	1361	84	0.00	0	27,768.36	114	1768.00

conclude that there is a strong positive relationship between repair time and units repaired.

Although cor(Y, X) is a useful quantity for measuring the direction and the strength of linear relationships, it cannot be used for prediction purposes, that is, we cannot use cor(Y, X) to predict the value of one variable given the value of the other. Furthermore, cor(Y, X) measures only pairwise relationships.

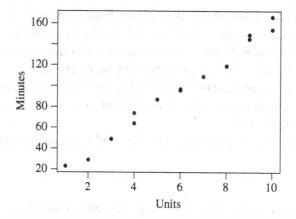

Figure 3.4 Computer repair data: scatter plot of minutes versus units.

Regression analysis, however, can be used to relate one or more response variable to one or more predictor variables. It can also be used in prediction. Regression analysis is an attractive extension to correlation analysis because it postulates a model that can be used not only to measure the direction and the strength of a relationship between the response and predictor variables, but also to numerically describe that relationship. We discuss simple linear regression models in the rest of this chapter. Chapter 4 is devoted to multiple regression models.

3.4 THE SIMPLE LINEAR REGRESSION MODEL

The relationship between a response variable Y and a predictor variable X is postulated as a linear model[3]

$$Y = \beta_0 + \beta_1 X + \varepsilon, \qquad (3.9)$$

where β_0 and β_1 are constants called the *model regression coefficients* or *parameters*, and ε is a random disturbance or error. It is assumed that in the range of the observations studied, the linear equation (3.9) provides an

---
[3] The adjective *linear* has a dual role here. It may be taken to describe the fact that the relationship between Y and X is linear. More generally, the word *linear* refers to the fact that the regression parameters, β_0 and β_1, enter (3.9) in a linear fashion. Thus, for example, $Y = \beta_0 + \beta_1 X^2 + \varepsilon$ is also a linear model even though the relationship between Y and X is quadratic.

acceptable approximation to the true relation between Y and X. In other words, Y is approximately a linear function of X, and ε measures the discrepancy in that approximation. In particular ε contains no systematic information for determining Y that is not already captured in X. The coefficient β_1, called the *slope*, may be interpreted as the change in Y for unit change in X. The coefficient β_0, called the *constant* coefficient or *intercept*, is the predicted value of Y when $X = 0$.

According to (3.9), each observation in Table 3.1 can be written as

$$y_i = \beta_0 + \beta_1 x_i + \varepsilon_i, \quad i = 1, 2, \ldots, n, \tag{3.10}$$

where y_i represents the ith value of the response variable Y, x_i represents the ith value of the predictor variable X, and ε_i represents the error in the approximation of y_i.

Regression analysis differs in an important way from correlation analysis. The correlation coefficient is symmetric in the sense that $\mathrm{cor}(Y, X)$ is the same as $\mathrm{cor}(X, Y)$. The variables X and Y are of equal importance. In regression analysis the response variable Y is of primary importance. The importance of the predictor X lies on its ability to account for the variability of the response variable Y and not in itself per se. Hence Y is of primary importance.

Returning to the Computer Repair Data example, suppose that the company wants to forecast the number of service engineers that will be required over the next few years. A linear model,

$$\text{Minutes} = \beta_0 + \beta_1 \text{ Units} + \varepsilon, \tag{3.11}$$

is assumed to represent the relationship between the length of service calls and the number of electronic components in the computer that must be repaired or replaced. To validate this assumption, we examine the graph of the response variable versus the explanatory variable. This graph, shown in Figure 3.4, suggests that the straight line relationship in (3.11) is a reasonable assumption.

3.5 PARAMETER ESTIMATION

Based on the available data, we wish to estimate the parameters β_0 and β_1. This is equivalent to finding the straight line that gives the *best fit* (representation) of the points in the scatter plot of the response versus the predictor variable (see Figure 3.4). We estimate the parameters using the popular *least squares method*, which gives the line that minimizes the sum of squares of the *vertical distances*[4] from each point to the line. The vertical distances represent the errors in the response variable. These errors can be obtained by rewriting (3.10) as

$$\varepsilon_i = y_i - \beta_0 - \beta_1 x_i, \quad i = 1, 2, \ldots, n. \tag{3.12}$$

The sum of squares of these distances can then be written as

$$S(\beta_0, \beta_1) = \sum_{i=1}^{n} \varepsilon_i^2 = \sum_{i=1}^{n} (y_i - \beta_0 - \beta_1 x_i)^2. \tag{3.13}$$

The values of $\hat{\beta}_0$ and $\hat{\beta}_1$ that minimize $S(\beta_0, \beta_1)$ are given by

$$\hat{\beta}_1 = \frac{\sum (y_i - \bar{y})(x_i - \bar{x})}{\sum (x_i - \bar{x})^2} \tag{3.14}$$

and

$$\hat{\beta}_0 = \bar{y} - \hat{\beta}_1 \bar{x}. \tag{3.15}$$

Note that we give the formula for $\hat{\beta}_1$ before the formula for $\hat{\beta}_0$ because $\hat{\beta}_0$ uses $\hat{\beta}_1$. The estimates $\hat{\beta}_0$ and $\hat{\beta}_1$ are called the least squares estimates of β_0 and β_1 because they are the solution to the *least squares method*, the intercept and the slope of the line that has the smallest possible sum of squares of the vertical distances from each point to the line. For this reason, the line is called the *least squares regression line*. The least squares regression line is given by

$$\hat{Y} = \hat{\beta}_0 + \hat{\beta}_1 X. \tag{3.16}$$

Note that a least squares line always exists because we can always find a line that gives the minimum sum of squares of the vertical distances. In fact, as we shall see later, in some cases a least squares line may not be unique. These cases are not common in practice.

For each observation in our data we can compute

$$\hat{y}_i = \hat{\beta}_0 + \hat{\beta}_1 x_i, \quad i = 1, 2, \ldots, n. \tag{3.17}$$

---

[4] An alternative to the vertical distance is the *perpendicular* (shortest) distance from each point to the line. The resultant line is called the *orthogonal regression* line.

These are called the *fitted* values. Thus, the ith fitted value, $\hat{y}_i$, is the point on the least squares regression line (3.16) corresponding to x_i. The vertical distance corresponding to the ith observation is

$$e_i = y_i - \hat{y}_i, \quad i = 1, 2, \ldots, n. \tag{3.18}$$

These vertical distances are called the *ordinary*[5] *least squares residuals*. One property of the residuals in (3.18) is that their sum is zero [see Exercise 3.5(a)]. This means that the sum of the distances above the line is equal to the sum of the distances below the line.

Using the Computer Repair data and the quantities in Table 3.5, we have

$$\hat{\beta}_1 = \frac{\sum(y_i - \bar{y})(x_i - \bar{x})}{\sum(x_i - \bar{x})^2} = \frac{1768}{114} = 15.509,$$

and

$$\hat{\beta}_0 = \bar{y} - \hat{\beta}_1 \bar{x} = 97.21 - 15.509 \times 6 = 4.162.$$

Then the equation of the least squares regression line is

$$\text{Minutes} = 4.162 + 15.509 \, \text{Units}. \tag{3.19}$$

This least squares line is shown together with the scatter plot of Minutes versus Units in Figure 3.5. The fitted values in (3.17) and the residuals in (3.18) are shown in Table 3.6.

The coefficients in (3.19) can be computed using the function lm(), where lm stands for Linear Models. The syntax for lm() is:

lm(Y ~ X, data = df).

This code assumes that both Y and X are in the dataset df. The ~ in this code means that Y is a function of X, in the same sense as $Y = f(X)$. In the current example, we use reg <- lm(Minutes ~ Units, data = df). This code indicates that the output of lm() is saved in an object named reg (short for regression). Typing print(reg) after this code will produce the following output:

```
reg <- lm(Minutes ~ Units, data = df)
print(reg)
Call:
lm(formula = Minutes ~ Units, data = df)
Coefficients:
(Intercept)        Units
      4.162       15.509
```

---

[5] To be distinguished from other types of residuals to be presented later.

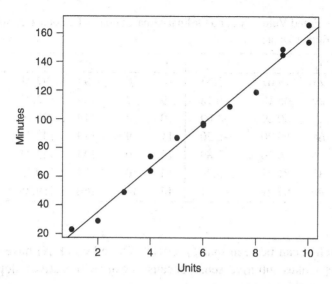

Figure 3.5 Computer repair data: plot of minutes versus uUnits with the fitted regression line.

The last line in the above output gives the regression coefficients $\hat{\beta}_0$ and $\hat{\beta}_1$, respectively. These coefficients can be interpreted in physical terms. The constant (intercept) term represents the setup or startup time for each repair and is approximately four minutes. The coefficient of Units represents the increase in the length of a service call for each additional component that has to be repaired. From the data given, we estimate that it takes about 15.5 minutes for each additional component that has to be repaired. For example, the length of a service call in which four components had to be repaired is obtained by substituting Units = 4 in the equation of the regression line (3.19) and obtaining $\hat{y} = 4.162 + 15.509 \times 4 = 66.20$. Since Units = 4, corresponds to two observations in our data set (observations 4 and 5), the value 66.198 is the fitted value for both observations 4 and 5, as can be seen from Table 3.6. Note, however, that since observations 4 and 5 have different values for the response variable Minutes, they have different residuals.

R knows that the object reg computed above is a lm() output. Accordingly, the fitted values ($\hat{y}_i$) and residuals (e_i) in Table 3.6 can be computed using the commands: fitted(reg) and residuals(reg), respectively.

We should note here that by comparing (3.2), (3.7), and (3.14), an alternative formula for $\hat{\beta}_1$ can be expressed as

$$\hat{\beta}_1 = \frac{\text{cov}(Y, X)}{\text{Var}(X)} = \text{cor}(Y, X)\frac{s_y}{s_x}, \qquad (3.20)$$

Table 3.6 Fitted Values, $\hat{y}_i$, and Ordinary Least Squares Residuals, e_i, for Computer Repair Data

i	x_i	y_i	$\hat{y}_i$	e_i	i	x_i	y_i	$\hat{y}_i$	e_i
1	1	23	19.67	3.33	8	6	97	97.21	−0.21
2	2	29	35.18	−6.18	9	7	109	112.72	−3.72
3	3	49	50.69	−1.69	10	8	119	128.23	−9.23
4	4	64	66.20	−2.20	11	9	149	143.74	5.26
5	4	74	66.20	7.80	12	9	145	143.74	1.26
6	5	87	81.71	5.29	13	10	154	159.25	−5.25
7	6	96	97.21	−1.21	14	10	166	159.25	6.75

from which it can be seen that $\hat{\beta}_1$, cov(Y, X), and cor(Y, X) have the same sign. This makes intuitive sense because positive (negative) slope means positive (negative) correlation.

So far in our analysis we have made only one assumption, namely, that Y and X are linearly related. This assumption is referred to as the *linearity* assumption. This is merely an assumption or a hypothesis about the relationship between the response and predictor variables. An early step in the analysis should always be the validation of this assumption. We wish to determine if the data at hand support the assumption that Y and X are linearly related. An informal way to check this assumption is to examine the scatter plot of the response versus the predictor variable, preferably drawn with the least squares line superimposed on the graph (see Figure 3.5). This graph can be obtained in R using:

```
attach(df)
plot(Units,Minutes,pch=19)
reg<- lm(Minutes ~ Units,data = df)
abline(reg)
detach(df)
```

If the scatter of points resembles a straight line, then we conclude that the linearity assumption is reasonable and continue with our analysis. But, if we observe a nonlinear pattern, we will have to take corrective action. For example, we may *reexpress* or *transform* the data before we continue the analysis. *Data transformation* is discussed in Chapter 7.

The least squares estimators have several desirable properties when some additional assumptions hold. The required assumptions are stated in Chapter 5. The validity of these assumptions must be checked before meaningful conclusions can be reached from the analysis. Chapter 5 also presents methods for the validation of these assumptions. Using the properties of least squares estimators, one can develop statistical inference procedures (e.g., confidence interval estimation, tests of hypothesis, and goodness-of-fit tests). These are presented in Sections 3.6–3.9.

3.6 TESTS OF HYPOTHESES

As stated earlier, the usefulness of X as a predictor of Y can be measured informally by examining the correlation coefficient and the corresponding scatter plot of Y versus X. A more formal way of measuring the usefulness of X as a predictor of Y is to conduct a test of hypothesis about the regression parameter β_1. Note that the hypothesis $\beta_1 = 0$ means that there is no linear relationship between Y and X. A test of this hypothesis requires the following assumption. For every fixed value of X, the ε's are assumed to be independent random quantities normally distributed with mean zero and a common variance σ^2. With these assumptions, the quantities $\hat{\beta}_0$ and $\hat{\beta}_1$ are unbiased[6] estimates of β_0 and β_1, respectively. Their variances are

$$\text{Var}(\hat{\beta}_0) = \sigma^2 \left[\frac{1}{n} + \frac{\bar{x}^2}{\sum (x_i - \bar{x})^2} \right], \qquad (3.21)$$

and

$$\text{Var}(\hat{\beta}_1) = \frac{\sigma^2}{\sum (x_i - \bar{x})^2}. \qquad (3.22)$$

Furthermore, the *sampling distributions* of the least squares estimates $\hat{\beta}_0$ and $\hat{\beta}_1$ are normal with means β_0 and β_1 and variance as given in (3.21) and (3.22), respectively.

The variances of $\hat{\beta}_0$ and $\hat{\beta}_1$ depend on the unknown parameter σ^2. So, we need to estimate σ^2 from the data. An unbiased estimate of σ^2 is given by

$$\hat{\sigma}^2 = \frac{\sum e_i^2}{n-2} = \frac{\sum (y_i - \hat{y}_i)^2}{n-2} = \frac{\text{SSE}}{n-2}, \qquad (3.23)$$

where SSE is the sum of squares of the residuals (errors). The square root of $\hat{\sigma}^2$, $\hat{\sigma}$, is called the *residual standard error*. The number $n - 2$ in the denominator of (3.23) is called the *degrees of freedom* (*df*). It is equal

---

[6] An estimate $\hat{\theta}$ is said to be an unbiased estimate of a parameter θ if the expected value of $\hat{\theta}$ is equal to θ.

SIMPLE LINEAR REGRESSION

to the number of observations minus the number of estimated regression coefficients.

Replacing σ^2 in (3.21) and (3.22) by $\hat{\sigma}^2$ in (3.23), we get unbiased estimates of the variances of $\hat{\beta}_0$ and $\hat{\beta}_1$. An *estimate* of the *standard deviation* is called the *standard error* (s.e.) of the estimate. Thus, the standard errors of $\hat{\beta}_0$ and $\hat{\beta}_1$ are

$$\text{s.e.}(\hat{\beta}_0) = \hat{\sigma}\sqrt{\frac{1}{n} + \frac{\bar{x}^2}{\sum(x_i - \bar{x})^2}} \qquad (3.24)$$

and

$$\text{s.e.}(\hat{\beta}_1) = \frac{\hat{\sigma}}{\sqrt{\sum(x_i - \bar{x})^2}}, \qquad (3.25)$$

respectively, where $\hat{\sigma}$ is the square root of $\hat{\sigma}^2$ in (3.23). The standard error of $\hat{\beta}_1$ is a measure of how precisely the slope has been estimated. The smaller the standard error, the more precise the estimator.

The R object reg from the lm() call actually includes the quantities in (3.23)–(3.25), among others. To see this, if we type summary(reg), we will get

```
summary(reg)
Call:
lm(formula = Minutes ~ Units, data = df)
Residuals:
    Min      1Q  Median      3Q     Max
-9.2318 -3.3415 -0.7143  4.7769  7.8033
Coefficients:
            Estimate Std. Error t value Pr(>|t|)
(Intercept)    4.162      3.355    1.24    0.239
Units         15.509      0.505   30.71 8.92e-13 ***
---
Signif. codes:  0 '***' 0.001 '**' 0.01 '*' 0.05 '.' 0.1 ' ' 1

Residual standard error: 5.392 on 12 degrees of freedom
Multiple R-squared:  0.9874,    Adjusted R-squared:  0.9864
F-statistic: 943.2 on 1 and 12 DF,  p-value: 8.916e-13
```

In the above output, after the call, a summary of the residuals is given followed by what is called the Coefficients Table. This table shows the estimated regression coefficients (4.162 and 15.509) and their standard errors (3.355 and 0.505), respectively. The first line below the Coefficients Table shows that the Residual standard error is 5.392 on 12 ($n - 2$) degrees of freedom. The other values in the above output will be explained shortly.

With the sampling distributions of $\hat{\beta}_0$ and $\hat{\beta}_1$, we are now in position to perform statistical analysis concerning the usefulness of X as a predictor of Y. Under the normality assumption, an appropriate test statistic for testing the null hypothesis $H_0 : \beta_1 = 0$ against the alternative $H_1 : \beta_1 \neq 0$ is the t-Test,

$$t_1 = \frac{\hat{\beta}_1}{\text{s.e.}(\hat{\beta}_1)}. \tag{3.26}$$

The statistic t_1 is distributed as a Student's t with $n - 2$ degrees of freedom. The test is carried out by comparing this observed value with the appropriate critical value obtained from the Student t-distribution, which is $t_{(n-2,\alpha/2)}$, where α is a specified significance level. Note that we divide α by 2 because we have a two-sided alternative hypothesis. The value of $t_{(n-2,\alpha/2)}$ can be obtained by R using the command qt($\alpha/2$, $n - 2$, lower.tail = FALSE). Accordingly, H_0 is to be rejected at the significance level α if

$$|t_1| \geq t_{(n-2,\alpha/2)}, \tag{3.27}$$

where $|t_1|$ denotes the absolute value of t_1.

A criterion equivalent to that in (3.27) is to compare the p-value for the t-Test with α and reject H_0 if

$$p(|t_1|) \leq \alpha, \tag{3.28}$$

where $p(|t_1|)$, called the *p-value*, is the probability that a random variable having a Student t distribution with $n - 2$ degrees of freedom is greater than $|t_1|$ (the absolute value of the observed value of the t-Test). Figure 3.6 is a graph of the density function of a t-distribution. The p-value is the sum of the two shaded areas under the curve. Letting abs.t = abs(t_1), the p-value can be obtained using R by

2 * pt(abs.t, $n - 2$, lower.tail = FALSE).

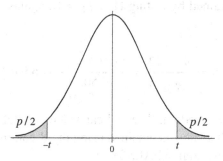

Figure 3.6 Graph of the probability density function of a t-distribution. The p-value for the t-Test is the shaded areas under the curve.

Note that we multiplied by 2 in the above line because the pt() gives only one side. However, the *p*-value is computed and supplied as part of the lm() output. In the summary(reg) above, the *t*-values and *p*-values are given for both the intercept and the slope under the columns labeled t value and Pr(>|t|), respectively. Note that the *p*-value for the slope is nearly zero (8.92e–13); hence, $H_0 : \beta_1 = 0$ is rejected at practically any small significant level. This is indicated in the summary(reg) output above by three stars. This would mean that β_1 is likely to be different from 0, and hence the predictor variable Units is a statistically significant predictor of the response variable Minutes.

To complete the picture of hypotheses testing regarding regression parameters, we give here tests for three other hypotheses that may arise in practice.

Testing $H_0: \beta_1 = \beta_1^0$

The above *t*-Test can be generalized to test the more general hypothesis $H_0 : \beta_1 = \beta_1^0$, where β_1^0 is a constant chosen by the investigator, against the two-sided alternative $H_1 : \beta_1 \neq \beta_1^0$. The appropriate test statistic in this case is the *t*-Test,

$$t_1 = \frac{\hat{\beta}_1 - \beta_1^0}{\text{s.e.}(\hat{\beta}_1)}. \tag{3.29}$$

Note that when $\beta_1^0 = 0$, the *t*-Test in (3.29) reduces to the *t*-Test in (3.26). The statistic t_1 in (3.29) is also distributed as a Student's *t* with $n - 2$ degrees of freedom. Thus, $H_0 : \beta_1 = \beta_1^0$ is rejected if (3.27) holds [or, equivalently, if (3.28) holds].

For illustration, using the Computer Repair data, let us suppose that the management expected the increase in service time for each additional unit to be repaired to be 12 minutes. Do the data support this conjecture? The answer may be obtained by testing $H_0 : \beta_1 = 12$ against $H_1 : \beta_1 \neq 12$. The appropriate statistic is

$$t_1 = \frac{\hat{\beta}_1 - 12}{\text{s.e.}(\hat{\beta}_1)} = \frac{15.509 - 12}{0.505} = 6.948,$$

with 12 degrees of freedom. The critical value for this test is $t_{(n-2, \alpha/2)} = t_{(12, 0.025)} = 2.18$. This value is obtained using the R command
   qt(0.05/2, 12, lower.tail = FALSE)
Since $t_1 = 6.948 > 2.18$, the result is highly significant, leading to the rejection of the null hypothesis. The management's estimate of the increase

in time for each additional component to be repaired is not supported by the data. Their estimate is too low.

Testing $H_0: \beta_0 = \beta_0^0$

The need for testing hypotheses regarding the regression parameter β_0 may also arise in practice. More specifically, suppose we wish to test $H_0 : \beta_0 = \beta_0^0$ against the alternative $H_1 : \beta_0 \neq \beta_0^0$, where β_0^0 is a constant chosen by the investigator. The appropriate test in this case is given by

$$t_0 = \frac{\hat{\beta}_0 - \beta_0^0}{\text{s.e.}(\hat{\beta}_0)}. \tag{3.30}$$

If we set $\beta_0^0 = 0$, a special case of this test is obtained as

$$t_0 = \frac{\hat{\beta}_0}{\text{s.e.}(\hat{\beta}_0)}, \tag{3.31}$$

which tests $H_0 : \beta_0 = 0$ against the alternative $H_1 : \beta_0 \neq 0$.

The least squares estimates of the regression coefficients, their standard errors, the t-Tests for testing that the corresponding coefficient are zero, and the p-values are usually given as part of the regression output by statistical packages. These values are usually displayed in a table such as the one in Table 3.7. This table is known as the *coefficients table*. To facilitate the connection between a value in the table and the formula used to obtain it, the equation number of the formula is given in parentheses.

As an illustrative example, Table 3.8 shows a part of the regression output for the Computer Repair data in Table 3.4. Thus, for example, $\hat{\beta}_1 = 15.509$, the s.e.$(\hat{\beta}_1) = 0.505$, and hence $t_1 = 15.509/0.505 = 30.71$. The critical value for this test using $\alpha = 0.05$, for example, is $t_{(12, 0.025)} = 2.18$. The $t_1 = 30.71$ is much larger than its critical value 2.18. Consequently, according to (3.27), $H_0 : \beta_1 = 0$ is rejected, which means that the predictor variable Units is a statistically significant predictor of the response variable Minutes. This

Table 3.7 Standard Regression Output

Variable	Coefficient (Formula)	s.e. (Formula)	t-Test (Formula)	p-Value
Constant	$\hat{\beta}_0$ (3.15)	s.e.$(\hat{\beta}_0)$ (3.24)	t_0 (3.31)	p_0
X	$\hat{\beta}_1$ (3.14)	s.e.$(\hat{\beta}_1)$ (3.25)	t_1 (3.26)	p_1

Equation number of the corresponding formulas are given in parentheses.

Table 3.8 Regression Output for Computer Repair Data

Variable	Coefficient	s.e.	t-Test	p-Value
Constant	4.162	3.355	1.24	0.2385
Units	15.509	0.505	30.71	0.0000

conclusion can also be reached using (3.28) by observing that the p-value ($p_1 < 0.0001$) is much less than $\alpha = 0.05$ indicating very high significance.

A Test Using Correlation Coefficient

As mentioned above, a test of $H_0 : \beta_1 = 0$ against $H_1 : \beta_1 \neq 0$ can be thought of as a test for determining whether the response and the predictor variables are linearly related. We used the t-Test in (3.26) to test this hypothesis. An alternative test, which involves the correlation coefficient between Y and X, can be developed. Suppose that the population correlation coefficient between Y and X is denoted by ρ. If $\rho \neq 0$, then Y and X are linearly related. An appropriate test for testing $H_0 : \rho = 0$ against $H_1 : \rho \neq 0$ is given by

$$t_1 = \frac{\operatorname{cor}(Y,X)\sqrt{n-2}}{\sqrt{1-[\operatorname{cor}(Y,X)]^2}}, \tag{3.32}$$

where $\operatorname{cor}(Y,X)$ is the sample correlation coefficient between Y and X, defined in (3.6), which is considered here to be an estimate of ρ. The t-Test in (3.32) is distributed as a Student's t with $n-2$ degrees of freedom. Thus, $H_0 : \rho = 0$ is rejected if (3.27) holds [or, equivalently, if (3.28) holds]. Again if $H_0 : \rho = 0$ is rejected, it means that there is a statistically significant linear relationship between Y and X.

It is clear that if no linear relationship exists between Y and X, then $\beta_1 = 0$. Consequently, the statistical tests for $H_0 : \beta_1 = 0$ and $H_0 : \rho = 0$ should be identical. Although the statistics for testing these hypotheses given in (3.26) and (3.32) look different, it can be demonstrated that they are indeed algebraically equivalent.

3.7 CONFIDENCE INTERVALS

To construct confidence intervals for the regression parameters, we also need to assume that the ϵ's have a normal distribution, which will enable us to conclude that the sampling distributions of $\hat{\beta}_0$ and $\hat{\beta}_1$ are normal, as discussed in Section 3.6. Consequently, the $(1 - \alpha) \times 100\%$ confidence interval for β_0 is given by

$$\hat{\beta}_0 \pm t_{(n-2,\alpha/2)} \times \text{s.e.}(\hat{\beta}_0), \qquad (3.33)$$

where $t_{(n-2,\alpha/2)}$ is the $(1 - \alpha/2)$ percentile of a t distribution with $n - 2$ degrees of freedom. Similarly, limits of the $(1 - \alpha) \times 100\%$ confidence interval for β_1 are given by

$$\hat{\beta}_1 \pm t_{(n-2,\alpha/2)} \times \text{s.e.}(\hat{\beta}_1). \qquad (3.34)$$

The confidence interval in (3.34) has the usual interpretation, namely, if we were to take repeated samples of the same size at the same values of X and construct, for example, 95% confidence intervals for the slope parameter for each sample, then 95% of these intervals would be expected to contain the true value of the slope.

From Table 3.8 we see that a 95% confidence interval for β_1 is

$$15.509 \pm 2.18 \times 0.505 = (14.408, 16.610). \qquad (3.35)$$

That is, the incremental time required for each broken unit is between 14 and 17 minutes. The calculation of confidence interval for β_0 in this example is left as an exercise for the reader.

Note that the confidence limits in (3.33) and (3.34) are constructed for each of the parameters β_0 and β_1, separately. This does not mean that a simultaneous (joint) confidence region for the two parameters is rectangular. Actually, the simultaneous confidence region is elliptical. This region is given for the general case of multiple regression in the Appendix to Chapter 4 in (A.15), of which the simultaneous confidence region for β_0 and β_1 is a special case.

3.8 PREDICTIONS

The fitted regression equation can be used for prediction. We distinguish between two types of predictions:

1. The prediction of the value of the response variable Y which corresponds to any chosen value, x_0, of the predictor variable.
2. The estimation of the mean response μ_0, when $X = x_0$.

For the first case, the predicted value $\hat{y}_0$ is

$$\hat{y}_0 = \hat{\beta}_0 + \hat{\beta}_1 x_0. \tag{3.36}$$

The standard error of this prediction is

$$\text{s.e.}(\hat{y}_0) = \hat{\sigma}\sqrt{1 + \frac{1}{n} + \frac{(x_0 - \bar{x})^2}{\sum(x_i - \bar{x})^2}}. \tag{3.37}$$

Hence, the confidence limits for the predicted value with confidence coefficient $(1 - \alpha)$ are given by

$$\hat{y}_0 \pm t_{(n-2,\alpha/2)} \text{ s.e.}(\hat{y}_0). \tag{3.38}$$

For the second case, the mean response μ_0 is estimated by

$$\hat{\mu}_0 = \hat{\beta}_0 + \hat{\beta}_1 x_0. \tag{3.39}$$

The standard error of this estimate is

$$\text{s.e.}(\hat{\mu}_0) = \hat{\sigma}\sqrt{\frac{1}{n} + \frac{(x_0 - \bar{x})^2}{\sum(x_i - \bar{x})^2}}, \tag{3.40}$$

from which it follows that the confidence limits for μ_0 with confidence coefficient $(1 - \alpha)$ are given by

$$\hat{\mu}_0 \pm t_{(n-2,\alpha/2)} \text{ s.e.}(\hat{\mu}_0). \tag{3.41}$$

Note that the point estimate of μ_0 is identical to the predicted response $\hat{y}_0$. This can be seen by comparing (3.36) with (3.39). The standard error of $\hat{\mu}_0$ is, however, smaller than the standard error of $\hat{y}_0$ and can be seen by comparing (3.37) with (3.40). Intuitively, this makes sense. There is greater uncertainty (variability) in predicting one observation (the next observation) than in estimating the mean response when $X = x_0$. The averaging that is implied in the mean response reduces the variability and uncertainty associated with the estimate.

To distinguish between the limits in (3.38) and (3.41), the limits in (3.38) are sometimes referred to as the *prediction* or *forecast* limits, whereas the limits given in (3.41) are called the *confidence limits*.

Suppose that we wish to predict the length of a service call in which four components had to be repaired. If $\hat{y}_4$ denotes the predicted value, then from (3.36) we get

$$\hat{y}_4 = 4.162 + 15.509 \times 4 = 66.20,$$

with a standard error that is obtained from (3.37) as

$$\text{s.e.}(\hat{y}_4) = 5.392\sqrt{1 + \frac{1}{14} + \frac{(4-6)^2}{114}} = 5.67.$$

On the other hand, if the service department wishes to estimate the expected (mean) service time for a call that needed four components repaired, we would use (3.39) and (3.40), respectively. Denoting by μ_4, the expected service time for a call that needed four components to be repaired, we have

$$\hat{\mu}_4 = 4.162 + 15.509 \times 4 = 66.20,$$

with a standard error

$$\text{s.e.}(\hat{\mu}_4) = 5.392\sqrt{\frac{1}{14} + \frac{(4-6)^2}{114}} = 1.76.$$

With these standard errors we can construct confidence intervals using (3.38) and (3.41), as appropriate.

As can be seen from (3.37), the standard error of prediction increases the farther the value of the predictor variable is from the center of the actual observations. Care should be taken when predicting the value of Minutes corresponding to a value for Units that does not lie close to the observed data. There are two dangers in such predictions. First, there is substantial uncertainty due to the large standard error. More important, the linear relationship that has been estimated may not hold outside the range of observations. Therefore, care should be taken in employing fitted regression lines for prediction far outside the range of observations. In our example we would not use the fitted equation to predict the service time for a service call which requires that 25 components be replaced or repaired. This value lies too far outside the existing range of observations.

3.9 MEASURING THE QUALITY OF FIT

After fitting a linear model relating Y to X, we are interested not only in knowing whether a linear relationship exists, but also in measuring the quality of the fit of the model to the data. The quality of the fit can be assessed by one of the following highly related (hence, somewhat redundant) ways:

1. When using the tests in (3.26) or (3.32), if H_0 is rejected, the magnitude of the values of the test (or the corresponding p-values) gives us information about the *strength* (not just the existence) of the linear relationship between Y and X. Basically, the larger the t (in absolute value) or the smaller the corresponding p-value, the stronger the linear relationship between Y and X. These tests are objective but they require all the assumptions stated earlier, specially the assumption of normality of the ε's.

2. The strength of the linear relationship between Y and X can also be assessed directly from the examination of the scatter plot of Y versus X together with the corresponding value of the correlation coefficient cor(Y, X) in (3.6). The closer the set of points to a straight line [the closer cor(Y, X) to 1 or -1], the stronger the linear relationship between Y and X. This approach is informal and subjective, but it requires only the linearity assumption.

3. Examine the scatter plot of Y versus $\hat{Y}$. The closer the set of points to a straight line, the stronger the linear relationship between Y and X. One can measure the strength of the linear relationship in this graph by computing the correlation coefficient between Y and $\hat{Y}$, which is given by

$$\text{cor}(Y, \hat{Y}) = \frac{\sum (y_i - \bar{y})(\hat{y}_i - \bar{\hat{y}})}{\sqrt{\sum (y_i - \bar{y})^2 \sum (\hat{y}_i - \bar{\hat{y}})^2}}, \quad (3.42)$$

where $\bar{y}$ is the mean of the response variable Y and $\bar{\hat{y}}$ is the mean of the fitted values. In fact, the scatter plot of Y versus X and the scatter plot of Y versus $\hat{Y}$ are redundant because the patterns of points in the two graphs are identical. The two corresponding values of the correlation coefficient are related by the following equation:

$$\text{cor}(Y, \hat{Y}) = |\text{cor}(Y, X)|. \quad (3.43)$$

Note that cor$(Y, \hat{Y})$ cannot be negative (why?), but cor(Y, X) can be positive or negative $[-1 \leq \text{cor}(Y, X) \leq 1]$. Therefore, in simple linear regression, the scatter plot of Y versus $\hat{Y}$ is redundant. However, in multiple regression, the scatter plot of Y versus $\hat{Y}$ is not redundant. The graph is very useful because, as we shall see in Chapter 4, it is used to assess the strength of the relationship between Y and the set of predictor variables $X_1, X_2, \ldots, X_p$.

4. Although scatter plots of Y versus $\hat{Y}$ and $\mathrm{cor}(Y, \hat{Y})$ are redundant in simple linear regression, they give us an indication of the quality of the fit in both simple and multiple regressions. Furthermore, in both simple and multiple regressions, $\mathrm{cor}(Y, \hat{Y})$ is related to another useful measure of the quality of fit of the linear model to the observed data. This measure is developed as follows. After we compute the least squares estimates of the parameters of a linear model, let us compute the following quantities:

$$\mathrm{SST} = \sum (y_i - \bar{y})^2,$$
$$\mathrm{SSR} = \sum (\hat{y}_i - \bar{y})^2, \qquad (3.44)$$
$$\mathrm{SSE} = \sum (y_i - \hat{y}_i)^2,$$

where SST stands for the total sum of squared deviations in Y from its mean $\bar{y}$, SSR denotes the sum of squares due to regression, and SSE represents the sum of squared residuals (errors). The quantities $(\hat{y}_i - \bar{y})$, $(\hat{y}_i - \bar{y})$, and $(y_i - \hat{y}_i)$ are depicted in Figure 3.7 for a typical point (x_i, y_i). The line $\hat{y}_i = \hat{\beta}_0 + \hat{\beta}_1 x_i$ is the fitted regression line based on all data points (not shown on the graph) and the horizontal line is drawn at $Y = \bar{y}$. Note that for every point (x_i, y_i), there are two points, $(x_i, \hat{y}_i)$, which lies on the fitted line, and $(x_i, \bar{y})$ which lies on the line $Y = \bar{y}$.

A fundamental equality, in both simple and multiple regressions, is given by

$$\mathrm{SST} = \mathrm{SSR} + \mathrm{SSE}. \qquad (3.45)$$

This equation arises from the description of an observation as

$$y_i \quad = \quad \hat{y}_i \quad + \quad (y_i - \hat{y}_i)$$
Observed = Fit + Deviation from fit.

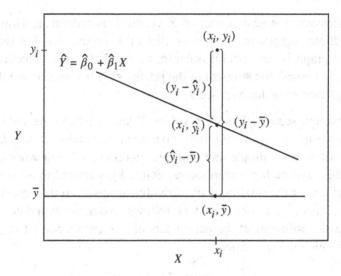

Figure 3.7 Graphical illustration of various quantities computed after fitting a regression line to data.

Subtracting $\bar{y}$ from both sides, we obtain

$$y_i - \bar{y} = (\hat{y}_i - \bar{y}) + (y_i - \hat{y}_i)$$
$$\text{Deviation from mean} = \text{Deviation due to fit} + \text{Residual.}$$

Accordingly, the total sum of squared deviations in Y can be decomposed into the sum of two quantities, the first, SSR, measures the quality of X as a predictor of Y, and the second, SSE, measures the error in this prediction. Therefore, the ratio $R^2 = \text{SSR/SST}$ can be interpreted as the proportion of the total variation in Y that is accounted for by the predictor variable X. Using (3.45), we can rewrite R^2 as

$$R^2 = \frac{\text{SSR}}{\text{SST}} = 1 - \frac{\text{SSE}}{\text{SST}}. \tag{3.46}$$

Additionally, it can be shown that

$$[\text{cor}(Y, X)]^2 = [\text{cor}(Y, \hat{Y})]^2 = R^2. \tag{3.47}$$

In simple linear regression, R^2 is equal to the square of the correlation coefficient between the response variable Y and the predictor X or to the square of the correlation coefficient between the response variable Y

and the fitted values $\hat{Y}$. The definition given in (3.46) provides us with an alternative interpretation of the squared correlation coefficients. The *goodness-of-fit index*, R^2, may be interpreted as the proportion of the total variability in the response variable Y that is accounted for by the predictor variable X. Note that $0 \leq R^2 \leq 1$ because SSE $\leq$ SST. If R^2 is near 1, then X accounts for a large part of the variation in Y. For this reason, R^2 is known as the *coefficient of determination* because it gives us an idea of how the predictor variable X accounts for (determines) the response variable Y. The same interpretation of R^2 will carry over to the case of multiple regression.

Using the Computer Repair data, the fitted values, and the residuals in Table 3.6, the reader can verify that $\text{cor}(Y, X) = \text{cor}(Y, \hat{Y}) = 0.994$, from which it follows that $R^2 = (0.994)^2 = 0.987$. The same value of R^2 can be computed using (3.46). Verify that SST $= 27{,}768.348$ and SSE $= 348.848$. So that

$$R^2 = 1 - \frac{\text{SSE}}{\text{SST}} = 1 - \frac{348.848}{27{,}768.348} = 0.987.$$

The value $R^2 = 0.987$ indicates that nearly 99% of the total variability in the response variable (Minutes) is accounted for by the predictor variable (Units). The high value of R^2 indicates a strong linear relationship between servicing time and the number of units repaired during a service call.

We reemphasize that the regression assumptions should be checked before drawing statistical conclusions from the analysis (e.g., conducting tests of hypothesis or constructing confidence or prediction intervals) because the validity of these statistical procedures hinges on the validity of the assumptions. Chapter 5 presents a collection of graphical displays that can be used for checking the validity of the assumptions. We have used these graphs for the computer repair data and found no evidence that the underlying assumptions of regression analysis are not in order. In summary, the 14 data points in the Computer Repair data have given us an informative view of the repair time problem. Within the range of observed data, we are confident of the validity of our inferences and predictions.

3.10 REGRESSION LINE THROUGH THE ORIGIN

We have considered fitting the model

$$Y = \beta_0 + \beta_1 X + \varepsilon, \tag{3.48}$$

which is a regression line with an intercept. Sometimes, it may be necessary to fit the model

$$Y = \beta_1 X + \varepsilon, \tag{3.49}$$

a line passing through the origin. This model is also called the *no-intercept* model. The line may be forced to go through the origin because of subject matter theory or other physical and material considerations. For example, distance traveled as a function of time should have no constant. Thus, in this case, the regression model in (3.49) is appropriate. Many other practical applications can be found where model (3.49) is more appropriate than (3.48). We shall see some of these examples in Chapter 8.

The least squares estimate of β_1 in (3.49) is

$$\hat{\beta}_1 = \frac{\sum y_i x_i}{\sum x_i^2}. \tag{3.50}$$

The *i*th fitted value is

$$\hat{y}_i = \hat{\beta}_1 x_i, \quad i = 1, 2, \ldots, n, \tag{3.51}$$

and the corresponding residual is

$$e_i = y_i - \hat{y}_i, \quad i = 1, 2, \ldots, n. \tag{3.52}$$

The standard error of the $\hat{\beta}_1$ is

$$\text{s.e.}(\hat{\beta}_1) = \frac{\hat{\sigma}}{\sqrt{\sum x_i^2}}, \tag{3.53}$$

where

$$\hat{\sigma} = \sqrt{\frac{\sum e_i^2}{n-1}} = \sqrt{\frac{\text{SSE}}{n-1}}. \tag{3.54}$$

Note that the degree of freedom for SSE is $n - 1$, not $n - 2$, as is the case for a model with an intercept.

Note that the residuals in (3.52) do not necessarily add up to zero as is the case for a model with an intercept [see Exercise 3.11(c)]. Also, the fundamental identity in (3.45) is no longer true in general. For this reason, some quality measures for models with an intercept such as R^2 in (3.46) are no longer appropriate for models with no intercept. The appropriate identity for the case of models with no intercept is obtained by replacing $\bar{y}$ in (3.45) by zero. Hence, the fundamental identity becomes

$$\sum_{i=1}^{n} y_i^2 = \sum_{i=1}^{n} \hat{y}_i^2 + \sum_{i=1}^{n} e_i^2, \qquad (3.55)$$

from which R^2 is redefined as

$$R^2 = \frac{\sum \hat{y}_i^2}{\sum y_i^2} = 1 - \frac{\sum e_i^2}{\sum y_i^2}. \qquad (3.56)$$

This is the appropriate form of R^2 for models with no intercept. Note, however, that the interpretations for the two formulas of R^2 are different. In the case of models with an intercept, R^2 can be interpreted as the proportion of the variation in Y that is accounted for by the predictor variable X after adjusting Y by its mean. For models without an intercept, no adjustment of Y is made. For example, if we fit (3.49) but use the formula for R^2 in (3.46), it is possible for R^2 to be negative in some cases [see Exercise 3.11(d)]. Therefore, the correct formula and the correct interpretation should be used.

The formula for the t-Test in (3.29) for testing $H_0 : \beta_1 = \beta_1^0$ against the two-sided alternative $H_1 : \beta_1 \neq \beta_1^0$ continues to hold but with the new definitions of $\hat{\beta}_1$ and s.e.($\hat{\beta}_1$) in (3.50) and (3.53), respectively.

As we mentioned earlier, models with no intercept should be used whenever they are consistent with the subject matter (domain) theory or other physical and material considerations. In some applications, however, one may not be certain as to which model should be used. In these cases, the choice between the models given in (3.48) and (3.49) has to be made with care. First, the goodness of fit should be judged by comparing the residual mean squares ($\hat{\sigma}^2$) produced by the two models because it measures the closeness of the observed and predicted values for the two models. Second, one can fit model (3.48) to the data and use the t-Test in (3.31) to test the significance of the intercept. If the test is significant, then use (3.48), otherwise use (3.49).

An excellent exposition of regression models through the origin is provided by Eisenhauer (2003) who also alerts the users of regression models through the origin to be careful when fitting these models using computer software programs because some of them give incorrect and confusing results for the case of regression models through the origin.

3.11 TRIVIAL REGRESSION MODELS

In this section we give two examples of trivial regression models, that is, regression equations that have no regression coefficients. The first example arises when we wish to test for the mean μ of a single variable Y based on a random sample of n observations $y_1, y_2, \ldots, y_n$. Here we have $H_0 : \mu = 0$ against $H_1 : \mu \neq 0$. Assuming that Y is normally distributed with mean μ and variance σ^2, the well-known *one-sample t-Test*

$$t = \frac{\bar{y} - 0}{\text{s.e.}(\bar{y})} = \frac{\bar{y}}{s_y/\sqrt{n}}, \qquad (3.57)$$

can be used to test H_0, where s_y is the sample standard deviation of Y. Alternatively, the above hypotheses can be formulated as

$$H_0(\text{Model 1}) : Y = \varepsilon \text{ against } H_1(\text{Model 2}) : Y = \beta_0 + \varepsilon, \qquad (3.58)$$

where $\beta_0 = \mu_0$. Thus, Model 1 indicates that $\mu = 0$ and Model 2 indicates that $\mu \neq 0$. The least squares estimate of β_0 in Model 2 is $\bar{y}$, the ith fitted value is $\hat{y}_i = \bar{y}$, and the ith residual is $e_i = y_i - \bar{y}$ (See Exercise. 3.15). It follows then that an estimate of σ^2 is

$$\hat{\sigma}^2 = \frac{\text{SSE}}{n-1} = \frac{\sum (y_i - \bar{y})^2}{n-1} = s_y^2, \qquad (3.59)$$

which is the *sample variance* of Y. The standard error of $\hat{\beta}_0$ is then $\hat{\sigma}/\sqrt{n} = s_y/\sqrt{n}$, which is the familiar standard error of the sample mean $\bar{y}$. The t-Test for testing Model 1 against Model 2 is

$$t_1 = \frac{\hat{\beta}_0 - 0}{\text{s.e.}(\hat{\beta}_0)} = \frac{\bar{y}}{s_y/\sqrt{n}}, \qquad (3.60)$$

which is the same as the one-sample t-Test in (3.57).

The second example occurs in connection with the *paired two-sample t-Test*. For example, to test whether a given diet is effective in weight reduction, a random sample of n people is chosen and each person in the sample follows the diet for a specified period of time. Each person's weight is measured at the beginning of the diet and at the end of the period. Let Y_1 and Y_2 denote the weight at the beginning and at the end of diet period, respectively. Let $Y = Y_1 - Y_2$ be the difference between the two weights. Then Y is a random variable with mean μ and variance σ^2. Consequently, testing whether or not the diet is effective is the same as testing $H_0 : \mu = 0$ against $H_1 : \mu > 0$. With the definition of Y and assuming that Y is normally distributed, the well-known paired two-sample t-Test is the same as the test in (3.57). This situation can be modeled as in (3.58) and the test in (3.60) can be used to test whether the diet is effective in weight reduction.

The above two examples show that the one-sample and the paired two-sample tests can be obtained as special cases using regression analysis.

3.12 BIBLIOGRAPHIC NOTES

The standard theory of regression analysis is developed in a number of good text books, some of which have been written to serve specific disciplines. Each provides a complete treatment of the standard results. The books by Snedecor and Cochran (1980), Fox (1984), and Kmenta (1986) develop the results using simple algebra and summation notation. The development in Searle (1971), Rao (1973), Seber (1977), Myers (1990), Sen and Srivastava (1990), Green (1993), Graybill and Iyer (1994), Draper and Smith (1998), Yan and Su (2009), and Montgomery et al. (2012) leans more heavily on theory and/or matrix algebra.

EXERCISES

3.1 Using the data in Table 3.5:
   (a) Compute $\text{Var}(Y)$ and $\text{Var}(X)$.
   (b) Prove or verify that $\sum_{i=1}^{n}(y_i - \bar{y}) = 0$.
   (c) Prove or verify that any standardized variable has a mean of 0 and a standard deviation of 1.
   (d) Prove or verify that the three formulas for $\text{cor}(Y, X)$ in (3.5), (3.6), and (3.7) are identical.
   (e) Prove or verify the inequalities in (3.8).
   (f) Prove or verify that the three formulas for $\hat{\beta}_1$ in (3.14) and (3.20) are identical.

3.2 Explain why you would or wouldn't agree with each of the following statements:
   (a) The $\text{cov}(Y, X)$ can take values between $-\infty$ and $+\infty$.
   (b) If $\text{cov}(Y, X) = 0$ or $\text{cor}(Y, X) = 0$, one can conclude that there is no relationship between Y and X.

(c) The least squares line fitted to the points in the scatter plot of Y versus $\hat{Y}$ has a zero intercept and a unit slope.

3.3 Using the regression output in Table 3.8, test the following hypotheses using $\alpha = 0.1$:
   (a) $H_0 : \beta_1 = 15$ versus $H_1 : \beta_1 \neq 15$
   (b) $H_0 : \beta_1 = 15$ versus $H_1 : \beta_1 > 15$
   (c) $H_0 : \beta_0 = 0$ versus $H_1 : \beta_0 \neq 0$
   (d) $H_0 : \beta_0 = 5$ versus $H_1 : \beta_0 \neq 5$

3.4 Using the regression output in Table 3.8, construct the 99% confidence interval for β_0.

3.5 When fitting the simple linear regression model $Y = \beta_0 + \beta_1 X + \varepsilon$ to a set of data using the least squares method, each of the following statements can be proven to be true. Prove each statement mathematically or demonstrate its correctness numerically (using the data in Table 3.4):
   (a) The sum of the ordinary least squares residuals is zero.
   (b) The two tests in (3.26) and (3.32) are equivalent.
   (c) The scatter plot of Y versus X and the scatter plot of Y versus $\hat{Y}$ have identical patterns.
   (d) The correlation coefficient between Y and $\hat{Y}$ must be nonnegative.

3.6 Using the data in Table 3.4, and the fitted values and the residuals in Table 3.6, verify that:
   (a) $\text{cor}(Y, X) = \text{cor}(Y, \hat{Y}) = 0.994$
   (b) SST $= 27{,}768.348$
   (c) SSE $= 348.848$

3.7 Verify that the four pairs in Anscombe's data give identical results for the following quantities:

   (a) $\hat{\beta}_0$ and $\hat{\beta}_1$ (b) $\text{cor}(Y, X)$
   (c) R^2 (d) The t-Test

3.8 When fitting a simple linear regression model $Y = \beta_0 + \beta_1 X + \varepsilon$ to a set of data using the least squares method, suppose that $H_0 : \beta_1 = 0$ was not rejected. This implies that the model can be written simply as:

Table 3.9 Regression Output When Y Is Regressed on X for Labor Force Participation Rate of Women

Variable	Coefficient	s.e.	t-Test	p-Value
Constant	0.203311	0.0976	2.08	0.0526
X	0.656040	0.1961	3.35	0.0038
$n = 19$	$R^2 = 0.397$	$R_a^2 = 0.362$	$\hat{\sigma} = 0.0566$	df = 17

$Y = \beta_0 + \varepsilon$. The least squares estimate of β_0 is $\hat{\beta}_0 = \bar{y}$. (Can you prove that?)

(a) What are the ordinary least squares residuals in this case?

(b) Show that the ordinary least squares residuals sum up to zero.

3.9 Let Y and X denote the labor force participation rate of women in 1972 and 1968, respectively, in each of 19 cities in the United States. The regression output for this data set is shown in Table 3.9. It was also found that SSR = 0.0358 and SSE = 0.0544. Suppose that the model $Y = \beta_0 + \beta_1 X + \varepsilon$ satisfies the usual regression assumptions.

(a) Compute Var(Y) and cor(Y, X).

(b) Suppose that the participation rate of women in 1968 in a given city is 45%. What is the estimated participation rate of women in 1972 for the same city?

(c) Suppose further that the mean and variance of the participation rate of women in 1968 are 0.5 and 0.005, respectively. Construct the 95% confidence interval for the estimate in (b).

(d) Construct the 95% confidence interval for the slope of the true regression line, β_1.

(e) Test the hypothesis: $H_0 : \beta_1 = 1$ versus $H_1 : \beta_1 > 1$ at the 5% significance level.

(f) If Y and X were reversed in the above regression, what would you expect R^2 to be?

3.10 One may wonder if people of similar heights tend to marry each other. For this purpose, a sample of newly married couples was selected. Let X be the height of the husband and Y be the height of the wife. The heights (in centimeters) of husbands and wives can be found at the Book's Website.

(a) Compute the covariance between the heights of the husbands and wives.

(b) What would the covariance be if heights were measured in inches rather than in centimeters?

(c) Compute the correlation coefficient between the heights of the husband and wife.

(d) What would the correlation be if heights were measured in inches rather than in centimeters?

(e) What would the correlation be if every man married a woman exactly 5 centimeters shorter than him?

(f) We wish to fit a regression model relating the heights of husbands and wives. Which one of the two variables would you choose as the response variable? Justify your answer.

(g) Using your choice of the response variable in Exercise 3.10(f), test the null hypothesis that the slope is zero.

(h) Using your choice of the response variable in 3.10(f), test the null hypothesis that the intercept is zero.

3.11 Consider fitting a simple linear regression model through the origin, $Y = \beta_1 X + \varepsilon$, to a set of data using the least squares method.

(a) Give an example of a situation where fitting the model (3.49) is justified by theoretical or other physical and material considerations.

(b) Show that least squares estimate of β_1 is as given in (3.50).

(c) Show that the residuals $e_1, e_2, \ldots, e_n$ will not necessarily add up to zero.

(d) Give an example of a data set Y and X in which R^2 in (3.46) but computed from fitting (3.49) to the data is negative.

(e) Which goodness of fit measures would you use to compare model (3.49) with model (3.48)?

3.12 The gold medal performance in the Men's Long Jump (measured in meters) for the Summer Olympic games can be found in the file Long.Jump.Men.csv in the Book's Website. The data are for the period from 1896 to 2022. Note that Olympic games were not played in some years due to wars.

(a) Compute the number of gold medals for each country.

(b) Construct the scatter plot of the Long.Jump versus Year.

(c) What is the minimum Long.Jump during these Olympic Series and in which year it occurred?

(d) How many Olympic games were skipped due to wars during this period and which years?
(e) Fit a simple regression model between the Long.Jump and Year and show the results.
(f) From the regression output, compute the variance of the variable Long.Jump.
(g) Which two years have the largest (in absolute value) residuals?
(h) What proportion of variability in the Long.Jump that can be explained by the variable Year?
(i) According to this model, what is your estimate of the Long.Jump in the years during which the Olympic games did not take place?
(j) What is the minimum Long.Jump during these Olympic Series and in which year it occurred?
(k) Construct the 94% confidence interval for the intercept of the regression line.
(l) At the 5% level, test the null hypothesis that there is no linear relationship between Long.Jump and Year. (Specify the two hypotheses, compute the test-statistics, and indicate your decision).
(m) At the 5% level, test the null hypothesis that slope of the true regression line is equal to 1. (Specify the two hypotheses, compute the test-statistic, compute the p-value, and indicate your decision).

3.13 The gold medal performance in the Women's Long Jump (measured in meters) for the Summer Olympic games can be found in the file Long.Jump.Women.csv in the Book's Website. The data are for the period from 1948 to 2022. Repeat Exercise 3.12 using the data for women.

3.14 In order to investigate the feasibility of starting a Sunday edition for a large metropolitan newspaper, information was obtained from a sample of 34 newspapers concerning their daily and Sunday circulations (in thousands) (*Source*: *Gale Directory of Publications*, 1994). The data can be found in the file Newspapers.csv at the Book's Website.
(a) Construct a scatter plot of Sunday circulation versus daily circulation. Does the plot suggest a linear relationship between daily and Sunday circulation? Do you think this is a plausible relationship?
(b) Fit a regression line predicting Sunday circulation from daily circulation.
(c) Obtain the 95% confidence intervals for β_0 and β_1.

(d) Is there a significant relationship between Sunday circulation and daily circulation? Justify your answer by a statistical test. Indicate what hypothesis you are testing and your conclusion.

(e) What proportion of the variability in Sunday circulation is accounted for by daily circulation?

(f) Provide an interval estimate (based on 95% level) for the average Sunday circulation of newspapers with daily circulation of 500,000.

(g) The particular newspaper that is considering a Sunday edition has a daily circulation of 500,000. Provide an interval estimate (based on 95% level) for the predicted Sunday circulation of this paper. How does this interval differ from that given in (f)?

(h) Another newspaper being considered as a candidate for a Sunday edition has a daily circulation of 2,000,000. Provide an interval estimate for the predicted Sunday circulation for this paper. How does this interval compare with the one given in (g)? Do you think it is likely to be accurate?

3.15 Let $y_1, y_2, \ldots, y_n$ be a sample drawn from a normal population with unknown mean μ and unknown variance σ^2. One way to estimate μ is to fit the linear model

$$y_i = \mu + \varepsilon; \quad i = 1, 2, \ldots, n, \quad (3.61)$$

and use the least squares (LS), that is, to minimize the sum of squares, $\sum_{i=1}^{n}(y_i - \mu)^2$. Another way is to use the least absolute value (LAV), that is, to minimize the sum of absolute value of the vertical distances, $\sum_{i=1}^{n}|y_i - \mu|$.

(a) Show that the least squares estimate of μ is the sample mean $\bar{y}$.

(b) Show that the LAV estimate of μ is the sample median.

(c) State one advantage and one disadvantage of the sample mean.

(d) State one advantage and one disadvantage of the sample median.

(e) Which of the above two estimates of μ would you choose? Why?

3.16 An alternative to the least squares method is the orthogonal regression method. According to the orthogonal regression method, the estimated regression coefficients in the simple regression model are obtained by minimizing the sum of squares of the perpendicular distances from each point to the regression line. Show that the intercept and the slope of the line that minimizes the sum of the squared orthogonal distances are obtained by finding β_0 and β_1 that minimize the function

$$g(\beta_0, \beta_1) = \frac{\sum_{i=1}^{n}(y_i - \beta_0 - \beta_1 x_i)^2}{1 + \beta_1^2}. \qquad (3.62)$$

Unlike the least squares criterion, there is no closed-form solution to the minimization problem in (3.62). A solution, however, can be obtained using iterative methods. This is one reason for the popularity of the least squares method.

CHAPTER 4

MULTIPLE LINEAR REGRESSION

4.1 INTRODUCTION

In this chapter the general multiple linear regression model is presented. The presentation serves as a review of the standard results on regression analysis. The standard theoretical results are given without mathematical derivations, but illustrated by numerical examples. Readers interested in mathematical derivations are referred to the bibliographic notes at the end of Chapter 3, where a number of books that contain a formal development of multiple linear regression theory are given.

4.2 DESCRIPTION OF THE DATA AND MODEL

The data consist of n observations on a dependent or response variable Y and p predictor or explanatory variables, $X_1, X_2, \ldots, X_p$. The observations are usually represented as in Table 4.1. The relationship between Y and $X_1, X_2, \ldots, X_p$ is formulated as a linear model

$$Y = \beta_0 + \beta_1 X_1 + \beta_2 X_2 + \cdots + \beta_p X_p + \varepsilon, \qquad (4.1)$$

Regression Analysis By Example Using R, Sixth Edition. Ali S. Hadi and Samprit Chatterjee
© 2024 John Wiley & Sons, Inc. Published 2024 by John Wiley & Sons, Inc.
Companion website: www.wiley.com/go/hadi/regression_analysis_6e

Table 4.1 Notation for Data Used in Multiple Regression Analysis

Observation Number	Response Y	X_1	Predictors X_2	...	X_p
1	y_1	x_{11}	x_{12}	...	x_{1p}
2	y_2	x_{21}	x_{22}	...	x_{2p}
3	y_3	x_{31}	x_{32}	...	x_{3p}
⋮	⋮	⋮	⋮	⋮	⋮
n	y_n	x_{n1}	x_{n2}	...	x_{np}

where $\beta_0, \beta_1, \beta_2, \ldots, \beta_p$ are constants referred to as the model *partial* regression coefficients (or simply as the *regression coefficients*) and ε is a random disturbance or error. It is assumed that for any set of fixed values of $X_1, X_2, \ldots, X_p$ that fall within the range of the data, the linear equation (4.1) provides an acceptable approximation of the true relationship between Y and the X's (Y is approximately a linear function of the X's, and ε measures the discrepancy in that approximation). In particular, ε contains no systematic information for determining Y that is not already captured by the X's.

According to (4.1), each observation in Table 4.1 can be written as

$$y_i = \beta_0 + \beta_1 x_{i1} + \cdots + \beta_p x_{ip} + \varepsilon_i, \quad i = 1, 2, \ldots, n, \quad (4.2)$$

where y_i represents the ith value of the response variable Y, $x_{i1}, x_{i2}, \ldots, x_{ip}$ represent values of the predictor variables for the ith unit (the ith row in Table 4.1), and ε_i represents the error in the approximation of y_i.

Multiple linear regression is an extension (generalization) of simple linear regression. Thus, the results given here are essentially extensions of the results given in Chapter 3. One can similarly think of simple regression as a special case of multiple regression because all simple regression results can be obtained using the multiple regression results when the number of predictor variables $p = 1$. For example, when $p = 1$, (4.1) and (4.2) reduce to (3.9) and (3.10), respectively.

4.3 EXAMPLE: SUPERVISOR PERFORMANCE DATA

Throughout this chapter we use data from a study in *industrial psychology* (management) to illustrate some of the standard regression results. A recent

Table 4.2 Description of Variables in Supervisor Performance Data

Variable	Description
Y	Overall rating of job being done by supervisor
X_1	Handles employee complaints
X_2	Does not allow special privileges
X_3	Opportunity to learn new things
X_4	Raises based on performance
X_5	Too critical of poor performance
X_6	Rate of advancing to better jobs

survey of the clerical employees of a large financial organization included questions related to employee satisfaction with their supervisors. There was a question designed to measure the overall performance of a supervisor, as well as questions that were related to specific activities involving interaction between supervisor and employee. An exploratory study was undertaken to try to explain the relationship between specific supervisor characteristics and overall satisfaction with supervisors as perceived by the employees. Initially, six questionnaire items were chosen as possible explanatory variables.

Table 4.2 gives the description of the variables in the study. As can be seen from the list, there are two broad types of variables included in the study. Variables X_1, X_2, and X_5 relate to direct interpersonal relationships between employee and supervisor, whereas variables X_3 and X_4 are of a less personal nature and relate to the job as a whole. Variable X_6 is not a direct evaluation of the supervisor but serves more as a general measure of how the employee perceives his or her own progress in the company.

The data for the analysis were generated from the individual employee response to the items on the survey questionnaire. The response on any item ranged from 1 through 5, indicating very satisfactory to very unsatisfactory, respectively. A dichotomous index was created for each item by collapsing the response scale to two categories: {1,2}, to be interpreted as a favorable response, and {3,4,5}, representing an unfavorable response. The data were collected in 30 departments selected at random from the organization. Each department had approximately 35 employees and one supervisor. The data set can be found in the file supervisor.csv at the Book's Website[1] and can be

---

[1] http://www.aucegypt.edu/faculty/hadi/RABE6

read into your R Workspace using the following commands:

url="http://www.aucegypt.edu/faculty/hadi/RABE6/Data6/Supervisor.csv"
download.file(url, "Supervisor.csv")
df <- read.csv("Supervisor.csv", header = TRUE)
head(df,2) # Display the first 2 lines of the dataset

The data were obtained by aggregating responses for departments to get the proportion of favorable responses for each item for each department. The resulting data therefore consist of 30 observations on seven variables, one observation for each department. We refer to this data set as the Supervisor Performance data.

A linear model of the form

$$Y = \beta_0 + \beta_1 X_1 + \beta_2 X_2 + \cdots + \beta_6 X_6 + \varepsilon, \qquad (4.3)$$

relating Y and the six explanatory variables, is assumed. Methods for the validation of this and other assumptions are presented in Chapter 5.

4.4 PARAMETER ESTIMATION

Based on the available data, we wish to estimate the parameters $\beta_0, \beta_1, \ldots, \beta_p$. As in the case of simple regression presented in Chapter 3, we use the least squares method, that is, we minimize the sum of squares of the errors. From (4.2), the errors can be written as

$$\varepsilon_i = y_i - \beta_0 - \beta_1 x_{i1} - \cdots - \beta_p x_{ip}, \quad i = 1, 2, \ldots, n. \qquad (4.4)$$

The sum of squares of these errors is

$$S(\beta_0, \beta_1, \ldots, \beta_p) = \sum_{i=1}^{n} \varepsilon_i^2 = \sum_{i=1}^{n} (y_i - \beta_0 - \beta_1 x_{i1} - \cdots - \beta_p x_{ip})^2. \qquad (4.5)$$

By a direct application of calculus, it can be shown that the least squares estimates $\hat{\beta}_0, \hat{\beta}_1, \ldots, \hat{\beta}_p$, which minimize $S(\beta_0, \beta_1, \ldots, \beta_p)$, are given by the solution of a system of linear equations known as the *normal equations*.[2]

---

[2] For readers who are familiar with matrix notation, the normal equations and the least squares estimates are given in the Appendix to this chapter as (A.2) and (A.3), respectively.

The estimate $\hat{\beta}_0$ is usually referred to as the *intercept* or *constant*, and $\hat{\beta}_j$ as the *estimate* of the (partial) regression coefficient of the predictor X_j.

We assume that the system of equations is solvable and has a unique solution. A closed-form formula for the solution is given in the Appendix at the end of this chapter for readers who are familiar with matrix notation. We shall not say anything more about the actual process of solving the normal equations. We assume the availability of computer software that gives a numerically accurate solution.

Using the estimated regression coefficients $\hat{\beta}_0, \hat{\beta}_1, \ldots, \hat{\beta}_p$, we write the fitted least squares regression equation as

$$\hat{Y} = \hat{\beta}_0 + \hat{\beta}_1 X_1 + \cdots + \hat{\beta}_p X_p. \tag{4.6}$$

For each observation in our data we can compute

$$\hat{y}_i = \hat{\beta}_0 + \hat{\beta}_1 x_{i1} + \cdots + \hat{\beta}_p x_{ip}, \quad i = 1, 2, \ldots, n. \tag{4.7}$$

These are called the *fitted* values. The corresponding *ordinary* least squares residuals are given by

$$e_i = y_i - \hat{y}_i, \quad i = 1, 2, \ldots, n. \tag{4.8}$$

An unbiased estimate of σ^2 is given by

$$\hat{\sigma}^2 = \frac{\text{SSE}}{n - p - 1}, \tag{4.9}$$

where

$$\text{SSE} = \sum_{i=1}^{n} (y_i - \hat{y}_i)^2 = \sum_{i=1}^{n} e_i^2, \tag{4.10}$$

is the *sum of squared residuals*. The number $n - p - 1$ in the denominator of (4.9) is called the *degrees of freedom*. It is equal to the number of observations minus the number of estimated regression coefficients.

We illustrate the computations of the above regression results using the Supervisor Performance dataset that we read earlier in Section 4.3 and stored it in an R object named df. The fitted regression equation in (4.6) is

$$\hat{Y} = -10.79 + 0.61 X_1 - 0.07 X_2 + 0.32 X_3 + 0.08 X_4 + 0.04 X_5 - 0.22 X_6. \tag{4.11}$$

114 MULTIPLE LINEAR REGRESSION

This is obtained in R using:

$$reg = lm(Y \sim ., data = df). \qquad (4.12)$$

The dot after $\sim$ in (4.12) indicates that all other variables in the object df will be used as predictor variables. If we wish to fit a subset of the predictor variables in df, say X1 and X2, we use lm(Y $\sim$ X1 + X2, data = df).

The object reg that we created in (4.12) is an object of class lm. As such, it includes a list of many of the regression results. For example,

reg$coef	# the estimated regression coefficients in (4.6)
reg$fitted	# the vector of fitted values in in (4.6)
reg$res	# the vector of residuals in (4.8)
summary(reg)	# gives several other regression results

Type ?lm to see the full list.

When certain assumptions hold, the least squares estimators have several desirable properties. Chapter 5 is devoted entirely to validation of the assumptions. We should note, however, that we have applied these validation procedures on the Supervisor Performance data that we use as illustrative numerical examples in this chapter and found no evidence for model misspecification. We will, therefore, continue with the presentation of multiple regression analysis in this chapter knowing that the required assumptions are valid for the Supervisor Performance data.

The properties of least squares estimators are presented in Section 4.7. Based on these properties, one can develop proper statistical inference procedures (e.g., confidence interval estimation, tests of hypothesis, and goodness-of-fit tests). These are presented in Sections 4.8–4.11.

4.5 INTERPRETATIONS OF REGRESSION COEFFICIENTS

The interpretation of the regression coefficients in a multiple regression equation is a source of common confusion. The simple regression equation represents a line, while the multiple regression equation represents a plane (in cases of two predictors) or a hyperplane (in cases of more than two predictors). In multiple regression, the coefficient β_0, called the *constant coefficient*, is the value of Y when $X_1 = X_2 = \cdots = X_p = 0$, as in simple regression. The regression coefficient $\beta_j, j = 1, 2, \ldots, p$, has several interpretations. It may be interpreted as the change in Y corresponding to a unit change in X_j

INTERPRETATIONS OF REGRESSION COEFFICIENTS

when all other predictor variables are held constant. Magnitude of the change is not dependent on the values at which the other predictor variables are fixed. In practice, however, the predictor variables may be inherently related, and holding some of them constant while varying the others may not be possible.

The regression coefficient β_j is also called the *partial regression coefficient* because β_j represents the contribution of X_j to the response variable Y after it has been adjusted for the other predictor variables. What does "adjusted for" mean in multiple regression? Without loss of generality, we address this question using the simplest multiple regression case where we have two predictor variables. When $p = 2$, the model is

$$Y = \beta_0 + \beta_1 X_1 + \beta_2 X_2 + \varepsilon. \tag{4.13}$$

We use the variables X_1 and X_2 from the Supervisor data to illustrate the concepts. We read this dataset earlier in Section 4.3 and stored it in an R object named df. The estimated regression equation is

$$\hat{Y} = 15.3276 + 0.7803 X_1 - 0.0502 X_2. \tag{4.14}$$

This equation is obtained by R using: reg=lm(Y ~ X1 + X2,data = df); reg$coef.

The coefficient of X_1 suggests that each unit of X_1 adds 0.7803 to Y when the value of X_2 is held fixed. As we show below, this is also the effect of X_1 after adjusting for X_2. Similarly, the coefficient of X_2 suggests that each unit of X_2 subtracts about 0.0502 from Y when the value of X_1 is held fixed. This is also the effect of X_2 after adjusting for X_1.

This interpretation can be easily understood when we consider the fact that the multiple regression equation can be obtained from a series of simple regression equations. For example, the coefficient of X_2 in (4.14) can be obtained as follows:

1. Fit the simple regression model that relates Y to X_1. Let the residuals from this regression be denoted by $e_{Y \cdot X_1}$. This notation indicates that the variable that comes before the dot is treated as a response variable and the variable that comes after the dot is considered as a predictor. The fitted regression equation is

$$\hat{Y} = 14.3763 + 0.754610 X_1. \tag{4.15}$$

This equation is obtained by R using: reg1=lm(Y ~ X1,data = df); reg1$coef.

2. Fit the simple regression model that relates X_2 (considered temporarily here as a response variable) to X_1. Let the residuals from this regression be denoted by $e_{X_2 \cdot X_1}$. The fitted regression equation is

$$\hat{X}_2 = 18.9654 + 0.513032 X_1. \qquad (4.16)$$

This equation is obtained by R using: reg2=lm(X2 ~ X1,data = df); reg2$coef.

The residuals $e_{Y \cdot X_1}$ and $e_{X_2 \cdot X_1}$ are given in Table 4.3. These two sets of residuals are obtained using the R command: cbind(reg1$res,reg2$res).

3. Fit the simple regression model that relates the above two residuals. In this regression, the response variable is $e_{Y \cdot X_1}$ and the predictor variable is $e_{X_2 \cdot X_1}$. The fitted regression equation is

$$\hat{e}_{Y \cdot X_1} = 0 - 0.0502 \, e_{X_2 \cdot X_1}. \qquad (4.17)$$

This equation is obtained by R using: lm(reg1$res ~ reg2$res)$coef.

The interesting result here is that the coefficient of $e_{X_2 \cdot X_1}$ in this last regression is the same as the multiple regression coefficient of X_2 in (4.14).

Table 4.3 Partial Residuals

Row	$e_{Y \cdot X_1}$	$e_{X_2 \cdot X_1}$	Row	$e_{Y \cdot X_1}$	$e_{X_2 \cdot X_1}$
1	−9.8614	−15.1300	16	−1.2912	−15.1383
2	0.3287	−0.7995	17	−4.5182	1.4269
3	3.8010	13.1224	18	5.3471	15.2527
4	−0.9167	−6.2864	19	−2.1990	−8.8776
5	7.7641	−2.9819	20	−8.1437	19.2787
6	−12.8799	1.8178	21	5.4393	−6.4867
7	−6.9352	−11.3385	22	3.5925	1.7397
8	0.0279	−7.4428	23	−11.1806	−0.8255
9	−4.2543	10.9660	24	−2.2969	4.0524
10	6.5925	−5.2604	25	7.8748	−4.6691
11	9.6294	6.8439	26	−6.4813	7.5311
12	7.3471	−2.7473	27	7.0279	0.5572
13	7.8379	6.2266	28	−9.3891	−4.2082
14	−9.0089	21.4529	29	6.4818	8.4269
15	4.5187	−4.4689	30	5.7457	−22.0340

The two coefficients are equal to −0.0502. In fact, their standard errors are also the same. What's the intuition here? In the first step, we found the linear relationship between Y and X_1. The residual from this regression is Y after taking or partialling out the linear effects of X_1. In other words, the residual is that part of Y that is not linearly related to X_1. In the second step we do the same thing, replacing Y by X_2, so the residual is the part of X_2 that is not linearly related to X_1. In the third step we look for the linear relationship between the Y residual and the X_2 residual. The resultant regression coefficient represents the effect of X_2 on Y after taking out the effects of X_1 from both Y and X_2.

The regression coefficient β_j is the partial regression coefficient because it represents the contribution of X_j to the response variable Y after both variables have been linearly adjusted for the other predictor variables (see also Exercise 4.7).

Note that the estimated intercept in the regression equation in (4.17) is zero because the two sets of residuals have a mean of zero (they sum up to zero). The same procedures can be applied to obtain the multiple regression coefficient of X_1 in (4.14). Simply interchange X_2 by X_1 in the above three steps. This is left as an exercise for the reader.

From the above discussion we see that the simple and the multiple regression coefficients are not the same unless the predictor variables are uncorrelated. In nonexperimental or observational data, the predictor variables are rarely uncorrelated. In an experimental setting, in contrast, the experimental design is often set up to produce uncorrelated explanatory variables because in an experiment the researcher sets the values of the predictor variables. So in samples derived from experiments it may be the case that the explanatory variables are uncorrelated and hence the simple and multiple regression coefficients in that sample would be the same.

4.6 CENTERING AND SCALING

The magnitudes of the regression coefficients in a regression equation depend on the unit of measurements of the variables. For example, if the regression coefficient of income, when measured in dollars, is 5.123, this coefficient will change to 5123 if income were measured in $1000 instead. To make the regression coefficients unitless, one may first *center* and/or *scale* the variables before performing the regression computations. There are other situations when centering and scaling the variables are desirable as in the case when dealing with the problem of collinearity in Chapters 10 and 11. This section describes the centering and scaling process.

We have been mainly dealing with regression models of the form

$$Y = \beta_0 + \beta_1 X_1 + \cdots + \beta_p X_p + \varepsilon, \qquad (4.18)$$

which are models with a constant term β_0. But there are also situations where fitting the *no-intercept* model

$$Y = \beta_1 X_1 + \cdots + \beta_p X_p + \varepsilon \qquad (4.19)$$

is necessary (see, e.g., Chapter 8). When dealing with constant term models, it is convenient to center and scale the variables, but when dealing with a no-intercept model, we need only to scale the variables.

4.6.1 Centering and Scaling in Intercept Models

If we are fitting an intercept model as in (4.18), we need to center and scale the variables. A *centered* variable is obtained by subtracting from each observation the mean of all observations. For example, the centered response variable is $(Y - \bar{y})$ and the centered jth predictor variable is $X_j - \bar{x}_j$. The mean of a centered variable is zero.

The centered variables can also be scaled. Two types of scaling are usually performed: *unit-length scaling* and *standardizing*. Unit length scaling of the response variable Y and the jth predictor variable X_j is obtained as follows:

$$\begin{aligned}\tilde{Z}_y &= (Y - \bar{y})/L_y, \\ \tilde{Z}_j &= (X_j - \bar{x}_j)/L_j, \quad j = 1, \ldots, p,\end{aligned} \qquad (4.20)$$

where $\bar{y}$ is the mean of Y, $\bar{x}_j$ is the mean of X_j, and

$$L_y = \sqrt{\sum_{i=1}^n (y_i - \bar{y})^2} \quad \text{and} \quad L_j = \sqrt{\sum_{i=1}^n (x_{ij} - \bar{x}_j)^2}, \quad j = 1, \ldots, p. \qquad (4.21)$$

The quantity L_y is referred to as the *length* of the centered variable $Y - \bar{y}$ because it measures the size or the magnitudes of the observations in $Y - \bar{y}$. Similarly, L_j measures the length of the variable $X_j - \bar{x}_j$. The variables $\tilde{Z}_y$ and $\tilde{Z}_j$ in (4.20) have zero means and unit lengths; hence, this type of scaling is called unit length scaling. In addition, unit length scaling has the following property:

$$\text{cor}(X_j, X_k) = \sum_{i=1}^{n} z_{ij} z_{ik}. \tag{4.22}$$

That is, the correlation coefficient between the original variables, X_j and X_k, can be computed easily as the sum of the products of the scaled versions Z_j and Z_k.

The unit length scaling of df can be done using the following R code:

```
z <- scale(df, scale = FALSE)      # Center the columns of df
nrm <- sqrt(apply(z*z, 2, sum))    # Compute the norm of each column
z = scale(z, scale = nrm)          # Unit length
```

The second type of scaling is called standardizing, which is defined by

$$\tilde{Y} = \frac{Y - \bar{y}}{s_y} \quad \text{and} \quad \tilde{X}_j = \frac{X_j - \bar{x}_j}{s_j}, \quad j = 1, \ldots, p, \tag{4.23}$$

where

$$s_y = \sqrt{\frac{\sum_{i=1}^{n}(y_i - \bar{y})^2}{n-1}} \quad \text{and} \quad s_j = \sqrt{\frac{\sum_{i=1}^{n}(x_{ij} - \bar{x}_j)^2}{n-1}}, \quad j = 1, \ldots, p, \tag{4.24}$$

are standard deviations of the response and jth predictor variable, respectively. The standardized variables $\tilde{Y}$ and $\tilde{X}_j$ in (4.23) have means zero and unit standard deviations. The data can be standardized using z = scale(df).

Since correlations are unaffected by centering and/or scaling the data, it is both sufficient and convenient to deal with either the unit length scaled or the standardized versions of the variables.

4.6.2 Scaling in No-Intercept Models

If we are fitting a no-intercept model as in (4.19), we do not center the data because centering has the effect of including a constant term in the model. This can be seen from

$$Y - \bar{y} = \beta_1(X_1 - \bar{x}_1) + \cdots + \beta_p(X_p - \bar{x}_p) + \varepsilon. \tag{4.25}$$

Rearranging terms, we obtain

$$Y = \bar{y} - (\beta_1\bar{x}_1 + \cdots + \beta_p\bar{x}_p) + \beta_1 X_1 + \cdots + \beta_p X_p + \varepsilon$$
$$= \beta_0 + \beta_1 X_1 + \cdots + \beta_p X_p + \varepsilon, \qquad (4.26)$$

where $\beta_0 = \bar{y} - (\beta_1\bar{x}_1 + \cdots + \beta_p\bar{x}_p)$. Although a constant term does not appear in an explicit form in (4.25), it is clearly seen in (4.26). Thus, when we deal with no-intercept models, we need only to scale the data. The scaled variables are defined by

$$\tilde{Z}_y = Y/L_y,$$
$$\tilde{Z}_j = X_j/L_j, \quad j = 1, \ldots, p, \qquad (4.27)$$

where

$$L_y = \sqrt{\sum_{i=1}^{n} y_i^2} \quad \text{and} \quad L_j = \sqrt{\sum_{i=1}^{n} x_{ij}^2}, \quad j = 1, 2, \ldots, p. \qquad (4.28)$$

The scaled variables in (4.27) have unit lengths but do not necessarily have means zero. Nor do they satisfy (4.22) unless the original variables have zero means.

We should mention here that centering (when appropriate) and/or scaling can be done without loss of generality because the regression coefficients of the original variables can be recovered from the regression coefficients of the transformed variables. For example, if we fit a regression model to centered data, the obtained regression coefficients $\hat{\beta}_1, \ldots, \hat{\beta}_p$ are the same as the estimates obtained from fitting the model to the original data. The estimate of the constant term when using the centered data will always be zero. The estimate of the constant term for an intercept model can be obtained from

$$\hat{\beta}_0 = \bar{y} - (\hat{\beta}_1\bar{x}_1 + \cdots + \hat{\beta}_p\bar{x}_p).$$

Scaling, however, will change the values of the estimated regression coefficients. For example, the relationship between the estimates, $\hat{\beta}_1, \ldots, \hat{\beta}_p$, obtained from using the original data and those obtained using the standardized data are given by

$$\hat{\beta}_j = (s_y/s_j)\hat{\theta}_j, \quad j = 1, 2, \ldots, p,$$
$$\hat{\beta}_0 = \bar{y} - \sum_{j=1}^{p} \hat{\beta}_j\bar{x}_j, \qquad (4.29)$$

where $\hat{\beta}_j$ and $\hat{\theta}_j$ are the jth estimated regression coefficients obtained when using the original and standardized data, respectively. Similar formulas can be obtained when using unit length scaling instead of standardizing.

The regression coefficients obtained using the standardized version of the variables are often referred to as the *beta coefficients*. They represent marginal effects of the predictor variables in standard deviation units. For example, θ_j measures the change in standardized units of Y corresponding to an increase of one standard deviation unit in X_j. We shall make use of the centered and/or scaled variables in Chapters 10 and 11.

4.7 PROPERTIES OF THE LEAST SQUARES ESTIMATORS

Under certain standard regression assumptions (to be stated in Chapter 5), the least squares estimators have the properties listed below. A reader familiar with matrix algebra will find concise statements of these properties employing matrix notation in the Appendix at the end of the chapter.

1. The estimator $\hat{\beta}_j, j = 0, 1, \ldots, p$, is an unbiased estimate of β_j and has a variance of $\sigma^2 c_{jj}$, where c_{jj} is the jth diagonal element of the inverse of the *variance–covariance* matrix in (A.7). The covariance between $\hat{\beta}_i$ and $\hat{\beta}_j$ is $\sigma^2 c_{ij}$, where c_{ij} is the element in the ith row and jth column of the inverse of the variance–covariance matrix in (A.7). For all unbiased estimates that are linear in the observations the least squares estimators have the smallest variance. Thus, the least squares estimators are said to be BLUE (*best linear unbiased estimators*).

2. The estimator $\hat{\beta}_j, j = 0, 1, \ldots, p$ is normally distributed with mean β_j and variance $\sigma^2 c_{jj}$.

3. $W = \text{SSE}/\sigma^2$ has a χ^2 distribution with $n - p - 1$ degrees of freedom, and $\hat{\beta}_j$'s and $\hat{\sigma}^2$ are distributed independently of each other.

4. The vector $\hat{\beta} = (\hat{\beta}_0, \hat{\beta}_1, \ldots, \hat{\beta}_p)$ has a $(p + 1)$-dimensional normal distribution with mean vector $\beta = (\beta_0, \beta_1, \ldots, \beta_p)$ and variance–covariance matrix with elements $\sigma^2 c_{ij}$.

The results above enable us to test various hypotheses about individual regression parameters and to construct confidence intervals. These are discussed in Section 4.9.

4.8 MULTIPLE CORRELATION COEFFICIENT

After fitting the linear model to a given data set, an assessment is made of the adequacy of fit. The discussion given in Section 3.9 applies here. All the material extend naturally to multiple regression and will not be repeated here.

The strength of the linear relationship between Y and the set of predictors $X_1, X_2, \ldots, X_p$ can be assessed through the examination of the scatter plot of Y versus $\hat{Y}$ and the correlation coefficient between Y and $\hat{Y}$, which is given by

$$\text{cor}(Y, \hat{Y}) = \frac{\sum (y_i - \bar{y})(\hat{y}_i - \bar{\hat{y}})}{\sqrt{\sum (y_i - \bar{y})^2 \sum (\hat{y}_i - \bar{\hat{y}})^2}}, \quad (4.30)$$

where $\bar{y}$ is the mean of the response variable Y and $\bar{\hat{y}}$ is the mean of the fitted values. As in the simple regression case, the coefficient of determination

$$[\text{cor}(Y, \hat{Y})]^2 = R^2, \quad (4.31)$$

where R^2 is numerically equivalent to

$$R^2 = \frac{\text{SSR}}{\text{SST}} = 1 - \frac{\text{SSE}}{\text{SST}} = 1 - \frac{\sum (y_i - \hat{y}_i)^2}{\sum (y_i - \bar{y})^2}, \quad (4.32)$$

as in (3.46). Thus, R^2 may be interpreted as the proportion of the total variability in the response variable Y that can be accounted for by the set of predictor variables $X_1, X_2, \ldots, X_p$. In multiple regression, $R = \sqrt{R^2}$ is called the *multiple correlation coefficient* because it measures the relationship between one variable Y and a set of variables $X_1, X_2, \ldots, X_p$.

The value of R^2 for the Supervisor Performance data is 0.73, showing that about 73% of the total variation in the overall rating of the job being done by the supervisor can be accounted for by the six variables.

When the model fits the data well, it is clear that the value of R^2 is close to unity. With a good fit, the observed and predicted values will be close to each other, and $\sum (y_i - \hat{y}_i)^2$ will be small. Then R^2 will be near unity. On the other hand, if there is no linear relationship between Y and the predictor variables, $X_1, \ldots, X_p$, the linear model gives a poor fit, the best predicted value for an observation y_i would be $\bar{y}$; that is, in the absence of any relationship with the predictors, the best estimate of any value of Y is the sample mean, because the sample mean minimizes the sum of squared deviations. So in the absence of any linear relationship between Y and the X's, R^2 will be near zero. The value of R^2 is used as a summary measure to judge the fit of the linear model to a given body of data. As pointed out in Chapter 3, a large value of R^2 does not necessarily mean that the model fits the data well. As we outline in Section 4.10, a more detailed analysis is needed to ensure that the model adequately describes the data.

A quantity related to R^2, known as the *adjusted R-squared*, R_a^2, is also used for judging the goodness of fit. It is defined as

$$R_a^2 = 1 - \frac{\text{SSE}/(n-p-1)}{\text{SST}/(n-1)}, \qquad (4.33)$$

which is obtained from R^2 in (4.32) after dividing SSE and SST by their respective degrees of freedom. From (4.33) and (4.32) it follows that

$$R_a^2 = 1 - \frac{n-1}{n-p-1}(1-R^2). \qquad (4.34)$$

R_a^2 is sometimes used to compare models having different numbers of predictor variables. (This is described in Chapter 12.) In comparing the goodness of fit of models with different numbers of explanatory variables, R_a^2 tries to "adjust" for the unequal number of variables in the different models. Unlike R^2, R_a^2 cannot be interpreted as the proportion of total variation in Y accounted for by the predictors. Many regression packages provide values for both R^2 and R_a^2.

4.9 INFERENCE FOR INDIVIDUAL REGRESSION COEFFICIENTS

Using the properties of the least squares estimators discussed in Section 4.7, one can make statistical inference regarding the regression coefficients. The statistic for testing $H_0 : \beta_j = \beta_j^0$ versus $H_1 : \beta_j \neq \beta_j^0$, where β_j^0 is a constant chosen by the investigator, is

$$t_j = \frac{\hat{\beta}_j - \beta_j^0}{\text{s.e.}(\hat{\beta}_j)}, \qquad (4.35)$$

which has a Student's t-distribution with $n-p-1$ degrees of freedom. The test is carried out by comparing the observed value with the appropriate critical value $t_{(n-p-1,\alpha/2)}$, where α is the significance level. This critical value is obtained using the R code:

qt($\alpha/2, n-p-1$, lower.tail = FALSE)

Note that we divide the significance level α by 2 because we have a two-sided alternative hypothesis. Accordingly, H_0 is to be rejected at the significance level α if

$$|t_j| \geq t_{(n-p-1,\alpha/2)}, \quad (4.36)$$

where $|t_j|$ denotes the absolute value of t_j. A criterion equivalent to that in (4.36) is to compare the *p*-value of the test with α and reject H_0 if

$$p(|t_j|) \leq \alpha, \quad (4.37)$$

where $p(|t_j|)$ is the *p-value* of the test, which is the probability that a random variable having a Student *t*-distribution, with $n - p - 1$, is greater than $|t_j|$ (the absolute value of the observed value of the *t*-Test); see Figure 3.6. The *p*-value is obtained using the R code:

2 * pt(abs(t$_1$), $n - 2$, lower.tail = FALSE).

Note that we multiplied by 2 in the above line because the pt() gives only one side.

The usual test is for $H_0 : \beta_j^0 = 0$, in which case the *t*-Test reduces to

$$t_j = \frac{\hat{\beta}_j}{\text{s.e.}(\hat{\beta}_j)}, \quad (4.38)$$

which is the ratio of $\hat{\beta}_j$ to its standard error, s.e.($\hat{\beta}_j$) given in the Appendix at the end of this chapter, in (A.10). The standard errors of the coefficients are computed by the statistical packages as part of their standard regression output.

Note that the rejection of H_0: $\beta_j = 0$ would mean that β_j is likely to be different from 0, and hence the predictor variable X_j is a statistically significant predictor of the response variable Y after adjusting for the other predictor variables.

As another example of statistical inference, the confidence limits for β_j with confidence coefficient α are given by

$$\hat{\beta}_j \pm t_{(n-p-1,\alpha/2)} \times \text{s.e.}(\hat{\beta}_j), \quad (4.39)$$

where $t_{(n-p-1,\alpha)}$ is the $1 - \alpha$ percentile point of the *t*-distribution with $n - p - 1$ degrees of freedom. The confidence interval in (4.39) is for the individual coefficient β_j. A joint confidence region of all regression coefficients is given in the Appendix at the end of this chapter in (A.15).

Note that when $p = 1$ (simple regression), the t-Test in (4.38) and the criteria in (4.36) and (4.37) reduce to the t-Test in (3.26) and the criteria in (3.27) and (3.28), respectively, illustrating the fact that simple regression results can be obtained from the multiple regression results by setting $p = 1$.

Many other statistical inference situations arise in practice in connection with multiple regression. These will be considered in the following sections.

Example: Supervisor Performance Data (Cont.)

Let us now illustrate the above t-Tests using the Supervisor Performance data set described earlier in this chapter. The results of fitting a linear regression model relating Y and the six explanatory variables are given in Table 4.4. The t-values in Table 4.4 test the null hypothesis $H_0 : \beta_j = 0, j = 0, 1, \ldots, p$, against an alternative $H_1 : \beta_j \neq 0$. From Table 4.4 it is seen that only the regression coefficient of X_1 is significantly different from zero and X_3 has a regression coefficient that approach being significantly different from zero at the 5% significance level. The other variables have insignificant t-Tests. The construction of confidence intervals for the individual parameters is left as an exercise for the reader.

It should be noted here that the constant in the above model is statistically not significant (t-value of 0.93 and p-value of 0.3616). In any regression model, unless there is strong theoretical reason, a constant should always be included even if the term is statistically not significant. The constant represents the base or background level of the response variable. Insignificant predictors should not be in general retained, but a constant should be retained.

Table 4.4 Regression Output for Supervisor Performance Data

Variable	Coefficient	s.e.	t-Test	p-value
Constant	10.787	11.5890	0.93	0.3616
X_1	0.613	0.1610	3.81	0.0009
X_2	−0.073	0.1357	−0.54	0.5956
X_3	0.320	0.1685	1.90	0.0699
X_4	0.081	0.2215	0.37	0.7155
X_5	0.038	0.1470	0.26	0.7963
X_6	−0.217	0.1782	−1.22	0.2356
$n = 30$	$R^2 = 0.73$	$R_a^2 = 0.66$	$\hat{\sigma} = 7.068$	df = 23

4.10 TESTS OF HYPOTHESES IN A LINEAR MODEL

In addition to looking at hypotheses about individual β's, several different hypotheses are considered in connection with the analysis of linear models. The most commonly investigated hypotheses are

1. All the regression coefficients associated with the predictor variables are zero.
2. Some of the regression coefficients are zero.
3. Some of the regression coefficients are equal to each other.
4. The regression parameters satisfy certain specified constraints.

The different hypotheses about the regression coefficients can all be tested in the same way by a unified approach. Rather than describing the individual tests, we first describe the general unified approach, then illustrate specific tests using the Supervisor Performance data.

The model given in (4.1) will be referred to as the *full model* (FM). The null hypothesis to be tested specifies values for some of the regression coefficients. When these values are substituted in the full model, the resulting model is called the *reduced model* (RM). The number of *distinct* parameters to be estimated in the reduced model is smaller than the number of parameters to be estimated in the full model. Accordingly, we wish to test

H_0 : Reduced model is adequate   against H_1 : Full model is adequate.

Note that the reduced model is *nested*. A set of models are said to be nested if they can be obtained from a larger model as special cases. The test for these nested hypotheses involves a comparison of the goodness of fit that is obtained when using the full model, to the goodness of fit that results using the reduced model specified by the null hypothesis. If the reduced model gives as good a fit as the full model, the null hypothesis, which defines the reduced model (by specifying some values of β_j), is not rejected. This procedure is described formally as follows.

Let $\hat{y}_i$ and $\hat{y}_i^*$ be the values predicted for y_i by the full model and the reduced model, respectively. The lack of fit in the data associated with the full model is the sum of the squared residuals obtained when fitting the full model to the data. We denote this by SSE(FM), the sum of squares due to error associated with the full model,

$$\text{SSE(FM)} = \sum (y_i - \hat{y}_i)^2. \qquad (4.40)$$

Similarly, the lack of fit in the data associated with the reduced model is the sum of the squared residuals obtained when fitting the reduced model to the data. This quantity is denoted by SSE(RM), the sum of squares due to error associated with the reduced model,

$$\text{SSE(RM)} = \sum (y_i - \hat{y}_i^*)^2. \tag{4.41}$$

In the full model there are $p+1$ regression parameters $(\beta_0, \beta_1, \beta_2, \ldots, \beta_p)$ to be estimated. Let us suppose that for the reduced model there are k distinct parameters. Note that SSE(RM) $\geq$ SSE(FM) because the additional parameters (variables) in the full model cannot increase the residual sum of squares. Note also that the difference SSE(RM) − SSE(FM) represents the increase in the residual sum of squares due to fitting the reduced model. If this difference is large, the reduced model is inadequate. To see whether the reduced model is adequate, we use the ratio

$$F = \frac{[\text{SSE(RM)} - \text{SSE(FM)}]/(p+1-k)}{\text{SSE(FM)}/(n-p-1)}. \tag{4.42}$$

This ratio is referred to as the F-Test. Note that we divide SSE(RM) − SSE(FM) and SSE(FM) in the above ratio by their respective degrees of freedom to compensate for the different number of parameters involved in the two models as well as to ensure that the resulting test statistic has a standard statistical distribution. The full model has $p+1$ parameters; hence, SSE(FM) has $n-p-1$ degrees of freedom. Similarly, the reduced model has k parameters and SSE(RM) has $n-k$ degrees of freedom. Consequently, the difference SSE(RM) − SSE(FM) has $(n-k) - (n-p-1) = p+1-k$ degrees of freedom. Therefore, the observed F-ratio in (4.42) has F-distribution with $p+1-k$ and $n-p-1$ degrees of freedom.

If the observed F-value is large in comparison to the tabulated value of F with $p+1-k$ and $n-p-1$ degrees of freedom, the result is significant at level α; that is, the reduced model is unsatisfactory and the null hypothesis, with its suggested values of β's in the full model is rejected. The reader interested in the proofs of the statements above is referred to Graybill (1976), Rao (1973), Searle (1971), or Seber and Lee (2003).

Accordingly, H_0 is rejected if

$$F \geq F_{(p+1-k, n-p-1; \alpha)}, \tag{4.43}$$

where F is the observed value of the F-Test in (4.42), $F_{(p+1-k, n-p-1; \alpha)}$ is the appropriate critical value, where α is the significance level. This critical value is obtained using the R code

qf(α, df1 = $p+1-k$, df2 = $n-p-1$, lower.tail = FALSE)

Another rule equivalent to (4.43) is to reject H_0 if

$$p(F) \leq \alpha, \qquad (4.44)$$

$p(F)$ is the *p-value* for the F-Test, which is the probability that a random variable having an F-distribution, with $p + 1 - k$ and $n - p - 1$ degrees of freedom, is greater than the observed F-Test in (4.42). The p-value can be computed using the R code:

pf(F, df1 = $p + 1 - k$, df2 = $n - p - 1$, lower.tail = FALSE)

In the rest of this section, we give several special cases of the general F-Test in (4.42) with illustrative numerical examples using the Supervisor Performance data.

4.10.1 Testing All Regression Coefficients Equal to Zero

An important special case of the F-Test in (4.42) is obtained when we test the hypothesis that all predictor variables under consideration have no explanatory power and that all their regression coefficients are zero. In this case, the reduced and full models become

$$\text{RM: } H_0 : Y = \beta_0 + \varepsilon, \qquad (4.45)$$

$$\text{FM: } H_1 : Y = \beta_0 + \beta_1 X_1 + \cdots + \beta_p X_p + \varepsilon. \qquad (4.46)$$

The residual sum of squares from the full model is SSE(FM) = SSE. Because the least squares estimate of β_0 in the reduced model is $\bar{y}$, the residual sum of squares from the reduced model is SSE(RM) = $\sum (y_i - \bar{y})^2$ = SST. The reduced model has one regression parameter and the full model has $p + 1$ regression parameters. Therefore, the F-Test in (4.42) reduces to

$$F = \frac{[\text{SSE(RM)} - \text{SSE(FM)}]/(p + 1 - k)}{\text{SSE(FM)}/(n - p - 1)}$$

$$= \frac{[\text{SST} - \text{SSE}]/p}{\text{SSE}/(n - p - 1)}. \qquad (4.47)$$

Because SST = SSR + SSE, we can replace SST − SSE in the above formula by SSR and obtain

$$F = \frac{\text{SSR}/p}{\text{SSE}/(n - p - 1)} = \frac{\text{MSR}}{\text{MSE}}, \qquad (4.48)$$

where MSR is the *mean square due to regression* and MSE is the *mean square due to error*. The F-Test in (4.48) can be used for testing the hypothesis that the regression coefficients of all predictor variables (excluding the constant) are zero.

The F-Test in (4.48) can also be expressed directly in terms of the sample multiple correlation coefficient. The null hypothesis which tests whether all the population regression coefficients are zero is equivalent to the hypothesis that states that the population multiple correlation coefficient is zero. Let R_p denote the sample multiple correlation coefficient, which is obtained from fitting a model to n observations in which there are p predictor variables (i.e., we estimate p regression coefficients and one intercept). The appropriate F for testing

$$H_0 : \beta_1 = \beta_2 = \cdots = \beta_p = 0$$

in terms of R_p is

$$F = \frac{R_p^2/p}{(1 - R_p^2)/(n - p - 1)}, \quad (4.49)$$

with p and $n - p - 1$ degrees of freedom.

The values involved in the above F-Test are customarily computed and compactly displayed in a table called the *analysis of variance* (ANOVA) table. The ANOVA table is given in Table 4.5. The first column indicates that there are two sources of variability in the response variable Y. The total variability in Y, SST $= \sum (y_i - \bar{y})^2$, can be decomposed into two sources: the *explained* variability, SSR $= \sum (\hat{y}_i - \bar{y})^2$, which is the variability in Y that can be accounted for by the predictor variables, and the *unexplained* variability, SSE $= \sum (y_i - \hat{y}_i)^2$. This is the same decomposition SST $=$ SSR $+$ SSE. This decomposition is given under the column heading Sum of Squares. The third column gives the degrees of freedom (df) associated with the sum of squares in the second column. The fourth column is the Mean Square (MS), which is obtained by dividing each sum of squares by its respective degrees of

Table 4.5 Analysis of Variance (ANOVA) Table in Multiple Regression

Source	Sum of Squares	df	Mean Square	F-Test
Regression	SSR	p	MSR $= \dfrac{\text{SSR}}{p}$	$F = \dfrac{\text{MSR}}{\text{MSE}}$
Residuals	SSE	$n - p - 1$	MSE $= \dfrac{\text{SSE}}{n - p - 1}$	

freedom. Finally, the F-Test in (4.48) is reported in the last column of the table. Some statistical packages also give an additional column containing the corresponding p-value, $p(F)$.

Returning now to the Supervisor Performance data, although the t-Tests for the regression coefficients have already indicated that some of the regression coefficients (β_1 and β_3) are significantly different from zero, we will, for illustrative purposes, test the hypothesis that all six predictor variables have no explanatory power, that is, $\beta_1 = \beta_2 = \cdots = \beta_6 = 0$. In this case, the reduced and full models in (4.45) and (4.46) become

$$\text{RM: } H_0 : Y = \beta_0 + \varepsilon, \qquad (4.50)$$

$$\text{FM: } H_1 : Y = \beta_0 + \beta_1 X_1 + \cdots + \beta_6 X_6 + \varepsilon. \qquad (4.51)$$

For the full model we have to estimate seven parameters, six regression coefficients and an intercept term β_0. The ANOVA table is given in Table 4.6.

Although the reg object we created in (4.12) using the function lm() contains information that can be used to construct the ANOVA tale, it does not compute the ANOVA table directly. The function ols_regress() in the package olsrr gives very neat displays of the regression results including the coefficients and ANOVA tables. To see what we mean, try out the following code:

```
if(!require("olsrr")) install.packages("olsrr")
ols=ols_regress(Y ~ . , data = df)
print(ols)
```

The object ols that we created in the above code is an object of class ols_regress, which is a list that includes many of the regression results. For example,

ols$betas	# the estimated regression coefficients
ols$std_errors	# the standard errors of the coefficients
ols$rsq	# Coefficient of determination R^2

Type ?ols_regress to see the entire list.

Table 4.6 Supervisor Performance Data: Analysis of Variance (ANOVA) Table

Source	Sum of Squares	df	Mean Square	F-Test
Regression	3147.97	6	524.661	10.5
Residuals	1149.00	23	49.9565	

The sum of squares due to error in the full model is SSE(FM) = SSE = 1149. Under the null hypothesis, where all the β's are zero, the number of parameters estimated for the reduced model is therefore 1 (β_0). Consequently, the sum of squares of the residuals in the reduced model is

$$\text{SSE(RM)} = \text{SST} = \text{SSR} + \text{SSE} = 3147.97 + 1149 = 4296.97.$$

Note that this is the same quantity obtained by $\sum (y_i - \bar{y})^2$. The observed F-ratio is 10.5. In our present example the numerical equivalence of (4.48) and (4.49) is easily seen for

$$F = \frac{R_p^2/p}{(1 - R_p^2)/(n - p - 1)} = \frac{0.7326/6}{(1 - 0.7326)/23} = 10.5.$$

This F-value has an F-distribution with 6 and 23 degrees of freedom. The 1% F-value with 6 and 23 degrees of freedom is found using

qf(0.01, 6, 23, lower.tail = FALSE)

to be 3.71. Since the observed F-value is larger than this value, the null hypothesis is rejected; not all the β's can be taken as zero. This, of course, comes as no surprise, because of the large values of some of the t-Tests.

If any of the t-Tests for the individual regression coefficients prove significant, the F for testing all the regression coefficients zero will usually be significant. A more puzzling case can, however, arise when none of the t-values for testing the regression coefficients are significant, but the F-Test given in (4.49) is significant. This implies that although none of the variables individually have significant explanatory power, the entire set of variables taken collectively explain a significant part of the variation in the dependent variable. This situation, when it occurs, should be looked at very carefully, for it may indicate a problem with the data analyzed, namely, that some of the explanatory variables may be highly correlated, a situation commonly called *collinearity*. We discuss this problem in Chapters 10 and 11.

4.10.2 Testing a Subset of Regression Coefficients Equal to Zero

We have so far attempted to explain Y in the Supervisor Performance data, in terms of six variables, $X_1, X_2, \ldots X_6$. The F-Test in (4.48) indicates that all the regression coefficients cannot be taken as zero; hence, one or more of the predictor variables is related to Y. The question of interest now is: Can Y be explained adequately by fewer variables? An important goal in regression analysis is to arrive at adequate descriptions of observed phenomenon in terms of as few meaningful variables as possible. This economy in description has two advantages. First, it enables us to isolate the most important variables, and second, it provides us with a simpler description of the process studied, thereby making it easier to understand the process. *Simplicity of description* or the *principle of parsimony*, as it is sometimes called, is one of the important guiding principles in regression analysis.

To examine whether the variable Y can be explained in terms of fewer variables, we look at a hypothesis that specifies that some of the regression coefficients are zero. If there are no overriding theoretical considerations as to which variables are to be included in the equation, preliminary t-Tests, like those given in Table 4.4, are used to suggest the variables. In our current example, suppose it was desired to explain the overall rating of the job being done by the supervisor by means of two variables, one taken from the group of personal employee interaction variables X_1, X_2, X_5, and another taken from the group of variables X_3, X_4, X_6, which are of a less personal nature. From this point of view X_1 and X_3 suggest themselves because they have significant t-Tests. Suppose then that we wish to determine whether Y can be explained by X_1 and X_3 as adequately as the full set of six variables. The reduced model in this case is

$$\text{RM: } Y = \beta_0 + \beta_1 X_1 + \beta_3 X_3 + \varepsilon. \tag{4.52}$$

This model corresponds to the null hypothesis

$$H_0 : \beta_2 = \beta_4 = \beta_5 = \beta_6 = 0. \tag{4.53}$$

The regression output from fitting this model is given in Table 4.7, which includes both the ANOVA and the coefficients tables.

The residual sum of squares in this output is the residual sum of squares for the reduced model, which is SSE(RM) = 1254.65. From Table 4.6, the residual sum of squares from the full model is SSE(FM) = 1149.00. Hence the F-Test in (4.42) is

$$F = \frac{[1254.65 - 1149]/4}{1149/23} = 0.528, \tag{4.54}$$

with 4 and 23 degrees of freedom.

The corresponding tabulated value for this test is $F_{(4,23,0.05)} = 2.8$. The value of F is not significant and the null hypothesis is not rejected. The variables X_1 and X_3 together explain the variation in Y as adequately as the

Table 4.7 Regression Output from the Regression of Y on X_1 and X_3

ANOVA Table

Source	Sum of Squares	df	Mean Square	F-Test
Regression	3042.32	2	1521.1600	32.7
Residuals	1254.65	27	46.4685	

Coefficients Table

Variable	Coefficient	s.e.	t-Test	p-value
Constant	9.8709	7.0610	1.40	0.1735
X_1	0.6435	0.1185	5.43	0.0000
X_3	0.2112	0.1344	1.57	0.1278
$n = 30$	$R^2 = 0.708$	$R_a^2 = 0.686$	$\hat{\sigma} = 6.817$	df = 27

full set of six variables. We conclude that the deletion of X_2, X_4, X_5, X_6 does not adversely affect the explanatory power of the model.

We conclude this section with a few remarks:

1. The F-Test in this case can also be expressed in terms of the sample multiple correlation coefficients. Let R_p denote the sample multiple correlation coefficient that is obtained when the full model with all the p variables in it is fitted to the data. Let R_q denote the sample multiple correlation coefficient when the model is fitted with q specific variables: that is, the null hypothesis states that $p - q$ specified variables have zero regression coefficients. The F-Test for testing the above hypothesis is

$$F = \frac{(R_p^2 - R_q^2)/(p - q)}{(1 - R_p^2)/(n - p - 1)}, \quad \text{df} = p - q \text{ and } n - p - 1. \quad (4.55)$$

In our present example, from Tables 4.6 and 4.7, we have $n = 30$, $p = 6, q = 2, R_6^2 = 0.7326$, and $R_2^2 = 0.7080$. Substituting these in (4.55) we get an F-value of 0.528, as before.

2. When the reduced model has only one coefficient (predictor variable) less than the full model, say β_j, then the F-Test in (4.42) has 1 and $n - p - 1$ degrees of freedom. In this case, it can be shown that the F-Test in (4.42) is equivalent to the t-Test in (4.36). More precisely, we have

$$F = t_j^2, \quad (4.56)$$

Table 4.8 Analysis of Variance (ANOVA) Table in Simple Regression

Source	Sum of Squares	df	Mean Square	F-Test
Regression	SSR	1	MSR = SSR	$F = \dfrac{\text{MSR}}{\text{MSE}}$
Residuals	SSE	$n - 2$	$\text{MSE} = \dfrac{\text{SSE}}{n - 2}$	

which indicates that an F-value with 1 and $n - p - 1$ degrees of freedom is equal to the square of a t-value with $n - p - 1$ degrees of freedom, a result which is well-known in statistical theory.

3. In simple regression the number of predictors is $p = 1$. Replacing p by one in the multiple regression ANOVA table (Table 4.5) we obtain the simple regression ANOVA table (Table 4.8). The F-Test in Table 4.8 tests the null hypothesis that the predictor variable X_1 has no explanatory power, that is, its regression coefficient is zero. But this is the same hypothesis tested by the t_1-Test introduced in Section 3.6 and defined in (3.26) as

$$t_1 = \frac{\hat{\beta}_1}{\text{s.e.}(\hat{\beta}_1)}. \tag{4.57}$$

Therefore in simple regression, the F and t_1 tests are equivalent, they are related by

$$F = t_1^2. \tag{4.58}$$

4.10.3 Testing the Equality of Regression Coefficients

It is possible to test the equality of two or more regression coefficients in the same model. In the present example we test whether the regression coefficient of the variables X_1 and X_3 can be treated as equal. The test is performed assuming that it has already been established that the regression coefficients for X_2, X_4, X_5, and X_6 are zero. The null hypothesis to be tested is

$$H_0 : \beta_1 = \beta_3 \mid (\beta_2 = \beta_4 = \beta_5 = \beta_6 = 0). \tag{4.59}$$

The full model assuming that $\beta_2 = \beta_4 = \beta_5 = \beta_6 = 0$ is

$$Y = \beta_0 + \beta_1 X_1 + \beta_3 X_3 + \varepsilon. \tag{4.60}$$

Under the null hypothesis, where $\beta_1 = \beta_3 = \beta_1'$, say, the reduced model is

$$Y = \beta_0' + \beta_1'(X_1 + X_3) + \varepsilon. \tag{4.61}$$

A simple way to carry out the test is to fit the model given by (4.60) to the data. The resulting regression output has been given in Table 4.7. We next fit the reduced model given in (4.61). This can be done quite simply by generating a new variable $W = X_1 + X_3$ and fitting the model

$$Y = \beta_0' + \beta_1' W + \varepsilon. \tag{4.62}$$

The least squares estimates of β_0', β_1' and the sample multiple correlation coefficient (in this case it is the simple correlation coefficient between Y and W since we have only one variable) are obtained. The fitted equation is

$$\hat{Y} = 9.988 + 0.444\, W \tag{4.63}$$

with $R_1^2 = 0.6685$. The appropriate F for testing the null hypothesis, defined in (4.55), becomes

$$F = \frac{(R_p^2 - R_q^2)/(p-q)}{(1 - R_p^2)/(n-p-1)} = \frac{(0.7080 - 0.6685)/(2-1)}{(1 - 0.7080)/(30-2-1)} = 3.65, \tag{4.64}$$

with 1 and 27 degrees of freedom. The tabulated value is $F_{(1,27,0.05)} = 4.21$. The resulting F is not significant; the null hypothesis is not rejected. The distribution of the residuals for this equation (not given here) was found satisfactory.

The equation

$$\hat{Y} = 9.988 + 0.444\,(X_1 + X_3)$$

is not inconsistent with the given data. We conclude then that X_1 and X_3 have the same incremental effect in determining employee satisfaction with a supervisor. This test could also be performed by using a t-Test, given by

$$t = \frac{\hat{\beta}_1 - \hat{\beta}_3}{\text{s.e.}(\hat{\beta}_1 - \hat{\beta}_3)}$$

with 27 degrees of freedom.[3] The conclusions are identical and follow from the fact that F with 1 and p degrees of freedom is equal to the square of t with p degrees of freedom.

---

[3] The $\text{s.e.}(\hat{\beta}_i - \hat{\beta}_j) = \sqrt{\text{Var}(\hat{\beta}_i) + \text{Var}(\hat{\beta}_j) - 2\text{cov}(\hat{\beta}_i, \hat{\beta}_j)}$. These quantities are defined in the Appendix to this chapter.

In this example we have discussed a sequential or step-by-step approach to model building. We have discussed the equality of β_1 and β_3 under the assumption that the other regression coefficients are equal to zero. We can, however, test a more complex null hypothesis which states that β_1 and β_3 are equal and $\beta_2, \beta_4, \beta_5,$ and β_6 are all equal to zero. This null hypothesis H'_0 is formally stated as

$$H'_0 : \beta_1 = \beta_3, \beta_2 = \beta_4 = \beta_5 = \beta_6 = 0. \tag{4.65}$$

The difference between (4.59) and (4.65) is that in (4.59), $\beta_2, \beta_4, \beta_5,$ and β_6 are assumed to be zero, whereas in (4.65) this is under test. The null hypothesis (4.65) can be tested quite easily. The reduced model under H'_0 is (4.61), but this model is not compared to the model of equation (4.60), as in the case of H_0, but with the full model with all six variables in the equation. The F-Test for testing H'_0 is, therefore,

$$F = \frac{(0.7326 - 0.6685)/5}{0.2674/23} = 1.10, \quad df = 5 \text{ and } 23.$$

The result is insignificant as before. The first test is more sensitive for detecting departures from equality of the regression coefficients than the second test. (Why?)

4.10.4 Estimating and Testing of Regression Parameters Under Constraints

Sometimes in fitting regression equations to a given body of data it is desired to impose some constraints on the values of the parameters. A common constraint is that the regression coefficients sum to a specified value, usually unity. The constraints often arise because of some theoretical or physical relationships that may connect the variables. Although no such relationships are obvious in our present example, we consider $\beta_1 + \beta_3 = 1$ for the purpose of demonstration. Assuming that the model in (4.60) has already been accepted, we may further argue that if each of X_1 and X_3 is increased by a fixed amount, Y should increase by that same amount. Formally, we are led to the null hypothesis H_0 which states that

$$H_0 : \beta_1 + \beta_3 = 1 \mid (\beta_2 = \beta_4 = \beta_5 = \beta_6 = 0). \tag{4.66}$$

Since $\beta_1 + \beta_3 = 1$, or equivalently, $\beta_3 = 1 - \beta_1$, then under H_0 the reduced model is

$$H_0 : Y = \beta_0 + \beta_1 X_1 + (1 - \beta_1)X_3 + \varepsilon.$$

Rearranging terms we obtain

$$H_0 : Y - X_3 = \beta_0 + \beta_1(X_1 - X_3) + \varepsilon,$$

which can be written as

$$H_0 : Y' = \beta_0 + \beta_1 V + \varepsilon,$$

where $Y' = Y - X_3$ and $V = X_1 - X_3$. The least squares estimates of the parameters β_1 and β_3 under the constraint are obtained by fitting a regression equation with Y' as response variable and V as the predictor variable. The fitted equation is

$$\hat{Y}' = 1.166 + 0.694\, V,$$

from which it follows that the fitted equation for the reduced model is

$$\hat{Y} = 1.166 + 0.694\, X_1 + 0.306\, X_3$$

with $R^2 = 0.7061$.

The test for H_0 is given by

$$F = \frac{(0.7080 - 0.6905)/1}{0.2920/27} = 1.62, \quad df = 1 \text{ and } 27,$$

which is not significant. The data support the proposition that the sum of the partial regression coefficients of X_1 and X_3 equals unity.

Recall that we have now tested two separate hypotheses about β_1 and β_3, one which states that they are equal and the other that they sum to unity. Since both hypotheses hold, it is implied that both coefficients can be taken to be 0.5. A test of this null hypothesis, $\beta_1 = \beta_3 = 0.5$, may be performed directly by applying the methods we have outlined.

The previous example, in which the equality of β_1 and β_3 was investigated, can be considered as a special case of a constrained problem in which the constraint is $\beta_1 - \beta_3 = 0$. The tests for the full set or subsets of regression coefficients being zero can also be thought of as examples of testing regression coefficients under constraints.

From the above discussion it is clear that several models may describe a given body of data adequately. Where several descriptions of the data are available, it is important that they all be considered. Some descriptions may be more meaningful than others (meaningful being judged in the context of the application and considerations of subject matter), and one of them may be finally adopted. Looking at alternative descriptions of the data provides insight that might be overlooked in focusing on a single description.

The question of which variables to include in a regression equation is very complex and is taken up in detail in Chapter 12. We make two remarks here that will be elaborated on in later chapters.

1. The estimates of regression coefficients that do not significantly differ from zero are most commonly replaced by zero in the equation. The replacement has two advantages: a simpler model and a smaller prediction variance.

2. A variable or a set of variables may sometimes be retained in an equation because of their theoretical importance in a given problem, even though the sample regression coefficients are statistically insignificant. That is, sample coefficients which are not significantly different from zero are not replaced by zero. The variables so retained should give a meaningful process description, and the coefficients help to assess the contributions of the X's to the value of the dependent variable Y.

4.11 PREDICTIONS

The fitted multiple regression equation can be used to predict the value of the response variable using a set of specific values of the predictor variables, $\mathbf{x}_0 = (x_{01}, x_{02}, \ldots, x_{0p})$. The predicted value, $\hat{y}_0$, corresponding to $\mathbf{x}_0$ is given by

$$\hat{y}_0 = \hat{\beta}_0 + \hat{\beta}_1 x_{01} + \hat{\beta}_2 x_{02} + \cdots + \hat{\beta}_p x_{0p}, \qquad (4.67)$$

and its standard error, s.e.($\hat{y}_0$), is given, in the Appendix to this chapter, in (4.12) for readers who are familiar with matrix notation. The standard error is usually computed by many statistical packages. Confidence limits for $\hat{y}_0$ with confidence coefficient α are

$$\hat{y}_0 \pm t_{(n-p-1, \alpha/2)} \text{ s.e.}(\hat{y}_0).$$

As already mentioned in connection with simple regression, instead of predicting the response Y corresponding to an observation $\mathbf{x}_0$ we may want to estimate the mean response corresponding to that observation. Let us denote the mean response at $\mathbf{x}_0$ by μ_0 and its estimate by $\hat{\mu}_0$. Then

$$\hat{\mu}_0 = \hat{\beta}_0 + \hat{\beta}_1 x_{01} + \hat{\beta}_2 x_{02} + \cdots + \hat{\beta}_p x_{0p},$$

as in (4.67), but its standard error, s.e.($\hat{\mu}_0$), is given, in the Appendix to this chapter, in (A.14) for readers who are familiar with matrix notation. Confidence limits for $\hat{\mu}_0$ with confidence coefficient α are

$$\hat{\mu}_0 \pm t_{(n-p-1, \alpha/2)} \text{s.e.}(\hat{\mu}_0).$$

4.12 SUMMARY

We have illustrated the testing of various hypotheses in connection with the linear model. Rather than describing individual tests we have outlined a general procedure by which they can be performed. It has been shown that the various tests can also be described in terms of the appropriate sample multiple correlation coefficients. It is to be emphasized here, that before starting on any testing procedure, the adequacy of the model assumptions should always be examined. As we shall see in Chapter 5, residual plots provide a very convenient graphical way of accomplishing this task. The test procedures are not valid if the assumptions on which the tests are based do not hold. If a new model is chosen on the basis of a statistical test, residuals from the new model should be examined before terminating the analysis. It is only by careful attention to detail that a satisfactory analysis of data can be carried out.

EXERCISES

4.1 Construct a small data set consisting of one response and two predictor variables so that the regression coefficient of X_1 in the following two fitted equations are equal: $\hat{Y} = \hat{\beta}_0 + \hat{\beta}_1 X_1$ and $\hat{Y} = \hat{\alpha}_0 + \hat{\alpha}_1 X_1 + \hat{\alpha}_2 X_2$. Hint: The two predictor variables should be uncorrelated.

4.2 Use the Supervisor.csv data that we read in Section 4.3 and stored it in an R object named df and write the R code to
  (a) Verify the equality in (4.22).
  (b) Verify that the formula for R^2 in (4.31) is the same as the one in (4.32).
  (c) Fit the reduced model in (4.50) to the data and verify that the SSE is 4296.97.
  (d) Fit the reduced model in (4.52) to the data and verify the F-test in (4.54).
  (e) reproduce the fitted model in (4.63) and verify the F-test in (4.64).

4.3 Use the Supervisor.csv data that we read in Section 4.3 and stored it in an R object named df and write the R code to verify that the coefficient of X_1 in the fitted equation in (4.14) can be obtained from a series of simple regression equations, as outlined in Section 4.5 for the coefficient of X_2.

4.4 The examination data consist of three variables: the scores in the final examination F and the scores in two preliminary examinations P_1 and P_2 for 22 students in a statistics course. The data can be found at the Book's Website.

(a) Fit each of the following models to the data:

$$\text{Model 1:} \quad F = \beta_0 + \beta_1 P_1 + \varepsilon$$
$$\text{Model 2:} \quad F = \beta_0 + \beta_2 P_2 + \varepsilon$$
$$\text{Model 3:} \quad F = \beta_0 + \beta_1 P_1 + \beta_2 P_2 + \varepsilon$$

(b) Test whether $\beta_0 = 0$ in each of the three models.

(c) Which variable individually, P_1 or P_2, is a better predictor of F?

(d) Which of the three models would you use to predict the final examination scores for a student who scored 78 and 85 on the first and second preliminary examinations, respectively? What is your prediction in this case?

4.5 Find or construct a simple or multiple regression data set such that the resulting adjusted R_a^2 a is negative.

4.6 Use the functions qt() and qf() to check the theoretical result given by (4.56).

4.7 The relationship between the simple and the multiple regression coefficients can be seen when we compare the following regression equations:

$$\hat{Y} = \hat{\beta}_0 + \hat{\beta}_1 X_1 + \hat{\beta}_2 X_2, \tag{4.68}$$

$$\hat{Y} = \hat{\beta}'_0 + \hat{\beta}'_1 X_1, \tag{4.69}$$

$$\hat{Y} = \hat{\beta}''_0 + \hat{\beta}'_2 X_2, \tag{4.70}$$

$$\hat{X}_1 = \hat{\alpha}_0 + \hat{\alpha}_2 X_2, \tag{4.71}$$

$$\hat{X}_2 = \hat{\alpha}'_0 + \hat{\alpha}_1 X_1. \tag{4.72}$$

Using the data in the file Examination.Data.csv at the Book's Website with $Y = F$, $X_1 = P_1$, and $X_2 = P_2$, verify that:

(a) $\hat{\beta}'_1 = \hat{\beta}_1 + \hat{\beta}_2 \hat{\alpha}_1$, that is, the simple regression coefficient of Y on X_1 is the multiple regression coefficient of X_1 plus the multiple regression coefficient of X_2 times the coefficient from the regression of X_2 on X_1.

(b) $\hat{\beta}'_2 = \hat{\beta}_2 + \hat{\beta}_1 \hat{\alpha}_2$, that is, the simple regression coefficient of Y on X_2 is the multiple regression coefficient of X_2 plus the multiple regression coefficient of X_1 times the coefficient from the regression of X_1 on X_2.

4.8 Table 4.9 shows the regression output, with some numbers erased, when a simple regression model relating a response variable Y to a predictor variable X_1 is fitted based on 20 observations. Complete the 13 missing numbers, then compute Var(Y) and Var(X_1).

4.9 Table 4.10 shows the regression output, with some numbers erased, when a simple regression model relating a response variable Y to a predictor variable X_1 is fitted based on 18 observations. Complete the 13 missing numbers, then compute Var(Y) and Var(X_1).

4.10 Construct the 95% confidence intervals for the individual parameters β_1 and β_2 using the regression output in Table 4.4.

4.11 Explain why the test for testing the hypothesis H_0 in (4.59) is more sensitive for detecting departures from equality of the regression coefficients than the test for testing the hypothesis H'_0 in (4.65).

Table 4.9 Regression Output When Y Is Regressed on X_1 for 20 Observations

	ANOVA Table			
Source	Sum of Squares	df	Mean Square	F-Test
Regression	1848.76	–	–	–
Residuals	–	–	–	

	Coefficients Table			
Variable	Coefficient	s.e.	t-Test	p-value
Constant	−23.4325	12.74	–	0.0824
X_1	–	0.1528	8.32	0.0000
$n = -$	$R^2 = -$	$R_a^2 = -$	$\hat{\sigma} = -$	df $= -$

Table 4.10 Regression Output When Y Is Regressed on X_1 for 18 Observations

ANOVA Table

Source	Sum of Squares	df	Mean Square	F-Test
Regression	–	–	–	–
Residuals	–	–	–	

Coefficients Table

Variable	Coefficient	s.e.	t-Test	p-value
Constant	3.43179	–	0.265	0.7941
X_1	–	0.1421	–	0.0000

$n = -$ $R^2 = 0.716$ $R_a^2 = -$ $\hat{\sigma} = 7.342$ $df = -$

4.12 Using the Supervisor Performance data, test the hypothesis $H_0 : \beta_1 = \beta_3 = 0.5$ in each of the following models:
  (a) $Y = \beta_0 + \beta_1 X_1 + \beta_3 X_3 + \varepsilon$.
  (b) $Y = \beta_0 + \beta_1 X_1 + \beta_2 X_2 + \beta_3 X_3 + \varepsilon$.

4.13 Refer to Exercise 3.10 and the data in the file Husband.Wife.csv, which can also be found at the Book's Website.
  (a) Using your choice of the response variable Exercise 3.10(f), test the null hypothesis that both the intercept and the slope are zero.
  (b) Which of the hypotheses and tests in Exercises 3.10(g), 3.10(h), and 4.13(a) would you choose to test whether people of similar heights tend to marry each other? What is your conclusion?
  (c) If none of the above tests is appropriate for testing the hypothesis that people of similar heights tend to marry each other, which test would you use? What is your conclusion based on this test?

4.14 To decide whether a company is discriminating against women, the following data were collected from the company's records: Salary is the annual salary in thousands of dollars, Qualification is an index of employee qualification, and Gender (1, if the employee is a man, and 0, if the employee is a woman). Two linear models were fit to the data and the regression outputs are shown in Table 4.11. Suppose that the usual regression assumptions hold.
  (a) Are men paid more than equally qualified women?
  (b) Are men less qualified than equally paid women?

Table 4.11 Regression Outputs for Salary Discriminating Data

Model 1: Dependent Variable Is Salary

Variable	Coefficient	s.e.	t-Test	p-value
Constant	20009.5	0.8244	24271	0.0000
Qualification	0.935253	0.0500	18.7	0.0000
Gender	0.224337	0.4681	0.479	0.6329

Model 2: Dependent Variable Is Qualification

Variable	Coefficient	s.e.	t-Test	p-value
Constant	−16744.4	896.4	−18.7	0.0000
Gender	0.850979	0.4349	1.96	0.0532
Salary	0.836991	0.0448	18.7	0.0000

Table 4.12 Regression Output When Salary Is Related to Four Predictor Variables

ANOVA Table

Source	Sum of Squares	df	Mean Square	F-Test
Regression	23,665,352	4	5,916,338	22.98
Residuals	22,657,938	88	257,477	

Coefficients Table

Variable	Coefficient	s.e.	t-Test	p-value
Constant	3526.4	327.7	10.76	0.000
Gender	722.5	117.8	6.13	0.000
Education	90.02	24.69	3.65	0.000
Experience	1.2690	0.5877	2.16	0.034
Months	23.406	5.201	4.50	0.000

| $n = 93$ | $R^2 = 0.515$ | $R_a^2 = 0.489$ | $\hat{\sigma} = 507.4$ | df = 88 |

(c) Do you detect any inconsistency in the above results? Explain.

(d) Which model would you advocate if you were the defense lawyer? Explain.

4.15 Table 4.12 shows the regression output of a multiple regression model relating the beginning salaries in dollars of employees in a given company to the following predictor variables:

Gender   An indicator variable (1 = man and 0 = woman)
Education   Years of schooling at the time of hire
Experience   Number of months of previous work experience
Months   Number of months with the company

In (a)–(b) below, specify the null and alternative hypotheses, the test used, and your conclusion using a 5% level of significance.

(a) Conduct the F-Test for the overall fit of the regression.

(b) Is there a *positive* linear relationship between Salary and Experience, after accounting for the effect of the variables Gender, Education, and Months?

(c) What salary would you forecast for a man with 12 years of education, 10 months of experience, and 15 months with the company?

(d) What salary would you forecast, on average, for men with 12 years of education, 10 months of experience, and 15 months with the company?

(e) What salary would you forecast, on average, for women with 12 years of education, 10 months of experience, and 15 months with the company?

4.16 Consider the regression model that generated the output in Table 4.12 to be a full model. Now consider the reduced model in which Salary is regressed on only Education. The ANOVA table obtained when fitting this model is shown in Table 4.13. Conduct a single test to compare the full and reduced models. What conclusion can be drawn from the result of the test? (Use $\alpha = 0.05$.)

4.17 Cigarette Consumption Data: A national insurance organization wanted to study the consumption pattern of cigarettes in all 50 states and

Table 4.13 ANOVA Table When the Beginning Salary Is Regressed on Education

ANOVA Table				
Source	Sum of Squares	df	Mean Square	F-Test
Regression	7,862,535	1	7,862,535	18.60
Residuals	38,460,756	91	422,646	

Table 4.14 Variables in the Cigarette Consumption Data

Variable	Definition
Age	Median age of a person living in a state
HS	Percentage of people over 25 years of age in a state who had completed high school
Income	Per capita personal income for a state (income in dollars)
Black	Percentage of blacks living in a state
Female	Percentage of females living in a state
Price	Weighted average price (in cents) of a pack of cigarettes in a state
Sales	Number of packs of cigarettes sold in a state on a per capita basis

the District of Columbia. The variables chosen for the study are given in Table 4.14. The data for 1970 can be found in the file Cigarette.Consumption.csv at the Book's Website. The states are given in alphabetical order.

In (a)–(b) below, specify the null and alternative hypotheses, the test used, and your conclusion using a 5% level of significance.

(a) Test the hypothesis that the variable Female is not needed in the regression equation relating Sales to the six predictor variables.

(b) Test the hypothesis that the variables Female and HS are not needed in the above regression equation.

(c) Compute the 95% confidence interval for the true regression coefficient of the variable Income.

(d) What percentage of the variation in Sales can be accounted for when Income is removed from the above regression equation? Explain.

(e) What percentage of the variation in Sales can be accounted for by the three variables: Price, Age, and Income? Explain.

(f) What percentage of the variation in Sales that can be accounted for by the variable Income, when Sales is regressed on only Income? Explain.

4.18 Consider the two models:

$$\text{RM: } H_0 : Y = \varepsilon,$$
$$\text{FM: } H_1 : Y = \beta_0 + \beta_1 X_1 + \cdots + \beta_p X_p + \varepsilon.$$

(a) Develop an F-Test for testing the above hypotheses.
(b) Let $p = 1$ (simple regression) and construct a data set Y and X_1 such that H_0 is not rejected at the 5% significance level.
(c) What does the null hypothesis indicate in this case?
(d) Compute an appropriate value of R^2 that relates the above two models.

Appendix: 4.A Multiple Regression in Matrix Notation

We present the standard results of multiple regression analysis in matrix notation. Let us define the following matrices:

$$Y = \begin{pmatrix} y_1 \\ y_2 \\ \vdots \\ y_n \end{pmatrix}, \quad X = \begin{pmatrix} x_{10} & x_{11} & \cdots & x_{1p} \\ x_{20} & x_{21} & \cdots & x_{2p} \\ \vdots & \vdots & & \vdots \\ x_{n0} & x_{n1} & \cdots & x_{np} \end{pmatrix}, \quad \beta = \begin{pmatrix} \beta_0 \\ \beta_1 \\ \vdots \\ \beta_p \end{pmatrix}, \quad \varepsilon = \begin{pmatrix} \varepsilon_1 \\ \varepsilon_2 \\ \vdots \\ \varepsilon_n \end{pmatrix}.$$

The linear model in (4.1) can be expressed in terms of the above matrices as

$$Y = X\beta + \varepsilon, \qquad (4.A.1)$$

where $x_{i0} = 1$ for all i. The assumptions made about ε for least squares estimation are

$$E(\varepsilon) = 0 \quad \text{and} \quad \text{Var}(\varepsilon) = E(\varepsilon\varepsilon^T) = \sigma^2 I_n,$$

where $E(\varepsilon)$ is the expected value (mean) of ε, I_n is the identity matrix of order n, and ε^T is the transpose of ε. Accordingly, ε_i's are independent and have zero mean and constant variance. This implies that

$$E(Y) = X\beta.$$

The least squares estimator $\hat{\beta}$ of β is obtained by minimizing the sum of squared deviations of the observations from their expected values. Hence the least squares estimators are obtained by minimizing $S(\beta)$, where

$$S(\beta) = \varepsilon^T \varepsilon = (Y - X\beta)^T (Y - X\beta).$$

Minimization of $S(\beta)$ leads to the system of equations

$$(\mathbf{X}^T\mathbf{X})\hat{\boldsymbol{\beta}} = \mathbf{X}^T\mathbf{Y}. \tag{4.A.2}$$

This is the system of *normal equations* referred to in Section (4.4). Assuming that $(\mathbf{X}^T\mathbf{X})$ has an inverse, the least squares estimates $\hat{\boldsymbol{\beta}}$ can be written explicitly as

$$\hat{\boldsymbol{\beta}} = (\mathbf{X}^T\mathbf{X})^{-1}\mathbf{X}^T\mathbf{Y}, \tag{4.A.3}$$

from which it can be seen that $\hat{\boldsymbol{\beta}}$ is a linear function of $\mathbf{Y}$. The vector of fitted values $\hat{\mathbf{Y}}$ corresponding to the observed $\mathbf{Y}$ is

$$\hat{\mathbf{Y}} = \mathbf{X}\hat{\boldsymbol{\beta}} = \mathbf{P}\mathbf{Y}, \tag{4.A.4}$$

where

$$\mathbf{P} = \mathbf{X}(\mathbf{X}^T\mathbf{X})^{-1}\mathbf{X}^T, \tag{4.A.5}$$

is known as the *hat* or *projection* matrix. The vector of residuals is given by

$$\mathbf{e} = \mathbf{Y} - \hat{\mathbf{Y}} = \mathbf{Y} - \mathbf{P}\mathbf{Y} = (\mathbf{I}_n - \mathbf{P})\mathbf{Y}. \tag{4.A.6}$$

The properties of the least squares estimators are

1. $\hat{\boldsymbol{\beta}}$ is an unbiased estimator of $\boldsymbol{\beta}$ (i.e., $E(\hat{\boldsymbol{\beta}}) = \boldsymbol{\beta}$) with variance-covariance matrix $\text{Var}(\hat{\boldsymbol{\beta}})$, which is

$$\text{Var}(\hat{\boldsymbol{\beta}}) = E(\hat{\boldsymbol{\beta}} - \boldsymbol{\beta})(\hat{\boldsymbol{\beta}} - \boldsymbol{\beta})^T = \sigma^2(\mathbf{X}^T\mathbf{X})^{-1} = \sigma^2\mathbf{C},$$

where

$$\mathbf{C} = (\mathbf{X}^T\mathbf{X})^{-1}. \tag{4.A.7}$$

Of all unbiased estimators of $\boldsymbol{\beta}$ that are linear in the observations, the least squares estimator has minimum variance. For this reason, $\hat{\boldsymbol{\beta}}$ is said to be the *best linear unbiased estimator* (BLUE) of $\boldsymbol{\beta}$.

2. The residual sum of squares can be expressed as

$$\mathbf{e}^T\mathbf{e} = \mathbf{Y}^T(\mathbf{I}_n - \mathbf{P})^T(\mathbf{I}_n - \mathbf{P})\mathbf{Y} = \mathbf{Y}^T(\mathbf{I}_n - \mathbf{P})\mathbf{Y}. \tag{4.A.8}$$

The last equality follows because $(\mathbf{I}_n - \mathbf{P})$ is a symmetric idempotent matrix.

3. An unbiased estimator of σ^2 is

$$\hat{\sigma}^2 = \frac{\mathbf{e}^T\mathbf{e}}{n-p-1} = \frac{\mathbf{Y}^T(\mathbf{I}_n - \mathbf{P})\mathbf{Y}}{n-p-1}. \qquad (4.\text{A}.9)$$

With the added assumption that the ε_i's are normally distributed we have the following additional results:

4. The vector $\hat{\boldsymbol{\beta}}$ has a $(p+1)$-dimensional normal distribution with mean vector $\boldsymbol{\beta}$ and variance–covariance matrix $\sigma^2\mathbf{C}$. The marginal distribution of $\hat{\beta}_j$ is normal with mean β_j and variance $\sigma^2 c_{jj}$, where c_{jj} is the jth diagonal element of $\mathbf{C}$ in (A.7). Accordingly, the standard error of $\hat{\beta}_j$ is

$$\text{s.e.}(\hat{\beta}_j) = \hat{\sigma}\sqrt{c_{jj}}, \qquad (4.\text{A}.10)$$

and the covariance of $\hat{\beta}_i$ and $\hat{\beta}_j$ is $\text{cov}(\hat{\beta}_i, \hat{\beta}_j) = \sigma^2 c_{ij}$.

5. The quantity $W = \mathbf{e}^T\mathbf{e}/\sigma^2$ has an χ^2 distribution with $n-p-1$ degrees of freedom.

6. $\hat{\boldsymbol{\beta}}$ and $\hat{\sigma}^2$ are distributed independently of one another.

7. The vector of fitted values $\hat{\mathbf{Y}}$ has a singular n-dimensional normal distribution with mean $E(\hat{\mathbf{Y}}) = \mathbf{X}\boldsymbol{\beta}$ and variance–covariance matrix $\text{Var}(\hat{\mathbf{Y}}) = \sigma^2\mathbf{P}$.

8. The residual vector $\mathbf{e}$ has a singular n-dimensional normal distribution with mean $E(\mathbf{e}) = \mathbf{0}$ and variance–covariance matrix $\text{Var}(\mathbf{e}) = \sigma^2(\mathbf{I}_n - \mathbf{P})$.

9. The predicted value $\hat{y}_0$ corresponding to an observation vector $\mathbf{x}_0 = (x_{00}, x_{01}, x_{02}, \ldots, x_{0p})^T$, with $x_{00} = 1$ is

$$\hat{y}_0 = \mathbf{x}_0^T\hat{\boldsymbol{\beta}} \qquad (4.\text{A}.11)$$

and its standard error is

$$\text{s.e.}(\hat{y}_0) = \hat{\sigma}\sqrt{1 + \mathbf{x}_0^T(\mathbf{X}^T\mathbf{X})^{-1}\mathbf{x}_0}. \qquad (4.\text{A}.12)$$

The mean response μ_0 corresponding to $\mathbf{x}_0^T$ is

$$\hat{\mu}_0 = \mathbf{x}_0^T\hat{\boldsymbol{\beta}} \qquad (4.\text{A}.13)$$

with a standard error

$$\text{s.e.}(\hat{\mu}_0) = \hat{\sigma}\sqrt{\mathbf{x}_0^T(\mathbf{X}^T\mathbf{X})^{-1}\mathbf{x}_0}. \qquad (4.A.14)$$

10. The $100(1-\alpha)\%$ joint confidence region for the regression parameters β is given by

$$\left\{\beta : \frac{(\beta - \hat{\beta})^T(\mathbf{X}^T\mathbf{X})(\beta - \hat{\beta})}{\hat{\sigma}^2(p+1)} \leq F_{(p+1,n-p-1,\alpha)}\right\}, \qquad (4.A.15)$$

which is an ellipsoid centered at $\hat{\beta}$.

CHAPTER 5

REGRESSION DIAGNOSTICS: DETECTION OF MODEL VIOLATIONS

5.1 INTRODUCTION

We have stated the basic results that are used for making inferences about simple and multiple linear regression models in Chapters 3 and 4. The results are based on summary statistics that are computed from the data. In fitting a model to a given body of data, we would like to ensure that the fit is not overly determined by one or a few observations. The distribution theory, confidence intervals, and tests of hypotheses outlined in Chapters 3 and 4 are valid and have meaning only if the standard regression assumptions are satisfied. These assumptions are stated in this chapter (Section 5.2). When these assumptions are violated, the standard results quoted previously do not hold, and an application of them may lead to serious error. We re-emphasize that the prime focus of this book is on the detection and correction of violations of the basic linear model assumptions as a means of achieving a thorough and informative analysis of the data. This chapter presents methods for checking these assumptions. We will rely mainly on graphical methods as opposed to applying rigid numerical rules to check for model violations.

Regression Analysis By Example Using R, Sixth Edition. Ali S. Hadi and Samprit Chatterjee.
© 2024 John Wiley & Sons, Inc. Published 2024 by John Wiley & Sons, Inc.
Companion website: www.wiley.com/go/hadi/regression_analysis_6e

5.2 THE STANDARD REGRESSION ASSUMPTIONS

In Chapters 4 and 5, we have given the least squares estimates of the regression parameters and stated their properties. The properties of least squares estimators and the statistical analysis presented in Chapters 3 and 4 are based on the following assumptions:

1. **Assumptions about the form of the model**: The model that relates the response Y to the predictors $X_1, X_2, \ldots, X_p$ is assumed to be linear in the regression parameters $\beta_0, \beta_1, \ldots, \beta_p$, namely,

$$Y = \beta_0 + \beta_1 X_1 + \cdots + \beta_p X_p + \varepsilon, \qquad (5.1)$$

which implies that the ith observation can be written as

$$y_i = \beta_0 + \beta_1 x_{i1} + \cdots + \beta_p x_{ip} + \varepsilon_i, \; i = 1, 2, \ldots, n. \qquad (5.2)$$

We refer to this as the *linearity* assumption. Checking the linearity assumption in simple regression is easy because the validity of this assumption can be determined by examining the scatter plot of Y versus X. A linear scatter plot ensures linearity. Checking the linearity in multiple regression is more difficult due to the high dimensionality of the data. Some graphs that can be used for checking the linearity assumption in multiple regression are given later in this chapter. When the linearity assumption does not hold, transformation of the data can sometimes lead to linearity. Data transformation is discussed in Chapter 7.

2. **Assumptions about the errors**: The errors $\varepsilon_1, \varepsilon_2, \ldots, \varepsilon_n$ in (5.2) are assumed to be *independently and identically distributed* (iid) normal random variables each with mean zero and a common variance σ^2. Note that this implies four assumptions:

   - The error $\varepsilon_i, i = 1, 2, \ldots, n$, has a normal distribution. We refer to this as the *normality assumption*. The normality assumption is not as easily validated especially when the values of the predictor variables are not replicated. The validity of the normality assumption can be assessed by examination of appropriate graphs of the residuals, as we describe later in this chapter.
   - The errors $\varepsilon_1, \varepsilon_2, \ldots, \varepsilon_n$ have mean zero.

- The errors $\varepsilon_1, \varepsilon_2, \ldots, \varepsilon_n$ have the same (but unknown) variance σ^2. This is the *constant variance assumption*. It is also known by other names such as the *homogeneity* or the *homoscedasticity* assumption. When this assumption does not hold, the problem is called the *heterogeneity* or the *heteroscedasticity* problem. This problem is considered in Chapter 8.

- The errors $\varepsilon_1, \varepsilon_2, \ldots, \varepsilon_n$ are independent of each other (their pairwise covariances are zero). We refer to this as the *independent-errors assumption*. When this assumption does not hold, we have the *autocorrelation* problem. This problem is considered in Chapter 9.

3. **Assumptions about the predictors**: There are three assumptions concerning the predictor variables:

    - The predictor variables $X_1, X_2, \ldots, X_p$ are nonrandom, that is, the values $x_{1j}, x_{2j}, \ldots, x_{nj}; j = 1, 2, \ldots, p$, are assumed fixed or selected in advance. This assumption is satisfied only when the experimenter can set the values of the predictor variables at predetermined levels. It is clear that under nonexperimental or observational situations this assumption will not be satisfied. The theoretical results that are presented in Chapters 3 and 4 will continue to hold, but their interpretation has to be modified. When the predictors are random variables, all inferences are conditional, conditioned on the observed data. It should be noted that this conditional aspect of the inference is consistent with the approach to data analysis presented in this book. Our main objective is to extract the maximum amount of information from the available data.

    - The values $x_{1j}, x_{2j}, \ldots, x_{nj}; j = 1, 2, \ldots, p$, are measured without error. This assumption is hardly ever satisfied. The errors in measurement will affect the residual variance, the multiple correlation coefficient, and the individual estimates of the regression coefficients. The exact magnitude of the effects will depend on several factors, the most important of which are the standard deviation of the errors of measurement and the correlation structure among the errors. The effect of the measurement errors will be to increase the residual variance and reduce the magnitude of the observed multiple correlation coefficient. The effects of measurement errors on individual regression coefficients are more difficult to assess. The estimate of the regression coefficient for a variable is affected not only by its own measurement errors, but also by the measurement errors of other variables included in the equation.

Correction for measurement errors on the estimated regression coefficients, even in the simplest case where all the measurement errors are uncorrelated, requires a knowledge of the ratio between the variances of the measurement errors for the variables and the variance of the random error. Since these quantities are seldom, if ever, known (particularly in the social sciences, where this problem is most acute), we can never hope to remove completely the effect of measurement errors from the estimated regression coefficients. If the measurement errors are not large compared to the random errors, the effect of measurement errors is slight. In interpreting the coefficients in such an analysis, this point should be remembered. Although there is some problem in the estimation of the regression coefficients when the variables are in error, the regression equation may still be used for prediction. However, the presence of errors in the predictors decreases the accuracy of predictions. For a more extensive discussion of this problem, the reader is referred to Fuller (1987), Chatterjee and Hadi (1988), and Chi-Lu and Van Ness (1999).

- The predictor variables $X_1, X_2, \ldots, X_p$ are assumed to be linearly independent of each other. This assumption is needed to guarantee the uniqueness of the least squares solution (the solution of the normal equations in (A.2) in the Appendix to Chapter 4). If this assumption is violated, the problem is referred to as the *collinearity* problem. This problem is considered in Chapters 10 and 11.

The first two of the above assumptions about the predictors cannot be validated, so they do not play a major role in the analysis. However, they do influence the interpretation of the regression results.

4. **Assumptions about the observations**: All observations are equally reliable and have an approximately equal role in determining the regression results and in influencing conclusions.

A feature of the method of least squares is that small or minor violations of the underlying assumptions do not invalidate the inferences or conclusions drawn from the analysis in a major way. Gross violations of the model assumptions can, however, seriously distort conclusions. Consequently, it is important to investigate the structure of the residuals and the data pattern through graphs.

5.3 VARIOUS TYPES OF RESIDUALS

A simple and effective method for detecting model deficiencies in regression analysis is the examination of residual plots. Residual plots will point to serious violations in one or more of the standard assumptions when they exist. Of more importance, the analysis of residuals may lead to suggestions of structure or point to information in the data that might be missed or overlooked if the analysis is based only on summary statistics. These suggestions or cues can lead to a better understanding and possibly a better model of the process under study. A careful graphical analysis of residuals may often prove to be the most important part of the regression analysis.

As we have seen in Chapters 3 and 4, when fitting the linear model in (5.1) to a set of data by least squares, we obtain the fitted values,

$$\hat{y}_i = \hat{\beta}_0 + \hat{\beta}_1 x_{i1} + \cdots + \hat{\beta}_p x_{ip}, \quad i = 1, 2, \ldots, n, \quad (5.3)$$

and the corresponding *ordinary* least squares residuals,

$$e_i = y_i - \hat{y}_i, \quad i = 1, 2, \ldots, n. \quad (5.4)$$

The fitted values in (5.3) can also be written in an alternative form as

$$\hat{y}_i = p_{i1} y_1 + p_{i2} y_2 + \cdots + p_{in} y_n, \quad i = 1, 2, \ldots, n, \quad (5.5)$$

where the p_{ij}'s are quantities that depend only on the values of the predictor variables (they do not involve the response variable). Equation (5.5) shows directly the relationship between the observed and predicted values. In simple regression, p_{ij} is given by

$$p_{ij} = \frac{1}{n} + \frac{(x_i - \bar{x})(x_j - \bar{x})}{\sum (x_i - \bar{x})^2}. \quad (5.6)$$

In multiple regression the p_{ij}'s are elements of a matrix known as the *hat* or *projection* matrix, which is defined in (A.5) in the Appendix to Chapter 4. When $i = j$, p_{ii} is the ith diagonal element of the projection matrix **P**. In simple regression,

$$p_{ii} = \frac{1}{n} + \frac{(x_i - \bar{x})^2}{\sum (x_i - \bar{x})^2}. \quad (5.7)$$

The value p_{ii} is called the *leverage* value for the ith observation because, as can be seen from (5.5), $\hat{y}_i$ is a weighted sum of all observations in Y and

p_{ii} is the weight (leverage) given to y_i in determining the ith fitted value $\hat{y}_i$ (Hoaglin and Welsch, 1978). Thus, we have n leverage values and they are denoted by

$$p_{11}, p_{22}, \ldots, p_{nn}. \tag{5.8}$$

The leverage values play an important role in regression analysis and we shall often encounter them.

When the assumptions stated in Section 5.2 hold, the ordinary residuals, $e_1, e_2, \ldots, e_n$, defined in (5.4), will sum to zero, but they will not have the same variance because

$$\text{Var}(e_i) = \sigma^2(1 - p_{ii}), \tag{5.9}$$

where p_{ii} is the ith leverage value in (5.8), which depends on $x_{i1}, x_{i2}, \ldots, x_{ip}$. To overcome the problem of unequal variances, we standardize the ith residual e_i by dividing it by its standard deviation and obtain

$$z_i = \frac{e_i}{\sigma\sqrt{1 - p_{ii}}}. \tag{5.10}$$

This is called the ith *standardized residual* because it has mean zero and standard deviation 1. The standardized residuals depend on σ, the unknown standard deviation of ε. An unbiased estimate of σ^2 is given by

$$\hat{\sigma}^2 = \frac{\sum e_i^2}{n - p - 1} = \frac{\sum (y_i - \hat{y}_i)^2}{n - p - 1} = \frac{\text{SSE}}{n - p - 1}, \tag{5.11}$$

where SSE is the sum of squares of the residuals. The number $n - p - 1$ in the denominator of (5.11) is called the *degrees of freedom* (df). It is equal to the number of observations, n, minus the number of estimated regression coefficients, $p + 1$.

An alternative unbiased estimate of σ^2 is given by

$$\hat{\sigma}_{(i)}^2 = \frac{\text{SSE}_{(i)}}{(n - 1) - p - 1} = \frac{\text{SSE}_{(i)}}{n - p - 2}, \tag{5.12}$$

where $\text{SSE}_{(i)}$ is the sum of squared residuals when we fit the model to the $n - 1$ observations obtained by omitting the ith observation. Both $\hat{\sigma}^2$ and $\hat{\sigma}_{(i)}^2$ are unbiased estimates of σ^2.

Using $\hat{\sigma}$ as an estimate of σ in (5.10), we obtain

$$r_i = \frac{e_i}{\hat{\sigma}\sqrt{1 - p_{ii}}}, \tag{5.13}$$

whereas using $\hat{\sigma}_{(i)}$ as an estimate of σ, we obtain

$$r_i^* = \frac{e_i}{\hat{\sigma}_{(i)}\sqrt{1-p_{ii}}}. \qquad (5.14)$$

The form of residual in (5.13) is called the *internally studentized residual*, and the residual in (5.14) is called the *externally studentized residual*, because e_i is not involved in (external to) $\hat{\sigma}_{(i)}$. For simplicity of terminology and presentation, however, we shall refer to the studentized residuals as the standardized residuals.

The standardized residuals do not sum to zero, but they all have the same variance. The externally standardized residuals follow a t-distribution with $n - p - 2$ degrees of freedom, but the internally standardized residuals do not. However, with a moderately large sample, these residuals should approximately have a standard normal distribution. The residuals are not strictly independently distributed, but with a large number of observations, the lack of independence may be ignored.

The two forms of residuals are related by

$$r_i^* = r_i \sqrt{\frac{n-p-2}{n-p-1-r_i^2}}, \qquad (5.15)$$

hence one is a monotone transformation of the other. Therefore, for the purpose of residual plots, it makes little difference as to which of the two forms of the standardized residuals is used. From here on, we shall use the internally standardized residuals in the graphs. We need not make any distinction between the internally and externally standardized residuals in our residual plots. Several graphs of the residuals are used for checking the regression assumptions.

5.4 GRAPHICAL METHODS

Graphical methods play an important role in data analysis. It is of particular importance in fitting linear models to data. As Chambers et al. (1983, p. 1) put it, "There is no single statistical tool that is as powerful as a well-chosen graph". Graphical methods can be regarded as exploratory tools. They are also an integral part of confirmatory analysis or statistical inference. Huber (1991, p. 121) says, "Eye-balling can give diagnostic insights no formal diagnostics will ever provide". One of the best examples that illustrates this is the Anscombe quartet data sets, which can be found in the file Anscombe.csv

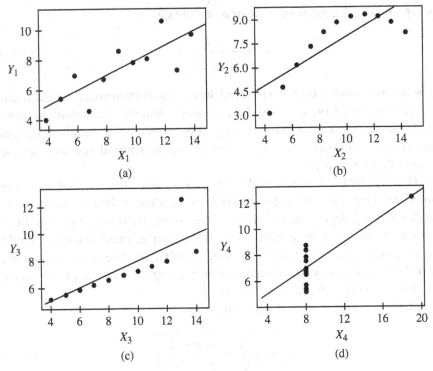

Figure 5.1 Plot of the data (X, Y) with the least squares fitted line for the Anscombe quartet.

at the Book's Website. The four data sets are constructed by Anscombe (1973) in such a way that all pairs (Y, X) have identical values of descriptive statistics (same correlation coefficients, same regression lines, same standard errors, etc.), yet their pairwise scatter plots (reproduced in Figure 5.1 for convenience)[1] give completely different scatters.

The scatter plot in Figure 5.1(a) indicates that a linear model may be reasonable, whereas the one in Figure 5.1(b) suggests a (possibly linearizable) nonlinear model. Figure 5.1(c) shows that the data follow a linear model closely except for one point which is clearly off the line. This point may be an outlier; hence, it should be examined before conclusions can be drawn from the data. Figure 5.1(d) indicates either a deficient experimental design or a bad sample. For the point at $X = 19$, the reader can verify that (a) the residual at this point is always zero (with a variance of zero) no matter how

---

[1] The R code for producing this and all other graphs in this book can be found at the Book's Website at http://www.aucegypt.edu/faculty/hadi/RABE6.

large or small its corresponding value of Y and (b) if the point is removed, the least squares estimates based on the remaining points are no longer unique (except the vertical line, any line that passes through the average of the remaining points is a least squares line!). Observations which unduly influence regression results are called *influential observations*. The point at $X = 19$ is therefore extremely influential because it alone determines both the intercept and the slope of the fitted line.

We have used the scatter plot here as an exploratory tool, but one can also use graphical methods to complement numerical methods in a confirmatory analysis. Suppose we wish to test whether there is a positive correlation between Y and X or, equivalently, if Y and X can be fitted by a positively sloped regression line. The reader can verify that the correlation coefficients are the same in all four data sets [cor$(Y, X) = 0.80$] and all four data sets also have the same regression line ($Y = 3 + 0.5 X$) with the same standard errors of the coefficients. Thus, based on these numerical summaries, one would reach the erroneous conclusion that all four data sets can be described by the same model. The underlying assumption here is that the relationship between Y and X is linear and this assumption does not hold here, for example, for the data set in Figure 5.1(b). Hence the test is invalid. The test for linear relationship, like other statistical methods, is based on certain underlying assumptions. Thus conclusions based on these methods are valid only when the underlying assumptions hold. It is clear from the above example that if analyses were solely based on numerical results, wrong conclusions will be reached.

Graphical methods can be useful in many ways. They can be used to:

1. Detect errors in the data (e.g., an outlying point may be a result of a typographical error)
2. Recognize patterns in the data (e.g., clusters, outliers, and gaps)
3. Explore relationships among variables
4. Discover new phenomena
5. Confirm or negate assumptions
6. Assess the adequacy of a fitted model
7. Suggest remedial actions (e.g., transform the data, redesign the experiment, and collect more data)
8. Enhance numerical analyses in general

This chapter presents some graphical displays useful in regression analysis. The graphical displays we discuss here can be classified into two (not mutually exclusive) classes:

- Graphs before fitting a model. These are useful, for example, in correcting errors in data and in selecting a model.
- Graphs after fitting a model. These are particularly useful for checking the assumptions and for assessing the goodness of the fit.

Our presentation draws heavily from Hadi (1993) and Hadi and Son (1997). Before examining a specific graph, consider what the graph should look like when the assumptions hold. Then examine the graph to see whether it is consistent with expectations. This will then confirm or disprove the assumption.

5.5 GRAPHS BEFORE FITTING A MODEL

The form of a model that represents the relationship between the response and predictor variables should be based on the theoretical background or the hypothesis to be tested. But if no prior information about the form of the model is available, the data may be used to suggest the model. The data should be examined thoroughly before a model is fitted. The graphs that one examines before fitting a model to the data serve as exploratory tools. Four possible groups of graphs are

1. One-dimensional graphs
2. Two-dimensional graphs
3. Rotating plots
4. Dynamic graphs.

5.5.1 One-Dimensional Graphs

Data analysis usually begins with the examination of each variable in the study. The purpose is to have a general idea about the distribution of each individual variable. One of the following graphs may be used for examining a variable:

- Histogram
- Stem-and-leaf display

- Dot plot
- Box plot.

The one-dimensional graphs serve two major functions. They indicate the distribution of a particular variable, whether the variable is symmetric or skewed. When a variable is very skewed, it should be transformed. For a highly skewed variable a logarithmic transformation is recommended. Univariate graphs provide guidance on the question as to whether one should work with the original or with the transformed variables.

Univariate graphs also point out the presence of outliers in the variables. Outliers should be checked to see if they are due to transcription errors. No observation should be deleted at this stage. They should be noted as they may show up as troublesome points later.

5.5.2 Two-Dimensional Graphs

Ideally, when we have multidimensional data, we should examine a graph of the same dimension as that of the data. Obviously, this is feasible only when the number of variables is small. However, we can take the variables in pairs and look at the scatter plots of each variable versus each other variable in the data set. The purposes of these pairwise scatter plots are to explore the relationships between each pair of variables and to identify general patterns.

When the number of variables is small, it may be possible to arrange these pairwise scatter plots in a matrix format, sometimes referred to as the *draftsman's* plot or the *plot* matrix. Figure 5.2 is an example of a plot matrix for one response and two predictor variables. The pairwise scatter plots are given in the upper triangular part of the plot matrix. We can also arrange the corresponding correlation coefficients in a matrix. The corresponding correlation coefficients are given in the lower triangular part of the plot matrix. These arrangements facilitate the examination of the plots. The pairwise correlation coefficients should always be interpreted in conjunction with the corresponding scatter plots. The reason for this is twofold: (a) the correlation coefficient measures only linear relationships, and (b) the correlation coefficient is nonrobust, that is, its value can be substantially influenced by one or two observations in the data.

What do we expect each of the graphs in the plot matrix to look like? In simple regression, the plot of Y versus X is expected to show a linear pattern. In multiple regression, however, the scatter plots of Y versus each predictor variable may or may not show linear patterns. Where the presence of a linear pattern is reassuring, the absence of such a pattern does not imply that our linear model is incorrect. An example is given below.

Example: Hamilton's Data

Hamilton (1987) generates sets of data in such a way that Y depends on the predictor variables collectively but not individually. One such data set is given in Table 5.1. It can be seen from the plot matrix of this data (Figure 5.2) that no linear relationships exist in the plot of Y versus X_1 ($R^2 = 0$) and Y versus X_2 ($R^2 = 0.19$). Yet, when Y is regressed on X_1 and X_2 simultaneously, we obtain an almost perfect fit. The reader can verify (e.g., using R) that the following fitted equations are obtained:

$\hat{Y} = 11.989 + 0.004X_1;$ t-Test $= 0.009;$ $R^2 = 0.0,$
$\hat{Y} = 10.632 + 0.195X_2;$ t-Test $= 1.74;$ $R^2 = 0.188,$
$\hat{Y} = -4.515 + 3.097X_1 + 1.032X_2;$ F-Test $= 39222;$ $R^2 = 1.0.$

The first two equations indicate that Y is related to neither X_1 nor X_2 individually, yet X_1 and X_2 predict Y almost perfectly. Incidentally, the first equation produces a negative value for the adjusted R^2, $R_a^2 = -0.08$.

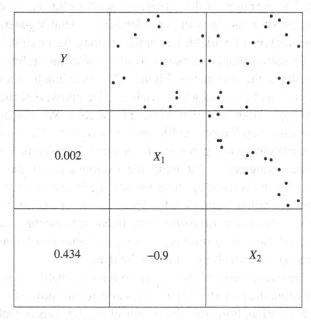

Figure 5.2 Plot matrix for Hamilton's data with the pairwise correlation coefficients.

Table 5.1 Hamilton's (1987) Data

Y	X_1	X_2	Y	X_1	X_2
12.37	2.23	9.66	12.86	3.04	7.71
12.66	2.57	8.94	10.84	3.26	5.11
12.00	3.87	4.40	11.20	3.39	5.05
11.93	3.10	6.64	11.56	2.35	8.51
11.06	3.39	4.91	10.83	2.76	6.59
13.03	2.83	8.52	12.63	3.90	4.90
13.13	3.02	8.04	12.46	3.16	6.96
11.44	2.14	9.05			

The scatter plots that should look linear in the plot matrix are the plots of Y versus each predictor variable after adjusting for all other predictor variables (i.e., taking the linear effects of all other predictor variables out). Two types of these graphs, known as the *added-variable plot* and the *residual plus component plot*, are presented in Section 5.13.1.

The pairwise scatter plot of the predictors should show no linear pattern (ideally, we should see no discernible pattern, linear or otherwise) because the predictors are assumed to be linearly independent. In Hamilton's data, this assumption does not hold because there is a clear linear pattern in the scatter plot of X_1 versus X_2 (Figure 5.2). We should caution here that the absence of linear relationships in these scatter plots does not imply that the entire set of predictors are linearly independent. The linear relationship may involve more than two predictor variables. Pairwise scatter plots will fail to detect such a multivariate relationship. This collinearity problem will be dealt with in Chapters 10 and 11.

5.5.3 Rotating Plots

Recent advances in computer hardware and software have made it possible to plot data of three or more dimensions. The simplest of these plots is the three-dimensional rotating plot. The rotating plot is a scatter plot of three variables in which the points can be rotated in various directions so that the three-dimensional structure becomes apparent. Describing rotating plots in words does not do them justice. The real power of rotation can be felt only when one watches a rotating plot in motion on a computer screen. The motion can be stopped when one sees an interesting view of the data. For example, in the Hamilton data we have seen that X_1 and X_2 predict Y almost perfectly.

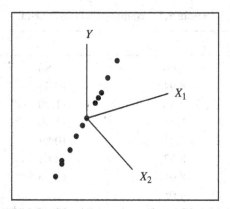

Figure 5.3 Rotating plot for Hamilton's data.

This finding is confirmed in the rotating plot of Y against X_1 and X_2. When this plot is rotated, the points fall on an almost perfect plane. The plot is rotated until an interesting direction is found. Figure 5.3 shows one such direction, where the plane is viewed from an angle that makes the scatter of points seem to fall on a straight line.

5.5.4 Dynamic Graphs

Dynamic graphics are an extraordinarily useful tool for exploring the structure and relationships in multivariate data. In a dynamic graphics environment the data analyst can go beyond just looking at a static graph. The graphs can be manipulated and the changes can be seen instantaneously on the computer screen. For example, one can make two or more three-dimensional rotating plots and then use dynamic graphical techniques to explore the structure and relationships in more than three dimensions. Articles and books have been written about the subject, and many statistical software programs include dynamic graphical tools (e.g., rotating, brushing, and linking). We refer the interested reader to Becker et al. (1987) and Velleman (1999).

5.6 GRAPHS AFTER FITTING A MODEL

The graphs presented in Section 5.5 are useful in data checking and the model formulation steps. The graphs after fitting a model to the data help in checking the assumptions and in assessing the adequacy of the fit of a given model. These graphs can be grouped into the following classes:

1. Graphs for checking the linearity and normality assumptions
2. Graphs for the detection of outliers and influential observations
3. Diagnostic plots for the effect of variables.

5.7 CHECKING LINEARITY AND NORMALITY ASSUMPTIONS

When the number of variables is small, the assumption of linearity can be checked by interactively and dynamically manipulating the plots discussed in Section 5.5. The task of checking the linearity assumption becomes difficult when the number of variables is large. However, one can check the linearity and normality assumptions by examining the residuals after fitting a given model to the data.

The following plots of the standardized residuals can be used to check the linearity and normality assumptions:

1. *Normal probability plot of the standardized residuals*: This is a plot of the ordered standardized residuals versus the so-called *normal scores*. The normal scores are what we would expect to obtain if we take a sample of size n from a standard normal distribution. If the residuals are normally distributed, the ordered residuals should be approximately the same as the ordered normal scores. Under normality assumption, this plot should resemble a (nearly) straight line with an intercept of zero and a slope of one (these are the mean and the standard deviation of the standardized residuals, respectively).

2. *Scatter plots of the standardized residual against each of the predictor variables*: Under the standard assumptions, the standardized residuals are uncorrelated with each of the predictor variables. If the assumptions hold, this plot should be a random scatter of points. Any discernible pattern in this plot may indicate violation of some assumptions. If the linearity assumption does not hold, one may observe a plot like the one given in Figure 5.4(a). In this case a transformation of the Y and/or the particular predictor variable may be necessary to achieve linearity. A plot that looks like Figure 5.4(b) may indicate heterogeneity of variance. In this case a transformation of the data that stabilizes the variance may be needed. Several types of transformations for the corrections of some model deficiencies are described in Chapter 7.

3. *Scatter plot of the standardized residual versus the fitted values*: Under the standard assumptions, the standardized residuals are also

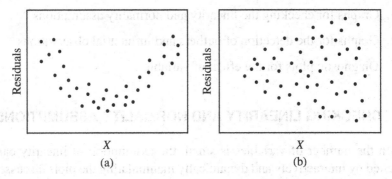

Figure 5.4 Two scatter plots of residuals versus X illustrating violations of model assumptions: (a) a pattern indicating nonlinearity and (b) a pattern indicating heterogeneity.

uncorrelated with the fitted values; therefore, this plot should also be a random scatter of points. In simple regression, the plots of standardized residuals against X and against the fitted values are identical.

4. *Index plot of the standardized residuals*: In this diagnostic plot we display the standardized residuals versus the observation number. If the order in which the observations were taken is immaterial, this plot is not needed. However, if the order is important (e.g., when the observations are taken over time or there is a spatial ordering), a plot of the residuals in serial order may be used to check the assumption of independence of the errors. Under the assumption of independent errors, the points should be scattered randomly within a horizontal band around zero.

5.8 LEVERAGE, INFLUENCE, AND OUTLIERS

In fitting a model to a given body of data, we would like to ensure that the fit is not overly determined by one or a few observations. Recall, for example, that in the Anscombe quartet data, the straight line for the data set in Figure 5.1(d) is determined entirely by one point. If the extreme point were to be removed, a very different line would result. When we have several variables, it is not possible to detect such a situation graphically. We would, however, like to know the existence of such points. It should be pointed out that looking at residuals in this case would be of no help, because the residual for this point is zero! The point is therefore not an outlier because it does not have a large residual, but it is a very influential point.

LEVERAGE, INFLUENCE, AND OUTLIERS

A point is an *influential* point if its deletion, singly or in combination with others (two or three), causes substantial changes in the fitted model (estimated coefficients, fitted values, *t*-Tests, etc.). Deletion of any point will in general cause changes in the fit. We are interested in detecting those points whose deletion cause large changes (i.e., they exercise undue influence). This point is illustrated by an example.

Example: New York Rivers Data

Consider the New York Rivers data described in Section 1.3.5 and given in the file NYRivers.csv at the Book's Website. Let us fit a linear model relating the mean nitrogen concentration, Y, and the four predictor variables representing land use:

$$Y = \beta_0 + \beta_1 X_1 + \beta_2 X_2 + \beta_3 X_3 + \beta_4 X_4 + \varepsilon. \qquad (5.16)$$

The data can be read into your R Workspace using the following commands:

```
url="http://www.aucegypt.edu/faculty/hadi/RABE6/Data6/NYRivers.csv"
download.file(url, "NYRivers.csv")
df <- read.csv("NYRivers.csv", header = TRUE, row.names = 1)
head(df,2) # Display the first 2 lines of the dataset
```

Note that in the above code row.names = 1 is an argument to tell R that the data file contains the names of the rivers (rows) in column 1. Table 5.2 shows the regression coefficients and the *t*-Tests for testing the significance of the coefficients for three subsets of the data. The second column in Table 5.2 gives the regression results based on all 20 observations (rivers). These results can be obtained using the R code:

```
if(!require("olsrr")) install.packages("olsrr") # Download the olsrr package
ols_regress(Nitrogen ~. , data = df)$tvalues # Display only the t-values
```

The third column in Table 5.2 gives the results after deleting the Neversink River (number 4). These results can be obtained using the R code:

```
df4 <- df[-4,] # df4 is the data without row number 4
ols_regress(Nitrogen ~. , data = df4)$tvalues # Display only the t-values
```

Table 5.2 New York Rivers Data: The t-Tests for the Individual Coefficients

	Observations Deleted		
Test	None	Neversink (River 4)	Hackensack (River 5)
t_0	1.40	1.21	2.08
t_1	0.39	0.92	0.25
t_2	−0.93	−0.74	−1.45
t_3	−0.21	−3.15	4.08
t_4	1.86	4.45	0.66

The fourth column gives the results after deleting the Hackensack River (number 5). It is left as an exercise for the reader to obtain these results.

Note the striking difference among the regression outputs of three data sets that differ from each other by only one observation! Observe, for example, the values of the t-Test for β_3. Based on all data, the test is insignificant, based on the data without the Neversink River, it is significantly negative, and based on the data without the Hackensack River, it is significantly positive. Only one observation can lead to substantially different results and conclusions! The Neversink and Hackensack Rivers are called influential observations because they influence the regression results substantially more than other observations in the data. Examining the raw data in the file NYRivers.csv, one can easily identify the Hackensack River because it has an unusually large value for X_3 (percentage of residential land) relative to the other values for X_3. The reason for this large value is that the Hackensack River is the only urban river in the data due to its geographic proximity to New York City with its high population density. The other rivers are in rural areas. Although the Neversink River is influential (as can be seen from Table 5.2), it is not obvious from the raw data that it is different from the other rivers in the data.

It is therefore important to identify influential observations if they exist in data. We describe methods for the detection of influential observations. Influential observations are usually outliers in either the response variable Y or the predictor variable (the X-space).

5.8.1 Outliers in the Response Variable

Observations with large standardized residuals are outliers in the response variable because they lie far from the fitted equation in the Y-direction. Since the standardized residuals are approximately normally distributed with mean zero and a standard deviation 1, points with standardized residuals larger than 2 or 3 standard deviations away from the mean (zero) are called *outliers*. Outliers may indicate a model failure for those points. They can be identified using formal testing procedures [see, e.g., Hawkins (1980), Barnett and Lewis (1994), Hadi and Simonoff (1993), and Hadi and Velleman (1997) and the references therein] or through appropriately chosen graphs of the residuals, the approach we adopt here. The pattern of the residuals is more important than their numeric values. Graphs of residuals will often expose gross model violations when they are present. Studying residual plots is one of the main tools in our analysis.

5.8.2 Outliers in the Predictors

Outliers can also occur in the predictor variables (the X-space). They can also affect the regression results. The leverage values p_{ii}, described earlier, can be used to measure outlyingness in the X-space. This can be seen from an examination of the formula for p_{ii} in the simple regression case given in (5.7), which shows that the farther a point is from $\bar{x}$, the larger the corresponding value of p_{ii}. This is also true in multiple regression. Therefore, p_{ii} can be used as a measure of outlyingness in the X-space because observations with large values of p_{ii} are outliers in the X-space (i.e., compared to other points in the space of the predictors). Observations that are outliers in the X-space [e.g., the point with the largest value of X_4 in Figure 5.1(d)] are known as *high-leverage* points to distinguish them from observations that are outliers in the response variable (those with large standardized residuals).

The leverage values possess several interesting properties [see Dodge and Hadi (1999) and Chatterjee and Hadi (1988), Chapter 2, for a comprehensive discussion]. For example, they lie between 0 and 1 and their average value is $(p + 1)/n$. Points with p_{ii} greater than $2(p + 1)/n$ (twice the average value) are generally regarded as points with high leverage (Hoaglin and Welsch, 1978).

In any analysis, points with high leverage should be flagged and then examined to see if they are also influential. A plot of the leverage values (e.g., index plot, dot plot, or a box plot) will reveal points with high leverage if they exist. Another interesting plot is proposed by Gray and Ling (1984).

5.8.3 Masking and Swamping Problems

The standardized residuals provide valuable information for validating linearity and normality assumptions and for the identification of outliers. However, analyses that are based on residuals alone may fail to detect outliers and influential observations for the following reasons:

1. *The presence of high-leverage points*: The ordinary residuals, e_i, and leverage values, p_{ii}, are related by

$$p_{ii} + \frac{e_i^2}{\text{SSE}} \leq 1, \qquad (5.17)$$

where SSE is the residual sum of squares. This inequality indicates that high-leverage points (points with large values of p_{ii}) tend to have small residuals. For example, the point at $X = 19$ in Figure 5.1(d) is extremely influential even though its residual is identically zero. Therefore, in addition to an examination of the standardized residuals for outliers, an examination of the leverage values is also recommended for the identification of troublesome points.

2. *The masking and swamping problems*: Masking occurs when the data contain outliers but we fail to detect them. This can happen because some of the outliers may be hidden by other outliers in the data. Swamping occurs when we wrongly declare some of the nonoutlying points as outliers. This can occur because outliers tend to pull the regression equation toward them, hence make other points lie far from the fitted equation. Thus, masking is a false negative decision whereas swamping is a false positive. An example of a data set in which masking and swamping problems are present is given below. Methods which are less susceptible to the masking and swamping problems than the standardized residuals and leverage values are given in Hadi and Simonoff (1993) and the references therein.

For the above reasons, additional measures of the influence of observations are needed. Before presenting these methods, we illustrate the above concepts using a real-life example.

Example: New York Rivers Data

Consider the New York Rivers data, but now for illustrative purpose, let us consider fitting the simple regression model

$$Y = \beta_0 + \beta_4 X_4 + \varepsilon, \qquad (5.18)$$

relating the mean nitrogen concentration, Y, to the percentage of land area in either industrial or commercial use, X_4. The scatter plot of Y versus X_4 together with the corresponding least squares fitted line are given in Figure 5.5. The corresponding standardized residuals, r_i, and the leverage values, p_{ii}, are given in Table 5.3 and their respective index plots are shown in Figure 5.6. In the index plot of the standardized residuals all the residuals are small indicating that there are no outliers in the data. This is a wrong conclusion because there are two clear outliers in the data as can be seen in the scatter plot in Figure 5.5. Thus masking has occurred! Because of the relationship between leverage and residual in (5.17), the Hackensack River with its large value of $p_{ii} = 0.67$ has a small residual. While a small value of the residual is desirable, the reason for the small value of the residual here is not due to a good fit; it is due to the fact that observation 5 is a high-leverage point and, in collaboration with observation 4, they pull the regression line toward them.

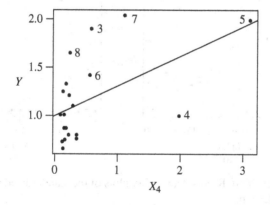

Figure 5.5 New York Rivers data: scatter plot of Y versus X_4.

A commonly used cutoff value for p_{ii} is $2(p+1)/n = 0.2$ (Hoaglin and Welsch, 1978). Accordingly, two points (Hackensack, $p_{ii} = 0.67$, and Neversink, $p_{ii} = 0.25$) that we have seen previously stand out in the scatter plot of points in Figure 5.5 are flagged as high-leverage points as can be seen in the index plot of p_{ii} in Figure 5.6(b), where the two points are far from the other points. This example shows clearly that looking solely at residual plots is inadequate.

Table 5.3 New York Rivers Data: Standardized Residuals, r_i, and Leverage Values, p_{ii}, from Fitting Model 5.18

Row	r_i	p_{ii}	Row	r_i	p_{ii}
1	0.03	0.05	11	0.75	0.06
2	−0.05	0.07	12	−0.81	0.06
3	1.95	0.05	13	−0.83	0.06
4	−1.85	0.25	14	−0.83	0.05
5	0.16	0.67	15	−0.94	0.05
6	0.67	0.05	16	−0.48	0.06
7	1.92	0.08	17	−0.72	0.06
8	1.57	0.06	18	−0.50	0.06
9	−0.10	0.06	19	−1.03	0.06
10	0.38	0.06	20	0.57	0.06

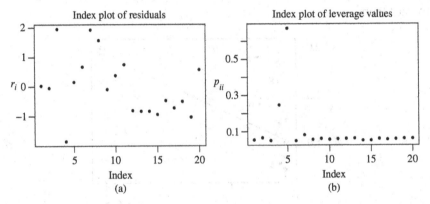

Figure 5.6 New York Rivers data: index plots of the standardized residuals, r_i, and the leverage values, p_{ii}.

5.9 MEASURES OF INFLUENCE

The influence of an observation may be measured by the effects it produces on the fit when it is omitted from the data in the fitting process. This deletion is almost always done one point at a time. Let $\hat{\beta}_{0(i)}, \hat{\beta}_{1(i)}, \ldots, \hat{\beta}_{p(i)}$ denote the regression coefficients obtained when the ith observation is deleted ($i = 1, 2, \ldots, n$). Similarly, let $\hat{y}_{1(i)}, \hat{y}_{2(i)}, \ldots, \hat{y}_{n(i)}$, and $\hat{\sigma}^2_{(i)}$ be the predicted values and residual mean square when we drop the ith observation. Note that

$$\hat{y}_{m(i)} = \hat{\beta}_{0(i)} + \hat{\beta}_{1(i)} x_{m1} + \cdots + \hat{\beta}_{p(i)} x_{mp} \qquad (5.19)$$

is the fitted value for observation m when the fitted equation is obtained with the ith observation deleted. Influence measures look at differences produced in quantities such as $\hat{\beta}_j - \hat{\beta}_{j(i)}$ or $\hat{y}_j - \hat{y}_{j(i)}$. There are numerous measures of influence in the literature, and the reader is referred to one of the books for details: Belsley et al. (1980), Cook and Weisberg (1982), Atkinson (1985), and Chatterjee and Hadi (1988). Here we give three of these measures.

5.9.1 Cook's Distance

An influence measure proposed by Cook (1977) is widely used. *Cook's distance* measures the difference between the regression coefficients obtained from the full data and the regression coefficients obtained by deleting the ith observation, or equivalently, the difference between the fitted values obtained from the full data and the fitted values obtained by deleting the ith observation. Accordingly, Cook's distance measures the influence of the ith observation by

$$C_i = \frac{\sum_{j=1}^{n} (\hat{y}_j - \hat{y}_{j(i)})^2}{\hat{\sigma}^2 (p+1)}, \quad i = 1, 2, \ldots, n. \qquad (5.20)$$

It can be shown that C_i can be expressed as

$$C_i = \frac{r_i^2}{p+1} \times \frac{p_{ii}}{1 - p_{ii}}, \quad i = 1, 2, \ldots, n. \qquad (5.21)$$

Thus, Cook's distance is a multiplicative function of two basic quantities. The first is the square of the standardized residual, r_i, defined in (5.13) and the second is the so-called *potential* function $p_{ii}/(1 - p_{ii})$, where p_{ii} is the leverage of the ith observation introduced previously. If a point is influential, its deletion causes large changes and the value of C_i will be large. Therefore, a large value of C_i indicates that the point is influential. It has been suggested that points with C_i values greater than the 50% point of the F-distribution with $p + 1$ and $n - p - 1$ degrees of freedom be classified as influential points. A practical operational rule is to classify points with C_i values greater than 1 as being influential. Rather than using a rigid cutoff rule, we suggest that all C_i values be examined graphically. A dot plot or an index plot of C_i is a useful graphical device. When the C_i values are all about the same, no action need be taken. On the other hand, if there are data points with C_i values that stand out from the rest, these points should be flagged and examined. The model may then be refitted without the offending points to see the effect of these points.

5.9.2 Welsch and Kuh Measure

A measure similar to Cook's distance has been proposed by Welsch and Kuh (1977) and named DFITS. It is defined as

$$\text{DFITS}_i = \frac{\hat{y}_i - \hat{y}_{i(i)}}{\hat{\sigma}_{(i)}\sqrt{p_{ii}}}, \quad i = 1, 2, \ldots, n. \quad (5.22)$$

Thus, DFITS_i is the scaled difference between the ith fitted value obtained from the full data and the ith fitted value obtained by deleting the ith observation. The difference is scaled by $\hat{\sigma}_{(i)}\sqrt{p_{ii}}$. It can be shown that DFITS_i can be written as

$$\text{DFITS}_i = r_i^* \sqrt{\frac{p_{ii}}{1 - p_{ii}}}, \quad i = 1, 2, \ldots, n, \quad (5.23)$$

where r_i^* is the standardized residual defined in (5.14). DFITS_i corresponds to $\sqrt{C_i}$ when the normalization is done by using $\hat{\sigma}_{(i)}$ instead of $\hat{\sigma}$. Points with $|\text{DFITS}_i|$ larger than $2\sqrt{(p+1)/(n-p-1)}$ are usually classified as influential points. Again, instead of having a strict cutoff value, we use the measure to sort out points of abnormally high influence relative to other points on a graph such as the index plot, the dot plot, or the box plot. There is not much to choose between C_i and DFITS_i – both give similar answers because they are functions of the residual and leverage values. Most computer software will give one or both of the measures, and it is sufficient to look at only one of them.

5.9.3 Hadi's Influence Measure

Hadi (1992) proposed a measure of the influence of the ith observation based on the fact that influential observations are outliers in either the response variable or in the predictors, or both. Accordingly, the influence of the ith observation can be measured by

$$H_i = \frac{p_{ii}}{1 - p_{ii}} + \frac{p+1}{1 - p_{ii}} \frac{d_i^2}{1 - d_i^2}, \quad i = 1, 2, \ldots, n, \quad (5.24)$$

where $d_i = e_i/\sqrt{SSE}$ is the so-called *normalized residual*. The first term on the right-hand side of (5.24) is the potential function which measures outlyingness in the X-space. The second term is a function of the residual, which measures outlyingness in the response variable. It can be seen that observations will have large values of H_i if they are outliers in the response and/or the predictor variables, that is, if they have large values of r_i, p_{ii}, or both. The measure H_i does not focus on a specific regression result, but it can be thought of as an overall general measure of influence which depicts observations that are influential on at least one regression result.

Note that C_i and $DFITS_i$ are multiplicative functions of the residuals and leverage values, whereas H_i is an additive function. The influence measure H_i can best be examined graphically in the same way as Cook's distance and Welsch and Kuh measure.

Example: New York Rivers Data

Consider again fitting the simple regression model in (5.18), which relates the mean nitrogen concentration, Y, to the percentage of land area in commercial/industrial use, X_4. The scatter plot of Y versus X_4 and the corresponding least squares regression are given in Figure 5.5. Observations 4 (the Neversink River) and 5 (the Hackensack River) are located far from the bulk of other data points in Figure 5.5. Also observations 7, 3, 8, and 6 are somewhat sparse in the upper-left region of the graph. The three influence measures discussed above which result from fitting model (5.18) are shown in Table 5.4, and the corresponding index plots are shown in Figure 5.7. No value of C_i exceeds its cutoff value of 1. However, the index plot of C_i in Figure 5.7(a) shows clearly that observation number 4 (Neversink) should be flagged as an influential observation. This observation also exceeds its $DFITS_i$ cutoff value of $2\sqrt{(p+1)/(n-p-1)} = 2/3$. As can be seen from Figure 5.7, observation number 5 (Hackensack) was not flagged by C_i or by $DFITS_i$. This is due to the small value of the residual because of its high leverage and to the multiplicative nature of the measure. The index plot of H_i in Figure 5.7(c) indicates that observation number 5 (Hackensack) is the most influential one, followed by observation number 4 (Neversink), which is consistent with the scatter plot in Figure 5.5.

Table 5.4 New York Rivers Data. Influence Measures from Fitting Model (5.18): Cook's Distance, C_i, Welsch and Kuh Measure, DFITS$_i$, and Hadi's Influence Measure H_i

Row	C_i	DFITS$_i$	H_i	Row	C_i	DFITS$_i$	H_i
1	0.00	0.01	0.06	11	0.02	0.19	0.13
2	0.00	−0.01	0.07	12	0.02	−0.21	0.14
3	0.10	0.49	0.58	13	0.02	−0.22	0.15
4	0.56	−1.14	0.77	14	0.02	−0.19	0.13
5	0.02	0.22	2.04	15	0.02	−0.22	0.16
6	0.01	0.15	0.10	16	0.01	−0.12	0.09
7	0.17	0.63	0.60	17	0.02	−0.18	0.12
8	0.07	0.40	0.37	18	0.01	−0.12	0.09
9	0.00	−0.02	0.07	19	0.04	−0.27	0.19
10	0.00	0.09	0.08	20	0.01	0.15	0.11

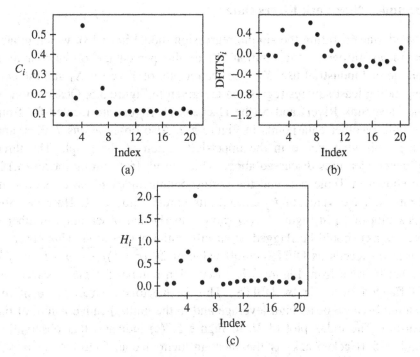

Figure 5.7 New York Rivers data: index plots of influence measures: (a) Cook's distance, C_i, (b) Welsch and Kuh measure, DFITS$_i$, and (c) Hadi's influence measure H_i.

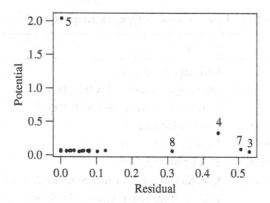

Figure 5.8 New York Rivers data: potential–residual plot.

5.10 THE POTENTIAL–RESIDUAL PLOT

The formula for H_i in (5.24) suggests a simple graph to aid in classifying unusual observations as high-leverage points, outliers, or a combination of both. The graph is called the *potential–residual* (P-R) plot (Hadi, 1992) because it is the scatter plot of

$$\text{Potential Function} \qquad \text{Residual Function}$$

$$\frac{p_{ii}}{1-p_{ii}} \qquad \text{versus} \qquad \frac{p+1}{1-p_{ii}}\frac{d_i^2}{1-d_i^2}. \qquad (5.25)$$

The P-R plot is related to the L-R (*leverage–residual*) plot suggested by Gray (1986) and McCulloch and Meeter (1983). The L-R plot is a scatter plot of p_{ii} versus d_i^2. For a comparison between the two plots, see Hadi (1992).

As an illustrative example, the P-R plot obtained from fitting model (5.18) is shown in Figure 5.8. Observation 5, which is a high-leverage point, is located by itself in the upper-left corner of the plot. Four outlying observations (3, 7, 4, and 8) are located in the lower-right area of the graph.

It is clear now that some individual data points may be flagged as outliers, leverage points, or influential points. The main usefulness of the leverage and influence measures is that they give the analyst a complete picture of the role played by different points in the entire fitting process. Any point falling in one of these categories should be carefully examined for accuracy (gross error, transcription error), relevancy (whether it belongs to the data set), and special significance (abnormal condition, unique situation). Outliers should always be scrutinized carefully. Points with high leverage that are not influential

Table 5.5 Functions for Computing Regression Diagnostics

Function	Output
residuals(reg)	Ordinary residuals, e_i
rstandard(reg)	Internally studentized residuals, r_i
rstudent(reg)	Externally studentized residuals, $r_i^\star$
cooks.distance(reg)	Cook's distance, C_i
ols_leverage(reg)	Leverage values, p_{ii}
ols_plot_dffits(reg)	Index plot of $DFITS_i$
ols_hadi(reg)	Computes Hadi's influence measure in (5.24). Output is a list of three components: $hadi, $potential, and $residual
ols_plot_hadi(reg)	Index plot of Hadi's influence, H_i
ols_plot_resid_pot(reg)	Potential–residual plot in (5.25)

do not cause problems. High-leverage points that are influential should be investigated because these points are outlying as far as the predictor variables are concerned and also influence the fit. To get an idea of the sensitivity of the analysis to these points, the model should be fitted without the offending points and the resulting coefficients examined.

5.11 REGRESSION DIAGNOSTICS IN R

R has a number of functions that provide the various regression diagnostics that we discussed in Sections 5.3 to 5.10. These functions are included in the packages stats (which is included in the base package) and olsrr. SO, we first need to download olsrr using if(!require("olsrr")) install.packages("olsrr"). Then we fit the regression model to the data stored in the object df using the lm() function: reg = lm(Y ~., data = df). The various diagnostics can then be computed as shown in Table 5.5. For example, one can use the functions in Table 5.5 to produce the results in Table 5.4 using the code: round(cbind(cooks.distance(reg),dffits(reg),ols_hadi(reg)$hadi),2).

5.12 WHAT TO DO WITH THE OUTLIERS?

Outliers and influential observations should not routinely be deleted or automatically down-weighted because they are not necessarily bad observations. On the contrary, if they are correct, they may be the most informative points in the data. For example, they may indicate that the data did not come from a normal population or that the model is not linear. To illustrate that outliers and influential observations can be the most informative points in the data, we use the exponential growth data described in the following example.

Example: Exponential Growth Data

Figure 5.9 is the scatter plot of two variables, the size of a certain population, Y, and time, X. As can be seen from the scatter of points, the majority of the points resemble a linear relationship between population size and time as indicated by the straight line in Figure 5.9. According to this model the two points 22 and 23 in the upper-right corner are outliers. If these points, however, are correct, they are the only observations in the data set that indicate that the data follow a nonlinear (e.g., exponential) model, such as the one shown in the graph. Think of this as a population of bacteria which increases very slowly over a period of time. After a critical point in time, however, the population explodes.

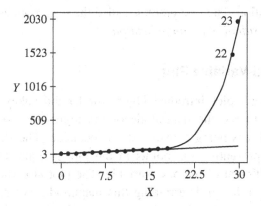

Figure 5.9 Scatter plot of population size, Y, versus time, X. The curve is obtained by fitting an exponential function to the full data. The straight line is the least squares line when observations 22 and 23 are deleted.

What to do with outliers and influential observations once they are identified? Because outliers and influential observations can be the most informative observations in the data set, they should not be automatically discarded without justification. Instead, they should be examined to determine why they are outlying or influential. Based on this examination, appropriate corrective actions can then be taken. These corrective actions include: correction of error in the data, deletion or down-weighing outliers, transforming the data, considering a different model, and redesigning the experiment or the sample survey, collecting more data.

5.13 ROLE OF VARIABLES IN A REGRESSION EQUATION

As we have indicated, successive variables are introduced sequentially into a regression equation. A question that arises frequently in practice is: Given a regression model which currently contains p predictor variables, what are the effects of deleting (or adding) one of the variables from (or to) the model? Frequently, the answer is to compute the t-Test for each variable in the model. If the t-Test is large in absolute value, the variable is retained; otherwise, the variable is omitted. This is valid only if the underlying assumptions hold. Therefore, the t-Test should be interpreted in conjunction with appropriate graphs of the data. Two plots have been proposed that give this information visually and are often very illuminating. They can be used to complement the t-Test in deciding whether one should retain or remove a variable in a regression equation. The first graph is called the *added-variable plot* and the second is the *residual plus component plot*.

5.13.1 Added-Variable Plot

The added-variable plot, introduced by Mosteller and Tukey (1977), enables us graphically to see the magnitude of the regression coefficient of the new variable that is being considered for inclusion. The slope of the least squares line representing the points in the plot is equal to the estimated regression coefficient of the new variable. The plot also shows data points which play key roles in determining this magnitude. We can construct an added-variable plot for each predictor variable X_j. The added-variable plot for X_j is essentially a graph of two different sets of residuals. The first is the residuals when Y is regressed on all predictor variables except X_j. We call this set the Y-residuals. The second set of residuals are obtained when we regress X_j (treated temporarily as a response variable) on all other predictor variables. We refer to this set as the X_j-residuals. Thus, the added-variable plot for X_j is simply a scatter plot of the

Y-residuals   versus X_j-residuals.

Therefore, if we have p predictor variables available, we can construct p added-variable plots, one for each predictor.

Note that the Y-residuals in the added-variable plot for X_j represent the part of Y not explained by all predictors other than X_j. Similarly, the X_j-residuals represent the part of X_j that is not explained by the other predictor variables. If a least squares regression line were fitted to the points in the added-variable plot for X_j, the slope of this line is equal to $\hat{\beta}_j$, the estimated regression coefficient of X_j when Y is regressed on all the predictor variables including X_j. This is an illuminating but equivalent interpretation of the partial regression coefficient as we have seen in Section 4.5.

The slope of the points in the plot gives the magnitude of the regression coefficient of the variable if it were brought into the equation. Thus, the stronger the linear relationship in the added-variable plot is, the more important the additional contribution of X_j to the regression equation already containing the other predictors. If the scatter of the points shows no marked slope, the variable is unlikely to be useful in the model. The scatter of the points will also indicate visually which of the data points are most influential in determining this slope and its corresponding t-Test. The added-variable plot is also known as the *partial regression plot*. We remark in passing that it is not actually necessary to carry out this fitting. These residuals can be obtained very simply from computations done in fitting Y on the full set of predictors. For a detailed discussion, see Velleman and Welsch (1981) and Chatterjee and Hadi (1988).

5.13.2 Residual Plus Component Plot

The residual plus component plot, introduced by Ezekiel (1924), is one of the earliest graphical procedures in regression analysis. It was revived by Larsen and McCleary (1972), who called it a *partial residual plot*. We are calling it a residual plus component plot, after Wood (1973), because this name is more self-explanatory.

The residual plus component plot for X_j is a scatter plot of

$$\mathbf{e} + \hat{\beta}_j X_j \quad \text{versus} \quad X_j,$$

where $\mathbf{e}$ is the ordinary least squares residuals when Y is regressed on all predictor variables and $\hat{\beta}_j$ is the coefficient of X_j in this regression. Note that $\hat{\beta}_j X_j$ is the contribution (component) of the jth predictor to the fitted values. As in the added-variable plot, the slope of the points in this plot is $\hat{\beta}_j$, the regression coefficient of X_j. Besides indicating the slope graphically, this plot indicates whether any nonlinearity is present in the relationship between Y and X_j. The plot can therefore suggest possible transformations for linearizing the data. The indication of nonlinearity is, however, not present in the added-variable plot because the horizontal scale in the plot is not the variable itself. Both plots are useful, but the residual plus component plot is more sensitive than the added-variable plot in detecting nonlinearities in the variable being considered for introduction in the model. The added-variable plot is, however, easier to interpret and points out the influential observations.

Example: The Scottish Hills Races Data

The Scottish hills races data consist of a response variable (record times, in seconds) and two explanatory variables (the distance in miles, and the climb in feet) for 35 races in Scotland in 1984. The data set can be found in the Book's Website. Since this data set is three dimensional, let us first examine a three-dimensional rotating plot of the data as an exploratory tool. An interesting direction in this rotating plot is shown in Figure 5.10. Five observations are marked in this plot. Clearly, observations 7 and 18 are outliers; they lie far away (in the direction of Time) from the plane suggested by the majority of other points. Observation 7 lies far away in the direction of Climb. Observations 33 and 31 are also outliers in the graph but to a lesser extent. While observations 11 and 31 are near the plane suggested by the majority of other points, they are located far from the rest of the points on the plane. (Observation 11 is far mainly in the direction of Distance and observation 31 is in the direction of Climb.) The rotating plot clearly shows

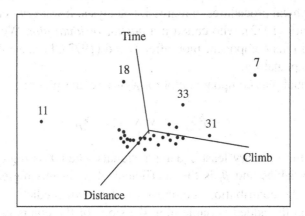

Figure 5.10 Rotating plot for the Scottish hills races data.

that the data contain unusual points (outliers, high-leverage points, and/or influential observations).

The fitted equation is

$$\text{Time} = -539.483 + 373.073 \, \text{Distance} + 0.662888 \, \text{Climb}. \quad (5.26)$$

We wish to address the question: Does each of the predictor variables contribute significantly when the other variable is included in the model? The t-Tests for the two predictors are 10.3 and 5.39, respectively, indicating very high significance. This implies that the answer to the above question is in the affirmative for both variables. The validity of this conclusion can be enhanced by examining the corresponding added-variable and residual plus component plots. These are given in Figures 5.11 and 5.12, respectively. For example, in the added-variable plot for Distance in Figure 5.11(a), the quantities plotted on the ordinate axis are the residuals obtained from the regression of Time on Climb (the other predictor variable), and the quantities plotted on the abscissa are the residuals obtained from the regression of Distance on Climb. Similarly for the added-variable plot for Climb, the quantities plotted are the residuals obtained from the regression of Time on Distance and the residuals obtained from the regression of Climb on Distance.

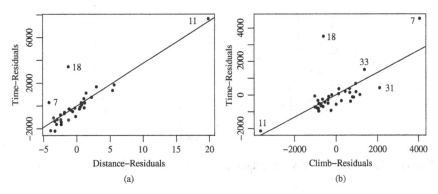

Figure 5.11 Scottish hills races data: added-variable plots for (a) Distance and (b) Climb.

It can be seen that there is a strong linear trend in all four graphs supporting the conclusions reached by the above t-Tests. The graphs, however, indicate the presence of some points that may influence our results and conclusions. Races 7, 11, and 18 clearly stand out. These points are marked on the graphs by their numbers. Races 31 and 33 are also suspects but to a lesser

184 REGRESSION DIAGNOSTICS: DETECTION OF MODEL VIOLATIONS

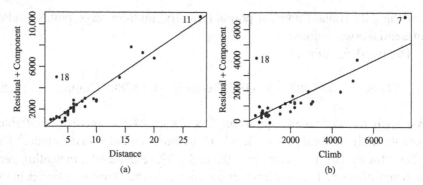

Figure 5.12 Scottish hills races data: residual plus component plots for (a) Distance and (b) Climb.

extent. An examination of the P-R plot obtained from the above fitted equation (Figure 5.13) classifies Race 11 as a high-leverage point, Race 18 as an outlier, and Race 7 as a combination of both. These points should be scrutinized carefully before continuing with further analysis.

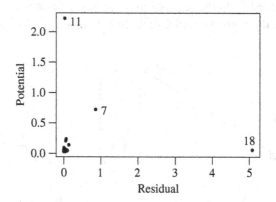

Figure 5.13 Scottish hills races data: potential–residual plot.

5.14 EFFECTS OF AN ADDITIONAL PREDICTOR

We discuss in general terms the effect of introducing a new variable in a regression equation. Two questions should be addressed: (a) Is the regression coefficient of the new variable significant? and (b) Does the introduction of the new variable substantially change the regression coefficients of the variables already in the regression equation? When a new variable is introduced in a regression equation, four possibilities result, depending on the answer to each of the above questions:

- **Case A:** The new variable has an insignificant regression coefficient and the remaining regression coefficients do not change substantially from their previous values. Under these conditions the new variable should not be included in the regression equation, unless some other external conditions (e.g., theory or subject matter considerations) dictate its inclusion.

- **Case B:** The new variable has a significant regression coefficient, and the regression coefficients for the previously introduced variables are changed in a substantial way. In this case the new variable should be retained, but an examination of collinearity[2] should be carried out. If there is no evidence of collinearity, the variable should be included in the equation and other additional variables should be examined for possible inclusion. On the other hand, if the variables show collinearity, corrective actions, as outlined in Chapter 11, should be taken.

- **Case C:** The new variable has a significant regression coefficient, and the coefficients of the previously introduced variables do not change in any substantial way. This is the ideal situation and arises when the new variable is uncorrelated with the previously introduced variables. Under these conditions the new variable should be retained in the equation.

- **Case D:** The new variable has an insignificant regression coefficient, but the regression coefficients of the previously introduced variables are substantially changed as a result of the introduction of the new variable. This is clear evidence of collinearity, and corrective actions have to be taken before the question of the inclusion or exclusion of the new variable in the regression equation can be resolved.

It is apparent from this discussion that the effect a variable has on the regression equation determines its suitability for being included in the fitted equation. The results presented in this chapter influence the formulation of different strategies devised for variable selection. Variable selection procedures are presented in Chapter 12.

---

[2] Collinearity occurs when the predictor variables are highly correlated. This problem is discussed in Chapters 10 and 11.

5.15 ROBUST REGRESSION

Another approach (not discussed here), useful for the identification of outliers and influential observations, is *robust regression*, a method of fitting that gives less weight to points with high leverage. There is a vast amount of literature on robust regression. The interested reader is referred, for example, to the books by Huber (1981), Hampel et al. (1986), Rousseeuw and Leroy (1987), Staudte and Sheather (1990), and Birkes and Dodge (1993). We must also mention the papers by Krasker and Welsch (1982), Coakley and Hettmansperger (1993), Chatterjee and Mächler (1997), and Billor et al. (2006), which incorporate ideas of bounding influence and leverage in fitting. In Section 14.5 we give a brief discussion of robust regression and present a numerical algorithm for robust fitting. Two examples are given as illustration.

EXERCISES

5.1 Check to see whether or not the standard regression assumptions are valid for each of the following data sets:
   (a) The Milk Production data described in Section 1.3.1 and given in the file Milk.Production.csv at the Book's Website.
   (b) The Right-To-Work Laws data described in Section 1.3.2 and given in the file RTWL.csv at the Book's Website.
   (c) The Egyptian Skulls data described in Section 1.3.4 and given in the file Egyptian.Skulls.csv at the Book's Website.
   (d) The Domestic Immigration data described in Section 1.3.3 and given in the file Domestic.Immigration.csv at the Book's Website.
   (e) The New York Rivers data described in Section 1.3.5 and given in the file NYRivers.csv at the Book's Website.

5.2 Find a data set where regression analysis can be used to answer a question of interest. Then:
   (a) Check to see whether or not the usual multiple regression assumptions are valid.
   (b) Analyze the data using the regression methods presented thus far, and answer the question of interest.

5.3 Consider the computer repair problem discussed in Section 3.3. In a second sampling period, 10 more observations on the variables Minutes and Units were obtained. Since all observations were collected by the

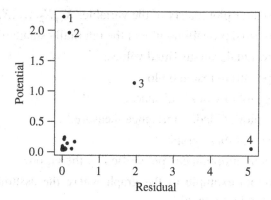

Figure 5.14 P-R plot used in Exercise 5.4.

same method from a fixed environment, all 24 observations were pooled to form one data set. The data can be found in the file Computer.Repair.Expanded.csv at the Book's Website.

(a) Fit a linear regression model relating Minutes to Units.

(b) Check each of the standard regression assumptions and indicate which assumption(s) seems to be violated.

5.4 In an attempt to find unusual points in a regression data set, a data analyst examines the P-R plot (shown in Figure 5.14). Classify each of the unusual points on this plot according to type.

5.5 Name one or more graphs that can be used to validate each of the following assumptions. For each graph, sketch an example where the corresponding assumption is valid and an example where the assumption is clearly invalid.

(a) There is a linear relationship between the response and predictor variables.

(b) The observations are independent of each other.

(c) The error terms have constant variance.

(d) The error terms are uncorrelated.

(e) The error terms are normally distributed.

(f) The observations are equally influential on least squares results.

5.6 The following graphs are used to verify some of the assumptions of the ordinary least squares regression of Y on $X_1, X_2, \ldots, X_p$:

1. The scatter plot of Y versus each predictor X_j.

2. The scatter plot matrix of the variables $X_1, X_2, \ldots, X_p$.
3. The normal probability plot of the internally standardized residuals.
4. The residuals versus fitted values.
5. The potential–residual plot.
6. Index plot of Cook's distance.
7. Index plot of Hadi's influence measure.

For each of these graphs:

(a) What assumption can be verified by the graph?
(b) Draw an example of the graph where the assumption does not seem to be violated.
(c) Draw an example of the graph which indicates the violation of the assumption.

5.7 Consider again the Cigarette Consumption data described in Exercise 4.17 and given in the file Cigarette.Consumption.csv at the Book's Website.

(a) What would you expect the relationship between Sales and each of the other explanatory variables to be (i.e., positive, negative)? Explain.
(b) Compute the pairwise correlation coefficients matrix and construct the corresponding scatter plot matrix.
(c) Are there any disagreements between the pairwise correlation coefficients and the corresponding scatter plot matrix?
(d) Is there any difference between your expectations in part (a) and what you see in the pairwise correlation coefficients matrix and the corresponding scatter plot matrix?
(e) Regress Sales on the six predictor variables. Is there any difference between your expectations in part (a) and what you see in the regression coefficients of the predictor variables? Explain inconsistencies if any.
(f) How would you explain the difference in the regression coefficients and the pairwise correlation coefficients between Sales and each of the six predictor variables?
(g) Is there anything wrong with the tests you made and the conclusions you reached in Exercise 4.17?

5.8 Consider again the Examination Data used in Exercise 4.4 and given in the file Examination.Data.csv at the Book's Website:

(a) For each of the three models, draw the P-R plot. Identify all unusual observations (by number) and classify as outlier, high-leverage point, and/or influential observation.

(b) What model would you use to predict the final score F?

5.9 Either prove each of the following statements mathematically or demonstrate its correctness numerically using the Cigarette Consumption data described in Exercise 4.17:

(a) The sum of the ordinary least squares residuals is zero.

(b) The relationship between $\hat{\sigma}^2$ and $\hat{\sigma}^2_{(i)}$ is

$$\hat{\sigma}^2_{(i)} = \hat{\sigma}^2 \left(\frac{n-p-1-r_i^2}{n-p-2} \right). \quad (5.27)$$

5.10 Figure 5.15 is the scatter plot of Y versus X, for n distinct observations, with the least squares line drawn on the plot. The plot also shows five points labeled as $A, B, C, D,$ and E.

(a) What is n?

(b) Classify each of the five points by putting check marks in the appropriate row in each of the columns in Table 5.6.

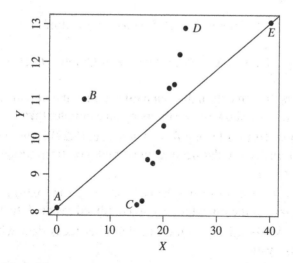

Figure 5.15 Plot of Y versus X, for n distinct observations with the least squares fitted line with five marked points.

Table 5.6 Classification of the Five Points in Figure 5.15

Point	Regression Outlier	High Leverage	Swamped Point	Masked Point	Cook's Influential	Hadi's Influential
A						
B						
C						
D						
E						

(c) Sketch the corresponding Potential–Residual Plot and label all the five points on your plot as well as the vertical and horizontal axes.

(d) Which of the five points have the smallest value(s) of Cook's Distance?

(e) Which of the five points have the smallest leverage value(s)?

5.11 Consider the Scottish hills races data in the file Scottish.Hills.csv in the Book's Website. Choose an observation index i (e.g., $i = 33$, which corresponds to the outlying observation number 33) and create an indicator (dummy) variable U_i, where all the values of U_i are zero except for its ith value which is one. Now consider comparing the following models:

$$H_0 : \text{Time} = \beta_0 + \beta_1 \text{ Distance} + \beta_2 \text{ Climb} + \varepsilon, \qquad (5.28)$$

$$H_1 : \text{Time} = \beta_0 + \beta_1 \text{ Distance} + \beta_2 \text{ Climb} + \beta_3 U_i + \varepsilon. \qquad (5.29)$$

Let r_i^* be the ith externally standardized residual obtained from fitting model (5.28). Show (or verify using an example) that

(a) The t-Test for testing $\beta_3 = 0$ in Model (5.29) is the same as the ith externally standardized residual obtained from Model (5.28), that is, $t_3 = r_i^*$.

(b) The F-Test for testing Model (5.28) versus (5.29) reduces to the square of the ith externally standardized residual, that is, $F = r_i^{*2}$.

(c) Fit Model (5.28) to the Scottish hills races data without the ith observation.

(d) Show that the estimates of β_0, β_1, and β_2 in Model (5.29) are the same as those obtained in (c). Hence adding an indicator variable for the ith observation is equivalent to deleting the corresponding observation!

5.12 Consider the data in the file Exercise5.12.csv in the Book's Website.[3] The data consist of a response variable Y and six predictor variables. Consider fitting a linear model relating Y to all six X-variables.

(a) What least squares assumptions (if any) seem to be violated?

(b) Compute r_i, C_i, DFITS$_i$, and H_i.

(c) Construct the index plots of r_i, C_i, DFITS$_i$, and H_i as well as the Potential–Residual plot.

(d) Identify all unusual observations in the data and classify each according to type (i.e., outliers, and leverage points).

5.13 Consider again the data set in Exercise 5.12. Suppose now that we fit a linear model relating Y to the first three X-variables. Justify your answer to each of the following questions with the appropriate added-variable plot:

(a) Should we add X_4 to the above model? If yes, keep X_4 in the model.

(b) Should we add X_5 to the above model? If yes, keep X_5 in the model.

(c) Should we add X_6 to the above model?

(d) Which model(s) would you recommend as the best possible description of Y? Use the above results and/or perform additional analysis if needed.

5.14 Consider fitting the model $Y = \beta_0 + \beta_1 X_1 + \beta_2 X_2 + \beta_3 X_3 + \varepsilon$ to the data set used in Exercise 5.12. Now let u be the residuals obtained from regressing Y on X_1 and X_2. Let v be the residuals obtained from regressing X_3 on X_1. Show (or verify using the data set in Exercise 5.12 as an example) that:

(a) $\hat{\beta}_3 = \sum_{i=1}^{n} u_i v_i / \sum_{i=1}^{n} v_i^2$

(b) The standard error of $\hat{\beta}_3$ is $\hat{\sigma} / \sqrt{\sum_{i=1}^{n} v_i^2}$.

---

[3] http://www.aucegypt.edu/faculty/hadi/RABE6

CHAPTER 6

QUALITATIVE VARIABLES AS PREDICTORS

6.1 INTRODUCTION

Qualitative or categorical variables can be very useful as predictor variables in regression analysis. Qualitative variables such as gender, marital status, or political affiliation can be represented by *indicator* or *dummy* variables. These variables take on only two values, usually 0 and 1. The two values signify that the observation belongs to one of two possible categories. The numerical values of indicator variables are not intended to reflect a quantitative ordering of the categories, but only serve to identify category or class membership. For example, an analysis of salaries earned by computer programmers may include variables such as education, years of experience, and gender as predictor variables. The gender variable could be quantified, say, as 1 for female and 0 for male. Indicator variables can also be used in a regression equation to distinguish among three or more groups as well as among classifications across various types of groups. For example, the regression described above may also include an indicator variable to distinguish whether the observation was for a systems or applications programmer.

Regression Analysis By Example Using R, Sixth Edition. Ali S. Hadi and Samprit Chatterjee
© 2024 John Wiley & Sons, Inc. Published 2024 by John Wiley & Sons, Inc.
Companion website: www.wiley.com/go/hadi/regression_analysis_6e

194 QUALITATIVE VARIABLES AS PREDICTORS

The four conditions determined by gender and type of programming can be represented by combining the two variables, as we shall see in this chapter.

Indicator variables can be used in a variety of ways and may be considered whenever there are qualitative variables affecting a relationship. We shall illustrate some of the applications with examples and suggest some additional applications. It is hoped that the reader will recognize the general applicability of the technique from the examples. In the first example, we look at data on a salary survey and use indicator variables to adjust for various categorical variables that affect the regression relationship. The second example uses indicator variables for analyzing and testing for equality of regression relationships in various subsets of a population.

We continue to assume that the response variable is a quantitative continuous variable, but the predictor variables can be quantitative and/or categorical. The case where the response variable is an indicator variable is dealt with in Chapter 13.

6.2 SALARY SURVEY DATA

The Salary Survey data set was developed from a salary survey of computer professionals in a large corporation. The objective of the survey was to identify and quantify those variables that determine salary differentials. In addition, the data could be used to determine if the corporation's salary administration guidelines were being followed. The data can be obtained from the Book's Website.[1] The data can be read into your R Workspace using the following commands:

url="http://www.aucegypt.edu/faculty/hadi/RABE6/Data6/Salary
.Survey.csv"
download.file(url, "Salary.Survey.csv")
df <- read.csv("Salary.Survey.csv", header = TRUE)
head(df, 2) # Display the first 2 lines of the dataset

The response variable is salary (S) and the predictors are: (a) experience (X), measured in years; (b) education (E), coded as 1 for completion of a high school (H.) diploma, 2 for completion of a bachelor degree (B.), and 3 for the completion of an advanced degree; and (c) management (M), which is coded as 1 for a person with management responsibility and 0 otherwise.

---

[1] http://www.aucegypt.edu/faculty/hadi/RABE6

We shall try to measure the effects of these three variables on salary using regression analysis.

A linear relationship will be used for salary and experience. We shall assume that each additional year of experience is worth a fixed salary increment. Education may also be treated in a linear fashion. If the education variable is used in the regression equation in raw form, we would be assuming that each step up in education is worth a fixed increment in salary. That is, with all other variables held constant, the relationship between salary and education is linear. That interpretation is possible but may be too restrictive. Instead, we shall view education as a categorical variable and define two indicator variables to represent the three categories. These two variables allow us to pick up the effect of education on salary whether or not it is linear. The management variable is also an indicator variable designating the two categories, 1 for management positions and 0 for regular staff positions.

Note that when using indicator variables to represent a set of categories, the number of these variables required is one less than the number of categories. For example, in the case of the education categories above, we create two indicator variables E_1 and E_2, where

$$E_{i1} = \begin{cases} 1, & \text{if the } i\text{th person is in the HS category,} \\ 0, & \text{otherwise,} \end{cases}$$

and

$$E_{i2} = \begin{cases} 1, & \text{if the } i\text{th person is in the BS category,} \\ 0, & \text{otherwise.} \end{cases}$$

These two indicator variable can be created and augmented to the data in df using the R code

```
E1 <- as.numeric(df$E == 1); E2 <- as.numeric(df$E == 2)
E3 <- as.numeric(df$E == 3); df <- cbind(df, E1, E2, E3)
```

As stated above, these two variables taken together uniquely represent the three groups. For HS, $E_1 = 1, E_2 = 0$; for BS, $E_1 = 0, E_2 = 1$; and for advanced degree, $E_1 = 0, E_2 = 0$. Furthermore, if there were a third variable, E_{i3}, defined to be 1 or 0 depending on whether or not the ith person is in the advanced degree category, then for each person we have $E_1 + E_2 + E_3 = 1$. Then $E_3 = 1 - E_1 - E_2$, showing clearly that one of the variables is superfluous.[2] Similarly, there is only one indicator variable required to distinguish

---

[2] Had E_1, E_2, and E_3 been used, there would have been a perfect linear relationship among the predictors, which is an extreme case of collinearity, a problem described in Chapters 10 and 11.

the two management categories. The category that is not represented by an indicator variable is referred to as the *base category* or the *control group* because the regression coefficients of the indicator variables are interpreted relative to the control group.

In terms of the indicator variables described above, the regression model is

$$S = \beta_0 + \beta_1 X + \gamma_1 E_1 + \gamma_2 E_2 + \delta_1 M + \varepsilon. \qquad (6.1)$$

This model can be fitted to the data using the following R code

```
reg <- lm(S ~ X + E1 + E2 + M, data = df)
```

By evaluating (6.1) for the different values of the indicator variables, it follows that there is a different regression equation for each of the six (three education and two management) categories as shown in Table 6.1. According to the proposed model, we may say that the indicator variables help to determine the base salary level as a function of education and management status after adjustment for years of experience.

The results of the regression computations for the model given in (6.1) appear in Table 6.2. The proportion of salary variation accounted for by the model is quite high ($R^2 = 0.957$). At this point in the analysis we should investigate the pattern of residuals to check on model specification. We shall postpone that investigation for now and assume that the model is satisfactory so that we can discuss the interpretation of the regression results. Later we shall return to analyze the residuals and find that the model must be altered.

We see that the coefficient of X is 546.18. That is, each additional year of experience is estimated to be worth an annual salary increment of $546. The other coefficients may be interpreted by looking into Table 6.1. The coefficient of the management indicator variable, δ_1, is estimated to

Table 6.1 Regression Equations for the Six Categories of Education and Management

Category	E	M	Regression Equation
1	1	0	$S = (\beta_0 + \gamma_1) \quad + \beta_1 X + \varepsilon$
2	1	1	$S = (\beta_0 + \gamma_1 + \delta_1) + \beta_1 X + \varepsilon$
3	2	0	$S = (\beta_0 + \gamma_2) \quad + \beta_1 X + \varepsilon$
4	2	1	$S = (\beta_0 + \gamma_2 + \delta_1) + \beta_1 X + \varepsilon$
5	3	0	$S = \beta_0 \quad + \beta_1 X + \varepsilon$
6	3	1	$S = (\beta_0 + \delta_1) \quad + \beta_1 X + \varepsilon$

Table 6.2 Regression Analysis of Salary Survey Data

Variable	Coefficient	s.e.	t-Test	p-Value
Constant	11031.81	383.22	28.79	0.0000
X	546.18	30.52	17.90	0.0000
E_1	−2996.21	411.75	−7.28	0.0000
E_2	147.82	387.66	0.38	0.7049
M	6883.53	313.92	21.93	0.0000
n = 46	$R^2 = 0.957$	$R_a^2 = 0.953$	$\hat{\sigma} = 1027$	df = 41

be 6883.53. From Table 6.1 we interpret this amount to be the average incremental value in annual salary associated with a management position. For the education variables, γ_1 measures the salary differential for the HS category relative to the advanced degree category and γ_2 measures the differential for the BS category relative to the advanced degree category. The difference, $\gamma_2 - \gamma_1$, measures the differential salary for the HS category relative to the BS category. From the regression results, in terms of salary for computer professionals, we see that an advanced degree is worth $2996 more than a high school diploma, a BS is worth $148 more than an advanced degree (this differential is not statistically significant, $t = 0.38$), and a BS is worth about $3144 more than a high school diploma. These salary differentials hold for every fixed level of experience.

6.3 INTERACTION VARIABLES

Returning now to the question of model specification, consider Figure 6.1,[3] where the residuals are plotted against X. The plot suggests that there may be three or more specific levels of residuals. Possibly the indicator variables that have been defined are not adequate for explaining the effects of education and management status. Actually, each residual is identified with one of the six education-management combinations. To see this we plot the residuals against Category (a new categorical variable that takes a separate value for each of the six combinations). This graph is, in effect, a plot of residuals versus a potential predictor variable that has not yet been used in

---
[3] The R code for producing this and all other graphs in this book can be found at the Book's Website at http://www.aucegypt.edu/faculty/hadi/RABE6.

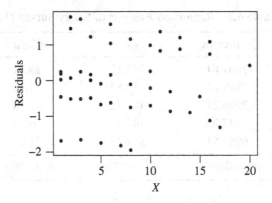

Figure 6.1 Standardized residuals versus years of experience (X).

the equation. The graph is given in Figure 6.2. It can be seen from the graph that the residuals cluster by size according to their education-management category. The combinations of education and management have not been satisfactorily treated in the model. Within each of the six groups, the residuals are either almost totally positive or totally negative. This behavior implies that the model given in (6.1) does not adequately explain the relationship between salary and experience, education, and management variables. The graph points to some hidden structure in the data that has not been explored.

The graphs strongly suggest that the effects of education and management status on salary determination are not additive. Note that in the model in (6.1) and its further exposition in Table 6.1, the incremental effects of both variables are determined by additive constants. For example, the effect

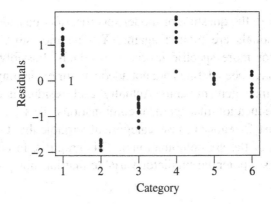

Figure 6.2 Standardized residuals versus education-management categorical variable.

of a management position is measured as δ_1, independently of the level of educational attainment. The nonadditive effects of these variables can be evaluated by constructing additional variables that are used to measure what may be referred to as *multiplicative* or *interaction effects*. Interaction variables are defined as products of the existing indicator variables $(E_1 \cdot M)$ and $(E_2 \cdot M)$. The inclusion of these two variables on the right-hand side of (6.1) leads to a model that is no longer additive in education and management, but recognizes the multiplicative effect of these two variables.

The expanded model is

$$S = \beta_0 + \beta_1 X + \gamma_1 E_1 + \gamma_2 E_2 + \delta_1 M$$
$$+ \alpha_1 (E_1 \cdot M) + \alpha_2 (E_2 \cdot M) + \varepsilon. \quad (6.2)$$

We can fit this model to the data using the following R code

reg <- lm(S ~ X + E1 + E2 + M + E1*M + E2*M, data = df);

The regression results are given in Table 6.3. The residuals from the regression of the expanded model are plotted against X in Figure 6.3. Note that observation 33 is an outlier. Salary is overpredicted by the model. Checking this observation in the listing of the raw data, it appears that this particular person seems to have fallen behind by a couple of hundred dollars in annual salary as compared to other persons with similar characteristics. To be sure that this single observation is not overly affecting the regression estimates, it has been deleted and the regression rerun. The new results are given in Table 6.4.

Table 6.3 Regression Analysis of Salary Data: Expanded Model

Variable	Coefficient	s.e.	t-Test	p-Value
Constant	11203.43	79.07	141.7	0.0000
X	496.99	5.57	89.3	0.0000
E_1	−1730.75	105.33	−16.4	0.0000
E_2	−349.08	97.57	−3.6	0.0009
M	7047.41	102.59	68.7	0.0000
$E_1 \cdot M$	−3066.04	149.33	−20.5	0.0000
$E_2 \cdot M$	1836.49	131.17	14.0	0.0000
$n = 46$	$R^2 = 0.999$	$R_a^2 = 0.999$	$\hat{\sigma} = 173.8$	df = 39

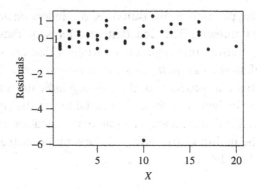

Figure 6.3 Standardized residuals versus years of experience: expanded model.

Table 6.4 Regression Analysis of Salary Data: Expanded Model, Observation 33 Deleted

Variable	Coefficient	s.e.	t-Test	p-Value
Constant	11199.71	30.53	366.8	0.0000
X	498.42	2.15	231.6	0.0000
E_1	−1741.34	40.68	−42.8	0.0000
E_2	−357.04	37.68	−9.5	0.0000
M	7040.58	39.62	177.7	0.0000
$E_1 \cdot M$	−3051.76	57.67	−52.9	0.0000
$E_2 \cdot M$	1997.53	51.79	38.6	0.0000
$n = 45$	$R^2 = 0.9998$	$R_a^2 = 0.998$	$\hat{\sigma} = 67.12$	df = 38

The regression coefficients are basically unchanged. However, the standard deviation of the residuals has been reduced to $67.12 and the proportion of explained variation has reached 0.9998. The plot of residuals versus X (Figure 6.4) appears to be satisfactory compared with the similar residual plot for the additive model. In addition, the plot of residuals for each education-management category (Figure 6.5) shows that each of these groups has residuals that appear to be symmetrically distributed about zero. Therefore the introduction of the interaction terms has produced an accurate representation of salary variations. The relationships between salary and experience, education, and management status appear to be adequately described by the model given in (6.2).

With the standard error of the residuals estimated to be $67.12, we can believe that we have uncovered the actual and very carefully administered

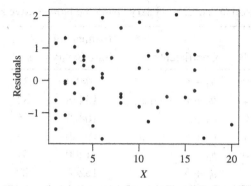

Figure 6.4 Standardized residuals versus years of experience: expanded model, observation 33 deleted.

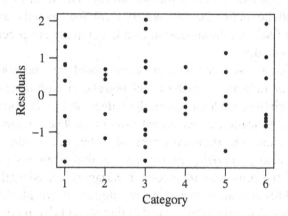

Figure 6.5 Standardized residuals versus education-management categorical variable: expanded model, observation 33 deleted.

salary formula. Using 95% confidence intervals, each year of experience is estimated to be worth between $494.06 and $502.77. These increments of approximately $500 are added to a starting salary that is specified for each of the six education-management groups. Since the final regression model is not additive, it is rather difficult to directly interpret the coefficients of the indicator variables. To see how the qualitative variables affect salary differentials, we use the coefficients to form estimates of the base salary for each of the six categories. These results are presented in Table 6.5 along with standard errors and confidence intervals. The standard errors are computed using (A.12) in the Appendix to Chapter 4.

Using a regression model with indicator variables and interaction terms, it has been possible to account for almost all the variation in salaries of

Table 6.5 Estimates of Base Salary Using the Nonadditive Model in (6.2)

Category	E	M	Coefficients	Estimate of Base Salary[a]	s.e.[a]	95% Confidence Interval
1	1	0	$\beta_0 + \gamma_1$	9459	31	(9398, 9520)
2	1	1	$\beta_0 + \gamma_1 + \delta + \alpha_1$	13448	32	(13385, 13511)
3	2	0	$\beta_0 + \gamma_2$	10843	26	(10792, 10894)
4	2	1	$\beta_0 + \gamma_2 + \delta + \alpha_2$	19880	33	(19815, 19945)
5	3	0	β_0	11200	31	(11139, 11261)
6	3	1	$\beta_0 + \delta$	18240	29	(18183, 18297)

[a] Recorded to the nearest dollar.

computer professionals selected for this survey. The level of accuracy with which the model explains the data is very rare! We can only conjecture that the methods of salary administration in this company are precisely defined and strictly applied.

In retrospect, we see that an equivalent model may be obtained with a different set of indicator variables and regression parameters. One could define five variables, each taking on the values of 1 or 0, corresponding to five of the six education-management categories. The numerical estimates of base salary and the standard errors of Table 6.5 would be the same. The advantage to proceeding as we have is that it allows us to separate the effects of the three sets of predictor variables, (a) education, (b) management, and (c) education-management interaction. Recall that interaction terms were included only after we found that an additive model did not satisfactorily explain salary variations. In general, we start with simple models and proceed sequentially to more complex models if necessary. We shall always hope to retain the simplest model that has an acceptable residual structure.

6.4 SYSTEMS OF REGRESSION EQUATIONS: COMPARING TWO GROUPS

A collection of data may consist of two or more distinct subsets, each of which may require a separate regression equation. Serious bias may be incurred if one regression relationship is used to represent the pooled data set. An analysis of this problem can be accomplished using indicator variables. An analysis of separate regression equations for subsets of the data may be applied to *cross-sectional* or *time series* data. The example discussed below treats cross-sectional data. Applications to time series data are discussed in Section 6.5.

The model for the two groups can be different in all aspects or in only some aspects. In this section we discuss three distinct cases:

1. Each group has a separate regression model.
2. The models have the same intercept but different slopes.
3. The models have the same slope but different intercepts.

We illustrate these cases below when we have only one quantitative predictor variable. These ideas can be extended straightforwardly to the cases where there are more than one quantitative predictor variable.

6.4.1 Models with Different Slopes and Different Intercepts

We illustrate this case with an important problem concerning equal opportunity in employment. Many large corporations and government agencies administer a preemployment test in an attempt to screen job applicants. The test is supposed to measure an applicant's aptitude for the job and the results are used as part of the information for making a hiring decision. The federal government has ruled[4] that these tests (a) must measure abilities that are directly related to the job under consideration and (b) must not discriminate on the basis of race or national origin. Operational definitions of requirements (a) and (b) are rather elusive. We shall not try to resolve these operational problems. We shall take one approach involving race represented as two groups, white and minority. The hypothesis that there are separate regressions relating test scores to job performance for the two groups will be examined. The implications of this hypothesis for discrimination in hiring are discussed.

Let Y represent job performance and let X be the score on the preemployment test. We want to compare

Model 1 (Pooled): $y_{ij} = \beta_0 + \beta_1 x_{ij} + \varepsilon_{ij}$, $j = 1, 2$; $i = 1, 2, \ldots, n_j$,

Model 2 (Minority): $y_{i1} = \beta_{01} + \beta_{11} x_{i1} + \varepsilon_{i1}$, (6.3)

Model 2 (White): $y_{i2} = \beta_{02} + \beta_{12} x_{i2} + \varepsilon_{i2}$.

Figure 6.6 depicts the two models. In Model 1, race distinction is ignored, the data are pooled, and there is one regression line. In Model 2, there is

---
[4] Tower amendment to Title VII, Civil Rights Act of 1964.

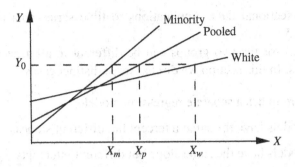

Figure 6.6 Requirements for employment on pretest.

a separate regression relationship for the two subgroups, each with distinct regression coefficients. We shall assume that the variances of the residual terms are the same in each subgroup.

Before analyzing the data, let us briefly consider the types of errors that could be present in interpreting and applying the results. If Y_0, as seen on the graph, has been set as the minimum required level of performance, then using Model 1, an acceptable score on the test is one that exceeds X_p. However, if Model 2 is in fact correct, the appropriate test score for whites is X_w and for minorities is X_m. Using X_p in place of X_m and X_w represents a relaxation of the pretest requirement for whites and a tightening of that requirement for minorities. Since inequity can result in the selection procedure if the wrong model is used to set cutoff values, it is necessary to examine the data carefully. It must be determined whether there are two distinct relationships or whether the relationship is the same for both groups and a single equation estimated from the pooled data is adequate. Note that whether Model 1 or Model 2 is chosen, the values X_m, X_w, and X_p are estimates subject to sampling errors and should only be used in conjunction with appropriate confidence intervals. (Construction of confidence intervals is discussed in the following paragraphs.)

Data were collected for this analysis using a special employment program. Twenty applicants were hired on a trial basis for six weeks. One week was spent in a training class. The remaining five weeks were spent on the job. The participants were selected from a pool of applicants by a method that was not related to the preemployment test scores. A test was given at the end of the training period and a work performance evaluation was developed at the end of the six-week period. These two scores were combined to form an index of job performance. (Those employees with unsatisfactory performance at the end of the six-week period were dropped.) The data appear in Table 6.6 and can be obtained from the Book's Website. We refer to this data set as the

SYSTEMS OF REGRESSION EQUATIONS: COMPARING TWO GROUPS

Table 6.6 Data on Preemployment Testing Program

Row	TEST	RACE	JPERF	Row	TEST	RACE	JPERF
1	0.28	1	1.83	11	2.36	0	3.25
2	0.97	1	4.59	12	2.11	0	5.30
3	1.25	1	2.97	13	0.45	0	1.39
4	2.46	1	8.14	14	1.76	0	4.69
5	2.51	1	8.00	15	2.09	0	6.56
6	1.17	1	3.30	16	1.50	0	3.00
7	1.78	1	7.53	17	1.25	0	5.85
8	1.21	1	2.03	18	0.72	0	1.90
9	1.63	1	5.00	19	0.42	0	3.85
10	1.98	1	8.04	20	1.53	0	2.95

Preemployment data. The data can be read into your R Workspace using the following commands:

url="http://www.aucegypt.edu/faculty/hadi/RABE6/Data6/Preemployment.csv"
download.file(url, "Preemployment.csv")
df <- read.csv("Preemployment.csv", header = TRUE)
head(df, 2) # Display the first 2 lines of the dataset

Formally, we want to test the null hypothesis $H_0 : \beta_{11} = \beta_{12}, \beta_{01} = \beta_{02}$ against the alternative that there are substantial differences in these parameters. The test can be performed using indicator variables. Let z_{ij} be defined to take the value 1 if $j = 1$ and to take the value 0 if $j = 2$. That is, Z is a new variable that has the value 1 for a minority applicant and the value 0 for a white applicant. We consider the two models

$$\text{Model 1: } y_{ij} = \beta_0 + \beta_1 x_{ij} + \varepsilon_{ij}$$
$$\text{Model 3: } y_{ij} = \beta_0 + \beta_1 x_{ij} + \gamma z_{ij} + \delta(z_{ij} \cdot x_{ij}) + \varepsilon_{ij}. \quad (6.4)$$

The variable $(z_{ij} \cdot x_{ij})$ represents the interaction between the group (race) variable Z and the preemployment test X. Note that Model 3 is equivalent to Model 2. This can be seen if we observe that for the minority group, $x_{ij} = x_{i1}$ and $z_{ij} = 1$; hence, Model 3 becomes

$$y_{i1} = \beta_0 + \beta_1 x_{i1} + \gamma + \delta x_{i1} + \varepsilon_{i1}$$
$$= (\beta_0 + \gamma) + (\beta_1 + \delta) x_{i1} + \varepsilon_{i1}$$
$$= \beta_{01} + \beta_{11} x_{i1} + \varepsilon_{i1},$$

which is the same as Model 2 for minority with $\beta_{01} = \beta_0 + \gamma$ and $\beta_{11} = \beta_1 + \delta$. Similarly, for the white group, we have $x_{ij} = x_{i2}$, $z_{ij} = 0$, and Model 3 becomes

$$y_{i2} = \beta_0 + \beta_1 x_{i2} + \varepsilon_{i2},$$

which is the same as Model 2 for white with $\beta_{02} = \beta_0$ and $\beta_{12} = \beta_1$. Therefore, a comparison between Models 1 and 2 is equivalent to a comparison between Models 1 and 3. Note that Model 3 can be viewed as a full model (FM) and Model 1 as a restricted model (RM) because Model 1 is obtained from Model 3 by setting $\gamma = \delta = 0$. Thus, our null hypothesis H_0 now becomes $H_0 : \gamma = \delta = 0$. The hypothesis is tested by constructing an F-Test for the comparison of two models as described in Chapter 4. In this case, the test statistics is

$$F = \frac{[\text{SSE(RM)} - \text{SSE(FM)}]/2}{\text{SSE(FM)}/16}, \qquad (6.5)$$

which has 2 and 16 degrees of freedom. (Why?) Proceeding with the analysis of the data, the regression results for Models 1 and 3 are given in Tables 6.7 and 6.8. These results ca be obtained using the following R code:

reg1 <- lm(JPERF ~ TEST, data = df); # Model 1
reg3 <- lm(JPERF ~ TEST + RACE + TEST * RACE, data = df) # Model 3

The plots of residuals against the predictor variable (Figures 6.7 and 6.8) look acceptable in both cases. The one residual at the lower right in Model 1 may require further investigation.

To evaluate the formal hypothesis we compute the F-ratio specified 6.5), which is equal to

$$F = \frac{(45.568 - 31.655)/2}{31.655/16} = 3.52 \qquad (6.6)$$

Table 6.7 Regression Results, Preemployment Data: Model 1

Variable	Coefficient	s.e.	t-Test	p-Value
Constant	1.03	0.87	1.19	0.2486
TEST (X)	2.36	0.54	4.39	0.0004
$n = 20$	$R^2 = 0.52$	$R_a^2 = 0.49$	$\hat{\sigma} = 1.59$	df = 18

Table 6.8 Regression Results, Preemployment Data: Model 3

Variable	Coefficient	s.e.	t-Test	p-Value
Constant	2.01	1.05	1.91	0.0736
TEST (X)	1.31	0.67	1.96	0.0677
RACE (Z)	−1.91	1.54	−1.24	0.2321
RACE · TEST ($X \cdot Z$)	2.00	0.95	2.09	0.0527
$n = 20$	$R^2 = 0.664$	$R_a^2 = 0.601$	$\hat{\sigma} = 1.41$	df = 16

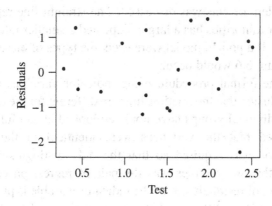

Figure 6.7 Standardized residuals versus test score: Model 1.

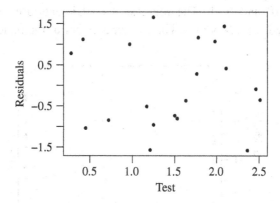

Figure 6.8 Standardized residuals versus test score: Model 3.

and is significant at a level slightly above 5%. Therefore, on the basis of this test we would conclude that the relationship is probably different for the two groups. Specifically, for minorities we have

$$Y_1 = 0.10 + 3.31 X_1$$

and for whites we have

$$Y_2 = 2.01 + 1.32 X_2.$$

The results are very similar to those that were described in Figure 6.5 when the problem of bias was discussed. The straight line representing the relationship for minorities has a larger slope and a smaller intercept than the line for whites. If a pooled model were used, the types of biases discussed in relation to Figure 6.6 would occur.

Although the formal procedure using indicator variables has led to the plausible conclusion that the relationships are different for the two groups, the data for the individual groups have not been looked at carefully. Recall that it was assumed that the variances were identical in the two groups. This assumption was required so that the only distinguishing characteristic between the two samples was the pair of regression coefficients. In Figure 6.9 a plot of residuals versus the indicator variable is presented. There does not appear to be a difference between the two sets of residuals. We shall now look more closely at each group. The regression coefficients for each sample taken separately are presented in Table 6.9. The residuals are shown in Figures 6.10 and 6.11. The regression coefficients are, of course, the values obtained from Model 3. The standard errors of the residuals are 1.29

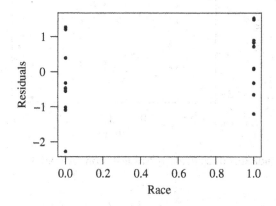

Figure 6.9 Standardized residuals versus race: Model 1.

Table 6.9 Separate Regression Results

Sample	$\hat{\beta}_0$	$\hat{\beta}_1$	t_1	p-Value	R^2	$\hat{\sigma}$	df
Minority	0.10	3.31	5.31	0.001	0.78	1.29	8
White	2.01	1.31	1.82	0.106	0.29	1.51	8

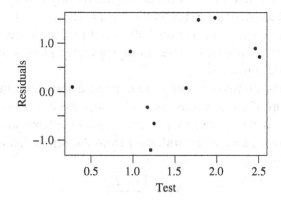

Figure 6.10 Standardized residuals versus test: Model 1, minority only.

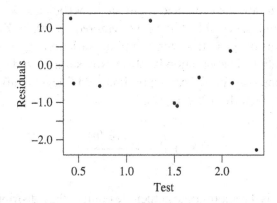

Figure 6.11 Standardized residuals versus test: Model 1, white only.

and 1.51 for the minority and white samples, respectively. The residual plots against the test score are acceptable in both cases. An interesting observation that was not available in the earlier analysis is that the preemployment test accounts for a major portion of the variation in the minority sample, but the test is only marginally useful in the white sample.

Our previous conclusion is still valid. The two regression equations are different. Not only are the regression coefficients different, but the residual mean squares also show slight differences. Of more importance, the values of R^2 are greatly different. For the white sample, $R^2 = 0.29$ is so small ($t = 1.82$; a critical value of 2.306 is required for significance) that the preemployment test score is not deemed an adequate predictor of job success. This finding has bearing on our original objective since it should be a prerequisite for comparing regressions in two samples that the relationships be valid in each of the samples when taken alone. Concerning the validity of the preemployment test, we conclude that if applied as the law prescribes, with indifference to race, it will give biased results for both racial groups. Moreover, based on these findings, we may be justified in saying that the test is of no value for screening white applicants.

We close the discussion with a note about determining the appropriate cutoff test score if the test were used. Consider the results for the minority sample. If Y_0 is designated as the minimum acceptable job performance value to be considered successful, then from the regression equation (also see Figure 6.6)

$$X_0 = \frac{Y_0 - \hat{\beta}_0}{\hat{\beta}_1},$$

where $\hat{\beta}_0$ and $\hat{\beta}_1$ are the estimated regression coefficients. X_0 is an estimate of the minimum acceptable test score required to attain Y_0. Since X_0 is defined in terms of quantities with sampling variation, X_0 is also subject to sampling variation. The variation is most easily summarized by constructing a confidence interval for X_0. An approximate 95% level confidence interval takes the form (Scheffé, 1959, p. 52)

$$X_0 \pm \frac{t_{(n-2,\alpha/2)}(\hat{\sigma}/n)}{\hat{\beta}_1},$$

where $t_{(n-2,\alpha/2)}$ is the appropriate percentile point of the t-distribution and $\hat{\sigma}^2$ is the least squares estimate of σ^2. If Y_0 is set at 4, then $X_0 = (4 - 0.10)/3.31 = 1.18$ and a 95% confidence interval for the test cutoff score is (1.09, 1.27).

6.4.2 Models with Same Slope and Different Intercepts

In Section 6.4.1 we dealt with the case where the two groups have distinct models with different sets of coefficients as given by Models 1 and 2 in (6.3) and as depicted in Figure 6.6. Suppose now that there is a reason to believe that the two groups have the same slope, β_1, and we wish to test the hypothesis that the two groups also have the same intercept, that is, $H_0 : \beta_{01} = \beta_{02}$. In this case we compare

Model 1 (Pooled): $\quad y_{ij} = \beta_0 + \beta_1 x_{ij} + \varepsilon_{ij}, \quad j = 1, 2; \quad i = 1, 2, \ldots, n_j,$

Model 2 (Minority): $\quad y_{i1} = \beta_{01} + \beta_1 x_{i1} + \varepsilon_{i1},$ \hfill (6.7)

Model 2 (White): $\quad y_{i2} = \beta_{02} + \beta_1 x_{i2} + \varepsilon_{i2}.$

Notice that the two models have the same value of the slope β_1 but different values of the intercepts β_{01} and β_{02}. Using the indicator variable Z defined earlier, we can write Model 2 as

$$\text{Model 3:} \quad y_{ij} = \beta_0 + \beta_1 x_{ij} + \gamma z_{ij} + \varepsilon_{ij}. \tag{6.8}$$

Note the absence of the interaction variable $(z_{ij} \cdot x_{ij})$ from Model 3 in (6.8). If it is present, as it is in (6.4), the two groups would have two models with different slopes and different intercepts.

The equivalence of Models 2 and 3 can be seen by noting that for the minority group, where $x_{ij} = x_{i1}$ and $z_{ij} = 1$, Model 3 becomes

$$y_{i1} = \beta_0 + \beta_1 x_{i1} + \gamma + \varepsilon_{i1}$$
$$= (\beta_0 + \gamma) + \beta_1 x_{i1} + \varepsilon_{i1}$$
$$= \beta_{01} + \beta_1 x_{i1} + \varepsilon_{i1},$$

which is the same as Model 2 for minority with $\beta_{01} = \beta_0 + \gamma$. Similarly, Model 3 for the white group becomes

$$y_{i2} = \beta_0 + \beta_1 x_{i2} + \varepsilon_{i2}.$$

Thus, Model 2 (or equivalently, Model 3) represents two parallel lines[5] (same slope) with intercepts $\beta_0 + \gamma$ and β_0. Therefore, our null hypothesis implies a restriction on γ in Model 3, namely, $H_0 : \gamma = 0$. To test this hypothesis, we use the F-Test

$$F = \frac{[\text{SSE(RM)} - \text{SSE(FM)}]/1}{\text{SSE(FM)}/17}, \tag{6.9}$$

---

[5] In the general case where the model contains $X_1, X_2, \ldots, X_p$ plus one indicator variable Z, Model 3 represents two parallel (hyper-) planes that differ only in the intercept.

which has 1 and 17 degrees of freedom. Equivalently, we can use the t-Test for testing $\gamma = 0$ in Model 3, which is

$$t = \frac{\hat{\gamma}}{\text{s.e.}(\hat{\gamma})}, \qquad (6.10)$$

which has 17 degrees of freedom. Again, the validation of the assumptions of Model 3 should be done before any conclusions are drawn from these tests. For the current example, we leave the computations of the above tests and the conclusions based on them as an exercise for the reader.

6.4.3 Models with Same Intercept and Different Slopes

Now we deal with the third case where the two groups have the same intercept, β_0, and we wish to test the hypothesis that the two groups also have the same slope, that is, $H_0 : \beta_{11} = \beta_{12}$. In this case we compare

Model 1 (Pooled): $\quad y_{ij} = \beta_0 + \beta_1 x_{ij} + \varepsilon_{ij}, \ j = 1, 2; \ i = 1, 2, \ldots, n_j,$

Model 2 (Minority): $\quad y_{i1} = \beta_0 + \beta_{11} x_{i1} + \varepsilon_{i1},$ \hfill (6.11)

Model 2 (White): $\quad y_{i2} = \beta_0 + \beta_{12} x_{i2} + \varepsilon_{i2}.$

Note that the two models have the same value of the intercept β_0 but different values of the slopes β_{11} and β_{12}. Using the indicator variable Z defined earlier, we can write Model 2 as

$$\text{Model 3:} \quad y_{ij} = \beta_0 + \beta_1 x_{ij} + \delta(z_{ij} \cdot x_{ij}) + \varepsilon_{ij}. \qquad (6.12)$$

Observe the presence of the interaction variable $(z_{ij} \cdot x_{ij})$ but the absence of the individual contribution of the variable Z. The equivalence of Models 2 and 3 can be seen by observing that for the minority group, where $x_{ij} = x_{i1}$ and $z_{ij} = 1$, Model 3 becomes

$$y_{i1} = \beta_0 + \beta_1 x_{i1} + \delta x_{i1} + \varepsilon_{i1}$$
$$= \beta_0 + (\beta_1 + \delta) x_{i1} + \varepsilon_{i1}$$
$$= \beta_0 + \beta_{11} x_{i1} + \varepsilon_{i1},$$

which is the same as Model 2 for minority with $\beta_{11} = \beta_1 + \delta$. Similarly, Model 3 for the white group becomes

$$y_{i2} = \beta_0 + \beta_{12} x_{i2} + \varepsilon_{i2}.$$

Therefore, our null hypothesis implies a restriction on δ in Model 3, namely, $H_0 : \delta = 0$. To test this hypothesis, we use the F-Test

$$F = \frac{[\text{SSE(RM)} - \text{SSE(FM)}]/1}{\text{SSE(FM)}/17},$$

which has 1 and 17 degrees of freedom. Equivalently, we can use the t-Test for testing $\delta = 0$ in Model 3, which is

$$t = \frac{\hat{\delta}}{\text{s.e.}(\hat{\delta})},$$

which has 17 degrees of freedom. Validation of the assumptions of Model 3, the computations of the above tests, and the conclusions based on them are left as an exercise for the reader.

6.5 OTHER APPLICATIONS OF INDICATOR VARIABLES

Applications of indicator variables such as those described in Section 6.4 can be extended to cover a variety of problems [see, e.g., Fox (1984), and Kmenta (1986) for a variety of applications]. Suppose, for example, that we wish to compare the means of $k \geq 2$ populations or groups. The techniques commonly used here are known as the *analysis of variance* (ANOVA). A random sample of size n_j is taken from the jth population, $j = 1, \ldots, k$. We have a total of $n = n_1 + \cdots + n_k$ observations on the response variable. Let y_{ij} be the ith response in the jth sample. Then y_{ij} can be modeled as

$$y_{ij} = \mu_0 + \mu_1 x_{i1} + \cdots + \mu_p x_{ip} + \varepsilon_{ij}. \tag{6.13}$$

In this model there are $p = k - 1$ indicator predictor variables $x_{i1}, \ldots, x_{ip}$. Each variable x_{ij} is 1 if the corresponding response is from population j, and zero otherwise. The population that is left out is usually known as the *control group*. All indicator variables for the control group are equal to zero. Thus, for the control group, (6.13) becomes

$$y_{ij} = \mu_0 + \varepsilon_{ij}. \tag{6.14}$$

In both (6.13) and (6.14), ε_{ij} are random errors assumed to be independent normal variables with zero means and constant variance σ^2. The constant μ_0 represents the mean of the control group and the regression coefficient μ_j can be interpreted as the difference between the means of the control and

jth groups. If $\mu_j = 0$, then the means of the control and jth groups are equal. The null hypothesis $H_0 : \mu_1 = \cdots = \mu_p = 0$ that all groups have the same mean can be represented by the model in (6.14). The alternate hypothesis that at least one of the μ_j's is different from zero can be represented by the model in (6.13). The models in (6.13) and (6.14) can be viewed as full and reduced models, respectively. Hence H_0 can be tested using the F-Test given in (4.47). Thus, the use of indicator variables allowed us to express ANOVA techniques as a special case of regression analysis. Both the number of quantitative predictor variables and the number of distinct groups represented in the data by indicator variables may be increased.

Note that the examples discussed above are based on cross-sectional data. Indicator variables can also be utilized with time series data. In addition, there are some models of growth processes where an indicator variable is used as the dependent variable. These models, known as *logistic regression models*, are discussed in Chapter 13.

In Sections 6.6 and 6.7 we discuss the use of indicator variables with time series data. In particular, notions of *seasonality* and *stability* of parameters over time are discussed. These problems are formulated and the data are provided. The analyses are left to the reader.

6.6 SEASONALITY

The data set we use as an example here, referred to as the Ski Sales data, can be found in the file Ski.Sales.csv at the Book's Website. The data can be read into your R Workspace using the following commands:

url="http://www.aucegypt.edu/faculty/hadi/RABE6/Data6/Ski.Sales.csv"
download.file(url, "Ski.Sales.csv")
df <- read.csv("Ski.Sales.csv", header = TRUE)
head(df, 2) # Display the first 2 lines of the dataset

The data consist of two variables: the sales, S, in millions for a firm that manufactures skis and related equipment for the years 1964–1973, and personal disposable income, PDI.[6] Each of these variables is measured quarterly. We use these data in Chapter 9 to illustrate the problem of *correlated errors*.

---
[6] Aggregate measure of purchasing potential.

The model is an equation that relates S to PDI, that is, $S_t = \beta_0 + \beta_1 \text{PDI}_t + \varepsilon_t$, where S_t is sales in millions in the tth period and PDI_t is the corresponding personal disposable income. Our approach here is to assume the existence of a seasonal effect on sales that is determined on a quarterly basis. To measure this effect we may define indicator variables to characterize the seasonality. Since we have four quarters, we define three indicator variables, Z_1, Z_2, and Z_3, where

$$z_{t1} = \begin{cases} 1, & \text{if the } t\text{th period is a first quarter,} \\ 0, & \text{otherwise,} \end{cases}$$

$$z_{t2} = \begin{cases} 1, & \text{if the } t\text{th period is a second quarter,} \\ 0, & \text{otherwise,} \end{cases}$$

$$z_{t3} = \begin{cases} 1, & \text{if the } t\text{th period is a third quarter,} \\ 0, & \text{otherwise.} \end{cases}$$

These variables can be created and augmented to the data in df using the following R code:

```
Q <- rep(1:4, 10); Z1 <- as.numeric(Q == 1); Z2 <- as.numeric
   (Q == 2);
Z3 <- as.numeric(Q == 3); df <- cbind(df[,-1], Z1, Z2, Z3)
```

The analysis and interpretation of this data set are left to the reader. The authors have analyzed these data and found that there are actually only two seasons. (See the discussion of these sales data in Chapter 9 for an analysis using only one indicator variable, two seasons.) See Kmenta (1986) for further discussion on using indicator variables for analyzing seasonality.

6.7 STABILITY OF REGRESSION PARAMETERS OVER TIME

Indicator variables may also be used to analyze the stability of regression coefficients over time or to test for structural change. We consider an extension of the system of regressions problem when data are available on a cross-section of observations and over time. Our objective is to analyze the constancy of the relationships over time. The methods described here are suitable for intertemporal and interspatial comparisons. To outline the method we use the Education Expenditure data for the years 1965, 1970, and 1975. The data can be found in the file Edu.Expend.csv at the Book's

216 QUALITATIVE VARIABLES AS PREDICTORS

Table 6.10 Variable Descriptions in the Education Expenditures Data

Variable	Description
State	Abbreviated name of the State
Y	Per capita expenditure on public education
Year	Year 1965, 1970, or 1975
X_1	Per capita personal income
X_2	Number of residents per thousand under 18 years of age
X_3	Number of people per thousand residing in urban areas

Website. The data can be read into your R Workspace using the following commands:

url="http://www.aucegypt.edu/faculty/hadi/RABE6/Data6/Edu.Expend.csv"
download.file(url, "Edu.Expend.csv")
df <- read.csv("Edu.Expend.csv", header = TRUE)
head(df,2) # Display the first 2 lines of the dataset

The description of the measured variables for the 50 states is given in Table 6.10. The variable Region is a categorical variable representing geographical regions:

1 = Northeast, 2 = North Central, 3 = South, 4 = West.

This data set is used in Chapter 8 to demonstrate methods of dealing with heteroscedasticity in multiple regression and to analyze the effects of regional characteristics on the regression relationships. Here we focus on the stability of the expenditure relationship with respect to time.

Data have been developed on the four variables described above for each state in 1965, 1970, and 1975. Assuming that the relationship can be identically specified in each of the three years,[7] the analysis of stability can be carried out by evaluating the variation in the estimated regression coefficients over time. Working with the pooled data set of 150 observations

---

[7] *Specification* as used here means that the same variables appear in each equation. Any transformations that are used apply to each equation. The assumption concerning identical specification should be empirically validated.

(50 states each in 3 years) we define two indicator variables, T_1 and T_2, where

$$T_{i1} = \begin{cases} 1, & \text{if the } i\text{th observation was from 1965,} \\ 0, & \text{otherwise,} \end{cases}$$

$$T_{i2} = \begin{cases} 1, & \text{if the } i\text{th observation was from 1970,} \\ 0, & \text{otherwise.} \end{cases}$$

These two indicator variables can be created and augmented to the data in df using the following code:

```
T1 <- as.numeric(df[, 2] == 1965); T2 <- as.numeric(df[, 2] == 1970)
df <- cbind(df,T1,T2)
```

Using Y to represent per capita expenditure on schools, the model takes the form

$$Y = \beta_0 + \beta_1 X_1 + \beta_2 X_2 + \beta_3 X_3 + \gamma_1 T_1 + \gamma_2 T_2 + \delta_1 T_1 \cdot X_1$$
$$+ \delta_2 T_1 \cdot X_2 + \delta_3 T_1 \cdot X_3 + \alpha_1 T_2 \cdot X_1 + \alpha_2 T_2 \cdot X_2$$
$$+ \alpha_3 T_2 \cdot X_3 + \varepsilon.$$

From the definitions of T_1 and T_2, the above model is equivalent to

For 1965: $Y = (\beta_0 + \gamma_1) + (\beta_1 + \delta_1)X_1 + (\beta_2 + \delta_2)X_2$
$\qquad\qquad + (\beta_3 + \delta_3)X_3 + \varepsilon,$

For 1970: $Y = (\beta_0 + \gamma_2) + (\beta_1 + \alpha_1)X_1 + (\beta_2 + \alpha_2)X_2$
$\qquad\qquad + (\beta_3 + \alpha_3)X_3 + \varepsilon,$

For 1975: $Y = \beta_0 + \beta_1 X_1 + \beta_2 X_2 + \beta_3 X_3 + \varepsilon.$

As noted earlier, this method of analysis necessarily implies that the variability about the regression function is assumed to be equal for all three years. One formal hypothesis of interest is

$$H_0 : \gamma_1 = \gamma_2 = \delta_1 = \delta_2 = \delta_3 = \alpha_1 = \alpha_2 = \alpha_3 = 0,$$

which implies that the regression system has remained unchanged throughout the period of investigation (1965–1975). The reader is invited to perform the analysis described above as an exercise.

EXERCISES

6.1 Using the model defined in (6.8):
  (a) Check to see if the usual least squares assumptions hold.
  (b) Test $H_0 : \gamma = 0$ using the F-Test.
  (c) Test $H_0 : \gamma = 0$ using the t-Test.
  (d) Verify the equivalence of the two tests above.

6.2 Compute the F and T tests in (6.9) and (6.10) and state your conclusions based on them.

6.3 Using the model defined in (6.12):
  (a) Check to see if the usual least squares assumptions hold.
  (b) Test $H_0 : \delta = 0$ using the F-Test.
  (c) Test $H_0 : \delta = 0$ using the t-Test.
  (d) Verify the equivalence of the two tests above.

6.4 Perform a thorough analysis of the Ski Sales data, given in the file Ski.Sales.csv at the Book's Website, using the ideas presented in Section 6.6.

6.5 Perform a thorough analysis of the Education Expenditures data in the file Edu.Expend.csv found at the Book's Website, using the ideas presented in Section 6.7.

6.6 Consider the Preemployment data and write the R code to produce
  (a) the value of the F-test in (6.6) and find its $p$value and critical value at the 5% level of significance.
  (b) the results in Table 6.9.

6.7 Table 6.11 shows a regression output obtained from fitting the model $Y = \beta_0 + \beta_1 X + \varepsilon$ to a set of data consisting of n workers in a given company, where Y is the weekly wages in $100 and X is the gender. The Gender variable is coded as 1 for Males and 0 for Females.
  (a) How many workers are there in this data set?
  (b) Compute the variance of Y.
  (c) Given that $\overline{X} = 0.52$, what is $\overline{Y}$?
  (d) Given that $\overline{X} = 0.52$, how many women are there in this data set?
  (e) What percentage of the variability in Y can be accounted for by X?
  (f) Compute the correlation coefficient between Y and X.

Table 6.11 Regression Output from the Regression of the Weekly Wages, Y, on X (Gender: 1 = Male, 0 = Female)

ANOVA Table

Source	Sum of Squares	df	Mean Square	F-Test
Regression	98.8313	1	98.8313	14
Residual	338.449	48	7.05101	

Coefficients Table

Variable	Coefficient	s.e.	t-Test	p-Value
Constant	15.58	0.54	28.8	0.0000
X	−2.81	0.75	−3.74	0.0005

(g) What is your interpretation of the estimated coefficient $\hat{\beta}_1$?

(h) What are the estimated weekly wages of a man chosen at random from the workers in the company?

(i) What are the estimated weekly wages of a woman chosen at random from the workers in the company?

(j) Construct a 95% confidence interval for β_1.

(k) Test the hypothesis that the average weekly wages of men are equal to those of women. [Specify (a) the null and alternative hypotheses, (b) the test statistics, (c) the critical value, and (d) your conclusion.]

6.8 The price of a car is thought to depend on the horsepower of the engine and the country where the car is made. The variable Country has four categories: USA, Japan, Germany, and Others. To include the variable Country in a regression equation, three indicator variables are created, one for USA, another for Japan, and the third for Germany. In addition, there are three interaction variables between the horsepower and each of the three Country categories (HP*USA, HP*Japan, and HP*Germany). Some regression outputs when fitting three models to the data are shown in Table 6.12. The usual regression assumptions hold.

(a) Compute the correlation coefficient between the price and the horsepower.

(b) What is the least squares estimated price of an American car with a 100 horsepower engine?

(c) Holding the horsepower fixed, which country has the least expensive car? Why?

220 QUALITATIVE VARIABLES AS PREDICTORS

Table 6.12 Some Regression Outputs When Fitting Three Models to the Car Data

Model 1

Source	Sum of Squares	df	Mean Square	F-Test
Regression	4604.7	1	4604.7	253
Residual	1604.44	88	18.2323	

Variable	Coefficient	s.e.	t-Test	p-Value
Constant	−6.107	1.487	−4.11	0.0001
Horsepower	0.169	0.011	15.9	0.0001

Model 2

Source	Sum of Squares	df	Mean Square	F-Test
Regression	4818.84	4	1204.71	73.7
Residual	1390.31	85	16.3566	

Variable	Coefficient	s.e.	t-Test	p-Value
Constant	−4.117	1.582	−2.6	0.0109
Horsepower	0.174	0.011	16.6	0.0001
USA	−3.162	1.351	−2.34	0.0216
Japan	−3.818	1.357	−2.81	0.0061
Germany	0.311	1.871	0.166	0.8682

Model 3

Source	Sum of Squares	df	Mean Square	F-Test
Regression	4889.3	7	698.471	43.4
Residual	1319.85	82	16.0957	

Variable	Coefficient	s.e.	t-Test	p-Value
Constant	−10.882	4.216	−2.58	0.0116
Horsepower	0.237	0.038	6.21	0.0001
USA	2.076	4.916	0.42	0.6740
Japan	4.755	4.685	1.01	0.3131
Germany	11.774	9.235	1.28	0.2059
HP*USA	−0.052	0.042	−1.23	0.2204
HP*Japan	−0.077	0.041	−1.88	0.0631
HP*Germany	−0.095	0.066	−1.43	0.1560

(d) Test whether there is an interaction between Country and horsepower. Specify the null and alternative hypotheses, test statistics, and conclusions.

(e) Given the horsepower of the car, test whether the Country is an important predictor of the price of a car. Specify the null and alternative hypotheses, test statistics, and conclusions.

(f) Would you recommend that the number of categories of Country be reduced? If so, which categories can be joined together to form one category?

(g) Holding the horsepower fixed, write down the formula for the test statistic for testing the equality of the price of American and Japanese cars.

6.9 Three types of fertilizer are to be tested to see which one yields more corn crop. Forty similar plots of land were available for testing purposes. The 40 plots are divided at random into four groups, 10 plots in each group. Fertilizer 1 was applied to each of the 10 corn plots in Group 1. Similarly, Fertilizers 2 and 3 were applied to the plots in Groups 2 and 3, respectively. The corn plants in Group 4 were not given any fertilizer; it will serve as the control group. Table 6.13 gives the corn yield y_{ij} for each of the 40 plots. The data are also available in the file Fertilizer.csv at the Book's Website.

(a) Create three indicator variables F_1, F_2, F_3, one for each of the three fertilizer groups.

(b) Fit the model $y_{ij} = \mu_0 + \mu_1 F_{i1} + \mu_2 F_{i2} + \mu_3 F_{i3} + \varepsilon_{ij}$.

Table 6.13 Corn Yields by Fertilizer Group

Fertilizer 1	Fertilizer 2	Fertilizer 3	Control Group
31	27	36	33
34	27	37	27
34	25	37	35
34	34	34	25
43	21	37	29
35	36	28	20
38	34	33	25
36	30	29	40
36	32	36	35
45	33	42	29

(c) Test the hypothesis that, on the average, none of the three types of fertilizer has an effect on corn crops. Specify the hypothesis to be tested, the test used, and your conclusions at the 5% significance level.

(d) Test the hypothesis that, on the average, the three types of fertilizer have equal effects on corn crop but different from those of the control group. Specify the hypothesis to be tested, the test used, and your conclusions at the 5% significance level.

(e) Which of the three fertilizers has the greatest effects on corn yield?

6.10 Presidential Election Data (1916–1996): These data were kindly provided by Professor Ray Fair of Yale University, who has found that the proportion of votes obtained by a presidential candidate in a U.S.A. presidential election can be predicted accurately by three macroeconomic variables, incumbency, and a variable which indicates whether the election was held during or just after a war. The variables considered are given in Table 6.14. The data can be found in the file Election.csv at the Book's Website. All growth rates are annual rates in percentage points. Consider fitting the following initial model

Table 6.14 Variables for the Presidential Election Data (1916–1996)

Variable	Definition
YEAR	Election year
V	Democratic share of the two-party presidential vote
I	Indicator variable (1 if there is a Democratic incumbent at the time of the election and -1 if there is a Republican incumbent)
D	Categorical variable (1 if a Democratic incumbent is running for election, -1 if a Republican incumbent is running for election, and 0 otherwise)
W	Indicator variable (1 for the elections of 1920, 1944, and 1948, and 0 otherwise)
G	Growth rate of real per capita GDP in the first three quarters of the election year
P	Absolute value of the growth rate of the GDP deflator in the first 15 quarters of the administration
N	Number of quarters in the first 15 quarters of the administration in which the growth rate of real per capita GDP is greater than 3.2%

to the data:

$$V = \beta_0 + \beta_1 I + \beta_2 D + \beta_3 W + \beta_4 (G \cdot I) + \beta_5 P + \beta_6 N + \varepsilon. \quad (6.15)$$

(a) Write the regression model corresponding to each of the three possible values of D in (6.15) and interpret the regression coefficient of D (β_2).
(b) Do we need to keep the variable I in the above model?
(c) Do we need to keep the interaction variable ($G \cdot I$) in the above model?
(d) Examine different models to produce the model or models that might be expected to perform best in predicting future presidential elections. Include interaction terms if needed.

6.11 Refer to the Presidential Election Data in Exercise 6.10, where the variable D is a categorical variable with three categories. Now, if we replace D by two indicator variables such as:

$D_1 = 1$ if $D = 1$ (Democratic incumbent is running) and 0 otherwise, and

$D_2 = 1$ if $D = -1$ (Republican incumbent is running) and 0 otherwise.

Then an alternative to the model in (6.15) is

$$V = \beta_0 + \beta_1 I + \alpha_1 D_1 + \alpha_2 D_2 + \beta_3 W + \beta_4 (G \cdot I) + \beta_5 P + \beta_6 N + \varepsilon. \quad (6.16)$$

(a) Write the regression model corresponding to each of the three possible values of D in (6.16) and interpret the regression coefficients of D_1 and D_2.
(b) Show that the model in (6.16) can be obtained as a special case of the model in (6.15) by assuming that $\alpha_1 = -\alpha_2$.
(c) Do these data support the assumption that $\alpha_1 = -\alpha_2$?

6.12 Use the Industrial Production data described in Section 1.3.6. The data can be found in the file Industrial.Production.csv at the Book' Website.

(a) Examine the relationship between polishing times, the diameters, and the product type. Does the relationship vary between the categories?

(b) Polishing time plays an important part in the cost. Construct a regression model which connects price with the product types, polishing time, and diameter.

6.13 In a statistics course personal information was collected on all the students for class analysis. Data on age (in years), height (in inches), and weight (in pounds) of the students can be found in the file Class.data.csv at the Book's Website. The gender of each student is also noted and coded as 1 for women and 0 for men. We want to study the relationship between the height and weight of students. Weight is taken as the response variable, and the height as the predictor variable.

(a) Do you agree or do you think the roles of the variables should be reversed?

(b) Is a single equation adequate to describe the relationship between height and weight for the two groups of students? Examine the standardized residual plot from the model fitted to the pooled data, distinguishing between the male and female students.

(c) Find the best model that describes the relationship between the weight and the height of students. Use interaction variables and the methodology described in this chapter.

(d) Do you think we should include age as a variable to predict weight? Give an intuitive justification for your answer.

CHAPTER 7

TRANSFORMATION OF VARIABLES

7.1 INTRODUCTION

Data do not always come in a form that is immediately suitable for analysis. We often have to transform the variables before carrying out the analysis. Transformations are applied to accomplish certain objectives such as to ensure linearity, to achieve normality, or to stabilize the variance. It often becomes necessary to fit a linear regression model to the transformed rather than the original variables. This is common practice. In this chapter, we discuss the situations where it is necessary to transform the data, the possible choices of transformation, and the analysis of transformed data.

We illustrate transformation mainly using simple regression. In multiple regression where there are several predictors, some may require transformation and others may not. Although the same technique can be applied to multiple regression, transformation in multiple regression requires more effort and care.

The necessity for transforming the data arises because the original variables, or the model in terms of the original variables, violates one or more of the standard regression assumptions. Two of the most commonly violated

---

Regression Analysis By Example Using R, Sixth Edition. Ali S. Hadi and Samprit Chatterjee
© 2024 John Wiley & Sons, Inc. Published 2024 by John Wiley & Sons, Inc.
Companion website: www.wiley.com/go/hadi/regression_analysis_6e

assumptions are the linearity of the model and the constancy of the error variance. As mentioned in Chapters 3 and 4, a regression model is linear when the parameters present in the model occur linearly even if the predictor variables occur nonlinearly. For example, each of the four following models is linear:

$$Y = \beta_0 + \beta_1 X + \varepsilon,$$
$$Y = \beta_0 + \beta_1 X + \beta_2 X^2 + \varepsilon,$$
$$Y = \beta_0 + \beta_1 \log X + \varepsilon,$$
$$Y = \beta_0 + \beta_1 \sqrt{X} + \varepsilon,$$

because the model parameters $\beta_0, \beta_1, \beta_2$ enter linearly. On the other hand,

$$Y = \beta_0 + e^{\beta_1 X} + \varepsilon$$

is a nonlinear model because the parameter β_1 does not enter the model linearly. To satisfy the assumptions of the standard regression model, instead of working with the original variables, we sometimes work with transformed variables. Transformations may be necessary for several reasons.

1. Theoretical considerations may specify that the relationship between two variables is nonlinear. An appropriate transformation of the variables can make the relationship between the transformed variables linear. Consider an example from learning theory (experimental psychology). A learning model that is widely used states that the time taken to perform a task on the ith occasion (T_i) is

$$T_i = \alpha \beta^i, \quad \alpha > 0, \quad 0 < \beta < 1. \tag{7.1}$$

The relationship between (T_i) and i as given in (7.1) is nonlinear, and we cannot directly apply techniques of linear regression. On the other hand, if we take logarithms of both sides, we get

$$\log T_i = \log \alpha + i \log \beta, \tag{7.2}$$

showing that $\log T_i$ and i are linearly related. The transformation enables us to use standard regression methods. Although the relationship between the original variables was nonlinear, the relationship between transformed variables is linear. A transformation is used to achieve the linearity of the fitted model.

2. The response variable Y, which is analyzed, may have a probability distribution whose variance is related to the mean. If the mean is related to the value of the predictor variable X, then the variance of Y will change with X and will not be constant. The distribution of Y will usually also be non-normal under these conditions. Non-normality invalidates the standard tests of significance (although not in a major way with large samples) since they are based on the normality assumption. The unequal variance of the error terms will produce estimates that are unbiased, but are no longer best in the sense of having the smallest variance. In these situations we often transform the data so as to ensure normality and constancy of error variance. In practice, the transformations are chosen to ensure the constancy of variance (*variance-stabilizing transformations*). It is a fortunate coincidence that the variance-stabilizing transformations are also good normalizing transforms.

3. There are neither prior theoretical nor probabilistic reasons to suspect that a transformation is required. The evidence comes from examining the residuals from the fit of a linear regression model in which the original variables are used.

Each of these cases where transformation is needed is illustrated in the following sections.

7.2 TRANSFORMATIONS TO ACHIEVE LINEARITY

One of the standard assumptions made in regression analysis is that the model which describes the data is linear. From theoretical considerations, or from an examination of scatter plot of Y against each predictor X_j, the relationship between Y and X_j may appear to be nonlinear. There are, however, several simple nonlinear regression models which by appropriate transformations can be made linear. We list some of these linearizable curves in Table 7.1. The corresponding graphs are given in Figures 7.1–7.4.

When curvature is observed in the scatter plot of Y against X, a linearizable curve from one of those given in Figures 7.1–7.4 may be chosen to represent the data. There are, however, many simple nonlinear models that cannot be linearized. Consider for example, $Y = \alpha + \beta \delta^X$, a modified exponential curve, or

$$Y = \alpha_1 e^{\theta_1 X} + \alpha_2 e^{\theta_2 X},$$

Table 7.1 Linearizable Simple Regression Functions with Corresponding Transformations

Function	Transformation	Linear Form	Graph
$Y = \alpha X^\beta$	$Y' = \log Y, X' = \log X$	$Y' = \log \alpha + \beta X'$	Figure 7.1
$Y = \alpha e^{\beta X}$	$Y' = \ln Y$	$Y' = \ln \alpha + \beta X$	Figure 7.2
$Y = \alpha + \beta \log X$	$X' = \log X$	$Y = \alpha + \beta X'$	Figure 7.3
$Y = \dfrac{X}{\alpha X - \beta}$	$Y' = \dfrac{1}{Y}, X' = \dfrac{1}{X}$	$Y' = \alpha - \beta X'$	Figure 7.4(a)
$Y = \dfrac{e^{\alpha+\beta X}}{1+e^{\alpha+\beta X}}$	$Y' = \ln \dfrac{Y}{1-Y}$	$Y' = \alpha + \beta X$	Figure 7.4(b)

In Chapter 13 we describe an application using the transformation in the last line of the table.

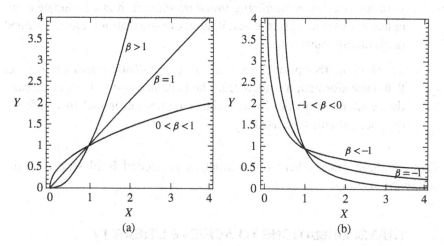

Figure 7.1 Graphs of the linearizable function $Y = \alpha X^\beta$.

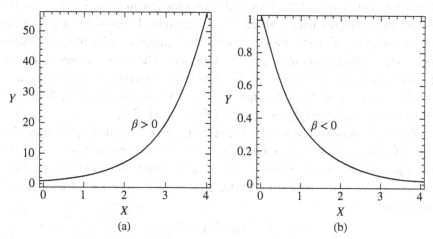

Figure 7.2 Graphs of the linearizable function $Y = \alpha e^{\beta X}$.

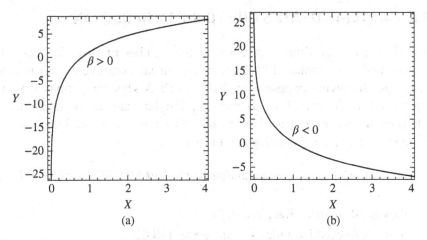

Figure 7.3 Graphs of the linearizable function $Y = \alpha + \beta \log X$.

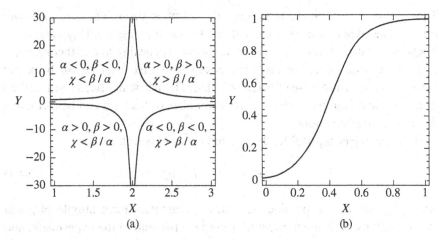

Figure 7.4 Graphs of the linearizable functions: (a) $Y = X/(\alpha X - \beta)$ and (b) $Y = (e^{\alpha + \beta X})/(1 + e^{\alpha + \beta X})$.

which is the sum of two exponential functions. The strictly nonlinear models (i.e., those not linearizable by variable transformation) require very different methods for fitting. We do not describe them in this book but refer the interested reader to Bates and Watts (1988) and Seber and Wild (1989), and Ratkowsky (1983, 1990).

In the following example, theoretical considerations lead to a model that is nonlinear. The model is, however, linearizable and we indicate the appropriate analysis.

7.3 BACTERIA DEATHS DUE TO X-RAY RADIATION

The data given in Table 7.2 represent the number of surviving bacteria (in hundreds) as estimated by plate counts in an experiment with marine bacterium following exposure to 200-kilovolt X-rays for periods ranging from $t = 1$ to 15 intervals of six minutes. The data can also be found in the file Bacteria.csv at the Book's Website.[1] The data can be read into your R Workspace using the following commands:

```
url="http://www.aucegypt.edu/faculty/hadi/RABE6/Data6/Bacteria.csv"
download.file(url, "Bacteria.csv")
df <- read.csv("Bacteria.csv", header = TRUE)
head(df, 2) # Display the first 2 lines of the dataset
```

The response variable n_t represents the number surviving after exposure time t. The experiment was carried out to test the single-hit hypothesis of X-ray action under constant field of radiation. According to this theory, there is a single vital center in each bacterium, and this must be hit by a ray before the bacterium is inactivated or killed. The particular bacterium studied does not form clumps or chains, so the number of bacteria can be estimated directly from plate counts.

If the theory is applicable, then n_t and t should be related by

$$n_t = n_0 e^{\beta_1 t}, \quad t \geq 0, \tag{7.3}$$

where n_0 and β_1 are parameters. These parameters have simple physical interpretations; n_0 is the number of bacteria at the start of the experiment, and

Table 7.2 Number of Surviving Bacteria (Units of 100)

t	n_t	t	n_t	t	n_t
1	355	6	106	11	36
2	211	7	104	12	32
3	197	8	60	13	21
4	166	9	56	14	19
5	142	10	38	15	15

---

[1] http://www.aucegypt.edu/faculty/hadi/RABE6

β_1 is the destruction (decay) rate. Taking logarithms of both sides of (7.3), we get

$$\ln n_t = \ln n_0 + \beta_1 t = \beta_0 + \beta_1 t, \qquad (7.4)$$

where $\beta_0 = \ln n_0$ and we have $\ln n_t$ as a linear function of t. If we introduce ε_t as the random error, our model becomes

$$\ln n_t = \beta_0 + \beta_1 t + \varepsilon_t \qquad (7.5)$$

and we can now apply standard least squares methods.

To get the error ε_t in the transformed model (7.5) to be additive, the error must occur in the multiplicative form in the original model (7.3). The correct representation of the model should be

$$n_t = n_0 e^{\beta_1 t} \varepsilon'_t, \qquad (7.6)$$

where ε'_t is the multiplicative random error. By comparing (7.5) and (7.6), it is seen that $\varepsilon_t = \ln \varepsilon'_t$. For standard least squares analysis ε_t should be normally distributed, which in turn implies that ε'_t has a log-normal distribution.[2] In practice, after fitting the transformed model we look at the residuals from the fitted model to see if the model assumptions hold. No attempt is usually made to investigate the random component, ε'_t, of the original model.

7.3.1 Inadequacy of a Linear Model

The first step in the analysis is to plot the raw data n_t versus t. The plot, shown in Figure 7.5,[3] suggests a nonlinear relationship between n_t and t. However, we proceed by fitting the simple linear model and investigate the consequences of misspecification. The model is

$$n_t = \beta_0 + \beta_1 t + \varepsilon_t, \qquad (7.7)$$

where β_0 and β_1 are constants; ε_t's are the random errors, with zero means and equal variances, and are uncorrelated with each other. Estimates of β_0, β_1, their standard errors, and the square of the correlation coefficient are given in Table 7.3. Despite the fact that the regression coefficient for the time variable is significant and we have a high value of R^2, the linear model is

---

[2] The random variable Y is said to have a log-normal distribution if $\ln Y$ has a normal distribution.
[3] The R code for producing this and all other graphs in this book can be found at the Book's Website at http://www.aucegypt.edu/faculty/hadi/RABE6.

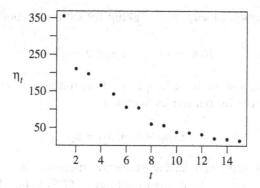

Figure 7.5 Plot of n_t against time t.

Table 7.3 Estimated Regression Coefficients from Model (7.7)

Variable	Coefficient	s.e.	t-Test	p-Value
Constant	259.58	22.73	11.42	0.0000
TIME (t)	−19.46	2.50	−7.79	0.0000
	$n = 15$	$R^2 = 0.823$	$\hat{\sigma} = 41.83$	df = 13

not appropriate. The plot of n_t against t shows departure from linearity for high values of t (Figure 7.5). We see this even more clearly if we look at a plot of the standardized residuals against time (Figure 7.6). The distribution of residuals has a distinct pattern. The residuals for $t = 2$ through 11 are all negative, for $t = 12$ through 15 are all positive, whereas the residual for $t = 1$ appears to be an outlier. This systematic pattern of deviation confirms that the linear model in (7.7) does not fit the data.

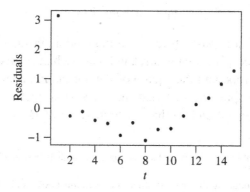

Figure 7.6 Plot of the standardized residuals from (7.7) against time t.

BACTERIA DEATHS DUE TO X-RAY RADIATION

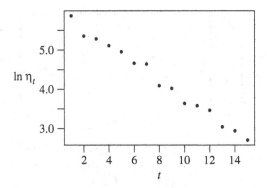

Figure 7.7 Plot of $\ln n_t$ against time t.

Table 7.4 Estimated Regression Coefficients When $\ln n_t$ Is Regressed on Time t

Variable	Coefficient	s.e.	t-Test	p-Value
Constant	5.973	0.0598	99.9	0.0000
TIME (t)	−0.218	0.0066	−33.2	0.0000
	$n = 15$	$R^2 = 0.988$	$\hat{\sigma} = 0.11$	df = 13

7.3.2 Logarithmic Transformation for Achieving Linearity

The relation between n_t and t appears distinctly nonlinear and we will work with the transformed variable $\ln n_t$, which is suggested from theoretical considerations as well as by Figure 7.7. The plot of $\ln n_t$ against t appears linear, indicating that the logarithmic transformation is appropriate. The results of fitting (7.5) appear in Table 7.4. The coefficients are highly significant, the standard errors are reasonable, and nearly 99% of the variation in the data is explained by the model. The standardized residuals are plotted against t in Figure 7.8. There are no systematic patterns to the distribution of the residuals and the plot is satisfactory. The single-hit hypothesis of X-ray action, which postulates that $\ln n_t$ should be linearly related to t, is confirmed by the data.

While working with transformed variables, careful attention must be paid to the estimates of the parameters of the model. In our example the point estimate of β_1 is −0.218 and the 95% confidence interval for the same parameter is (−0.232, −0.204). The estimate of the constant term in the equation is the best linear unbiased estimate of $\ln n_0$. If $\hat{\beta}_0$ denotes the estimate, $e^{\hat{\beta}_0}$ may be used as an estimate of n_0. With $\hat{\beta}_0 = 5.973$, the estimate of n_0 is $e^{\hat{\beta}_0} = 392.68$. This estimate is not an unbiased estimate of n_0; that

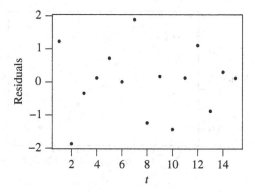

Figure 7.8 Plot of the standardized residuals against time t after transformation.

is, the true size of the bacteria population at the start of the experiment was probably somewhat smaller than 392.68. A correction can be made to reduce the bias in the estimate of n_0. The estimate $\exp[\hat{\beta}_0 - \frac{1}{2}\text{Var}(\hat{\beta}_0)]$ is nearly unbiased of n_0. In our present example, the modified estimate of n_0 is 391.98. Note that the bias in estimating n_0 has no effect on the test of the theory or the estimation of the decay rate.

In general, if nonlinearity is present, it will show up in a plot of the data. If the plot corresponds approximately to one of the graphs given in Figures 7.1–7.4, one of those curves can be fitted after transforming the data. The adequacy of the transformed model can then be investigated by methods outlined in Chapter 5.

7.4 TRANSFORMATIONS TO STABILIZE VARIANCE

We have discussed in the preceding section the use of transformations to achieve linearity of the regression function. Transformations are also used to stabilize the error variance, that is, to make the error variance constant for all the observations. The constancy of error variance is one of the standard assumptions of least squares theory. It is often referred to as the assumption of *homoscedasticity*. When the error variance is not constant over all the observations, the error is said to be *heteroscedastic*. *Heteroscedasticity* is usually detected by suitable graphs of the residuals such as the scatter plot of the standardized residuals against the fitted values or against each of the predictor variables. A plot with the characteristics of Figure 7.9 typifies the situation. The residuals tend to have a funnel-shaped distribution, either fanning out or closing in with the values of X.

TRANSFORMATIONS TO STABILIZE VARIANCE 235

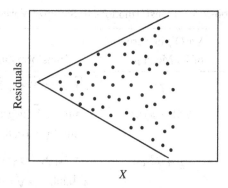

Figure 7.9 An example of heteroscedastic residuals.

If heteroscedasticity is present and no corrective action is taken, application of the ordinary least squares to the raw data will result in estimated coefficients which lack precision in a theoretical sense. The estimated standard errors of the regression coefficients are often understated, giving a false sense of accuracy.

Heteroscedasticity can be removed by means of a suitable transformation. We describe an approach for (a) detecting heteroscedasticity and its effects on the analysis and (b) removing heteroscedasticity from the data analyzed using transformations.

The response variable Y, in a regression problem, may follow a probability distribution whose variance is a function of the mean of that distribution. One property of the normal distribution, that many other probability distributions do not have, is that its mean and variance are independent in the sense that one is not a function of the other. The binomial and Poisson are but two examples of common probability distributions that have this characteristic. We know, for example, that a variable that is distributed binomially with parameters n and π has mean $n\pi$ and variance $n\pi(1 - \pi)$. It is also known that the mean and variance of a Poisson random variable are equal. When the relationship between the mean and variance of a random variable is known, it is possible to find a simple transformation of the variable, which makes the variance approximately constant (stabilizes the variance). We list in Table 7.5, for convenience and easy reference, transformations that stabilize the variance for some random variables with commonly occurring probability distributions whose variances are functions of their means. The transformations listed in Table 7.5 not only stabilize the variance, but also have the effect of making the distribution of the transformed variable closer to the normal distribution. Consequently, these transformations serve the

236 TRANSFORMATION OF VARIABLES

Table 7.5 Transformations to Stabilize Variance

Probability Distribution of Y	Var(Y) in Terms of Its Mean μ	Transformation	Resulting Variance
Poisson[a]	μ	$\sqrt{Y}$ or $(\sqrt{Y} + \sqrt{Y+1})$	0.25
Binomial[b]	$\mu(1-\mu)/n$	$\sin^{-1}\sqrt{Y}$ (degrees)	$821/n$
		$\sin^{-1}\sqrt{Y}$ (radians)	$0.25/n$
Negative Binomial[c]	$\mu + \lambda^2\mu^2$	$\lambda^{-1}\sinh^{-1}(\lambda\sqrt{Y})$ or	
		$\lambda^{-1}\sinh^{-1}(\lambda\sqrt{Y}+0.5)$	0.25

[a] For small values of Y, $\sqrt{Y+0.5}$ is sometimes recommended.
[b] The sample size is denoted by n; for $Y = r/n$ a slightly better transformation is $\sin^{-1}\sqrt{(r+3/8)/(n+3/4)}$.
[c] Note that the parameter $\lambda = 1/\sqrt{r}$.

dual purpose of normalizing the variable as well as making the variance functionally independent of the mean.

As an illustration, consider the following situation: Let Y be the number of accidents and X the speed of operating a lathe in a machine shop. We want to study the relationship between the number of accidents Y and the speed of lathe operation X. Suppose that a linear relationship is postulated between Y and X and is given by $Y = \beta_0 + \beta_1 X + \varepsilon$, where ε is the random error. The mean of Y is seen to increase with X. It is known from empirical observation that rare events (events with small probabilities of occurrence) often have a Poisson distribution. Let us assume that Y has a Poisson distribution. Since the mean and variance of Y are the same,[4] it follows that the variance of Y is a function of X, and consequently the assumption of homoscedasticity will not hold. From Table 7.5 we see that the square root of a Poisson variable ($\sqrt{Y}$) has a variance independent of the mean and is approximately equal to 0.25. To ensure homoscedasticity we, therefore, regress $\sqrt{Y}$ on X. Here the transformation is chosen to stabilize the variance, the specific form being suggested by the assumed probability distribution of the response variable. An analysis of data employing transformations suggested by probabilistic considerations is demonstrated in the following example.

---

[4] The probability mass function of a Poisson random variable Y is $\Pr(Y = y) = e^{-\lambda}\lambda^y/y!$; $y = 0, 1, \ldots$, where λ is a parameter. The mean and variance of a Poisson random variable are equal to λ.

TRANSFORMATIONS TO STABILIZE VARIANCE

Table 7.6 Number of Injury Incidents Y and Proportion of Total Flights N

Row	Y	N	Row	Y	N	Row	Y	N
1	11	0.0950	4	19	0.2078	7	3	0.1292
2	7	0.1920	5	9	0.1382	8	1	0.0503
3	7	0.0750	6	4	0.0540	9	3	0.0629

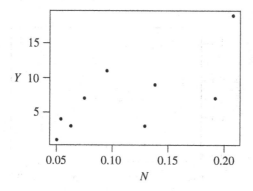

Figure 7.10 Plot of Y against N.

Injury Incidents in Airlines

The number of injury incidents and the proportion of total flights from New York for nine ($n = 9$) major U.S.A. airlines for a single year is given in Table 7.6 and plotted in Figure 7.10. Let f_i and y_i denote the total flights and the number of injury incidents for the ith airline that year. Then the proportion of total flights n_i made by the ith airline is $n_i = f_i / \sum f_j$. If all the airlines are equally safe, the injury incidents can be explained by the model

$$y_i = \beta_0 + \beta_1 n_i + \varepsilon_i, \tag{7.8}$$

where β_0 and β_1 are constants and ε_i is the random error.

The results of fitting the model are given in Table 7.7. The plot of residuals against n_i is given in Figure 7.11. The residuals are seen to increase with n_i in Figure 7.11 and, consequently, the assumption of homoscedasticity seems to be violated. This is not surprising, since the injury incidents may behave as a Poisson variable which has a variance proportional to its mean. To ensure the assumption of homoscedasticity, we make the square root transformation. Instead of working with Y we work with $\sqrt{Y}$, a variable which has an approximate variance of 0.25, and is more normally distributed than the original variable.

238 TRANSFORMATION OF VARIABLES

Table 7.7 Estimated Regression Coefficients (When Y Is Regressed on N)

Variable	Coefficient	s.e.	t-Test	p-Value
Constant	−0.14	3.14	−0.045	0.9657
N	64.98	25.20	2.580	0.0365
	$n = 9$	$R^2 = 0.487$	$\hat{\sigma} = 4.201$	df = 7

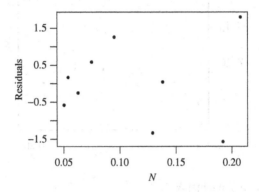

Figure 7.11 Plot of the standardized residuals versus N.

Consequently, the model we fit is

$$\sqrt{y_i} = \beta'_0 + \beta'_1 n_i + \varepsilon_i. \tag{7.9}$$

The result of fitting (7.9) is given in Table 7.8. The residuals from (7.9) when plotted against n_i are shown in Figure 7.12. The residuals for the transformed model do not seem to increase with n_i. This suggests that for the transformed model the homoscedastic assumption is not violated. The analysis of the model in terms of $\sqrt{y_i}$ and n_i can now proceed using standard techniques. The regression is significant here (as judged by the t statistic) but is not very strong. Only 48% of the total variability of the injury incidents of the

Table 7.8 Estimated Regression Coefficients When $\sqrt{Y}$ Is Regressed on N

Variable	Coefficient	s.e.	t-Test	p-Value
Constant	1.169	0.578	2.02	0.0829
N	11.856	4.638	2.56	0.0378
	$n = 9$	$R^2 = 0.483$	$\hat{\sigma} = 0.773$	df = 7

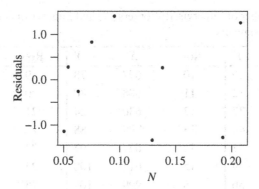

Figure 7.12 Plot of the standardized residuals from the regression of $\sqrt{y_i}$ on n_i.

airlines is explained by the variation in their number of flights. It appears that for a better explanation of injury incidents other factors have to be considered.

In the preceding example the nature of the response variable (injury incidents) suggested that the error variance was not constant about the fitted line. The square root transformation was considered based on the well-established empirical fact that the occurrence of accidents tends to follow the Poisson probability distribution. For Poisson variables, the square root is the appropriate transformation (Table 7.5). There are situations, however, when the error variance is not constant and there is no a priori reason to suspect that this would be the case. Empirical analysis will reveal the problem, and by making an appropriate transformation this effect can be eliminated. If the unequal error variance is not detected and eliminated, the resulting estimates will have large standard errors, but will be unbiased. This will have the effect of producing wide confidence intervals for the parameters and tests with low sensitivity. We illustrate the method of analysis for a model with this type of heteroscedasticity in the next example.

7.5 DETECTION OF HETEROSCEDASTIC ERRORS

In a study of 27 industrial establishments of varying size, the number of supervised workers (X) and the number of supervisors (Y) were recorded (Table 7.9). The data can also be found in the file Supervised.Workers.csv at the Book's Website. It was decided to study the relationship between the two variables, and as a start a linear model

$$y_i = \beta_0 + \beta_1 x_i + \epsilon_i \qquad (7.10)$$

Table 7.9 Number of Supervised Workers (X) and Supervisors (Y) in 27 Industrial Establishments

Row	X	Y	Row	X	Y	Row	X	Y
1	294	30	10	697	78	19	700	106
2	247	32	11	688	80	20	850	128
3	267	37	12	630	84	21	980	130
4	358	44	13	709	88	22	1025	160
5	423	47	14	627	97	23	1021	97
6	311	49	15	615	100	24	1200	180
7	450	56	16	999	109	25	1250	112
8	534	62	17	1022	114	26	1500	210
9	438	68	18	1015	117	27	1650	135

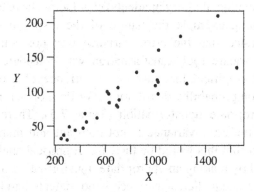

Figure 7.13 Number of supervisors (Y) versus number supervised (X).

was postulated. A plot of Y versus X suggests a simple linear model as a starting point (Figure 7.13). The results of fitting the linear model are given in Table 7.10.

The plot of the standardized residuals versus X (Figure 7.14) shows that the residual variance tends to increase with X. The residuals tend to lie in a band that diverges as one moves along the X axis. In general, if the band within which the residuals lie diverges (i.e., becomes wider) as X increases, the error variance is also increasing with X. On the other hand, if the band converges (i.e., becomes narrower), the error variance decreases with X. If the band that contains the residual plots consists of two lines parallel to the X axis, there is no evidence of heteroscedasticity. A plot of the standardized residuals against the predictor variable points up the presence

Table 7.10 Estimated Regression Coefficients When Number of Supervisors (Y) Is Regressed on the Number Supervised (X)

Variable	Coefficient	s.e.	t-Test	p-Value
Constant	14.448	9.562	1.51	0.1433
X	0.105	0.011	9.30	0.0000
	$n = 27$	$R^2 = 0.776$	$\hat{\sigma} = 21.73$	df = 25

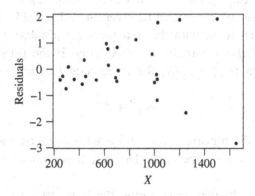

Figure 7.14 Plot of the standardized residuals against X when number of supervisors (Y) Is regressed on the number supervised (X).

of heteroscedastic errors. As can be seen in Figure 7.14, in our present example the residuals tend to increase with X.

7.6 REMOVAL OF HETEROSCEDASTICITY

In many industrial, economic, and biological applications, when unequal error variances are encountered, it is often found that the standard deviation of residuals tends to increase as the predictor variable increases. Based on this empirical observation, we will hypothesize in the present example that the standard deviation of the residuals is proportional to X (some indication of this is available from the plot of the residuals in Figure 7.14):

$$\text{Var}(\varepsilon_i) = k^2 x_i^2, \quad k > 0. \tag{7.11}$$

Dividing both sides of (7.10) by x_i, we obtain

$$\frac{y_i}{x_i} = \frac{\beta_0}{x_i} + \beta_1 + \frac{\varepsilon_i}{x_i}. \tag{7.12}$$

Now, define a new set of variables and coefficients,

$$Y' = \frac{Y}{X}, \quad X' = \frac{1}{X}, \quad \beta'_0 = \beta_1, \quad \beta'_1 = \beta_0, \quad \varepsilon' = \frac{\varepsilon}{X}.$$

In terms of the new variables (7.12) reduces to

$$y'_i = \beta'_0 + \beta'_1 x'_i + \varepsilon'_i. \tag{7.13}$$

Note that for the transformed model, Var(ε'_i) is constant and equals k^2. If our assumption about the error term as given in (7.11) holds, to fit the model properly we must work with the transformed variables: Y/X and $1/X$ as response and predictor variables, respectively. If the fitted model for the transformed data is $\hat{\beta}'_0 + \hat{\beta}'_1/X$, the fitted model in terms of the original variables is

$$\hat{Y} = \hat{\beta}'_1 + \hat{\beta}'_0 X. \tag{7.14}$$

The constant in the transformed model is the regression coefficient of X in the original model, and vice versa. This can be seen from comparing (7.12) and (7.13).

The residuals obtained after fitting the transformed model are plotted against the predictor variable in Figure 7.15. It is seen that the residuals are randomly distributed and lie roughly within a band parallel to the horizontal axis. There is no marked evidence of heteroscedasticity in the transformed model. The distribution of residuals shows no distinct pattern and we conclude that the transformed model is adequate. Our assumption about the error term appears to be correct; the transformed model has homoscedastic errors and

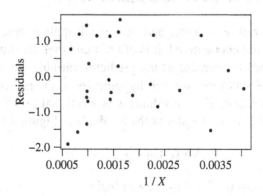

Figure 7.15 Plot of the standardized residuals against $1/X$ when Y/X is regressed on $1/X$.

Table 7.11 Estimated Regression Coefficients of the Original Equation When Fitted by the Transformed Variables Y/X and $1/X$

Variable	Coefficient	s.e.	t-Test	p-Value
Constant	3.803	4.570	0.832	0.4131
X	0.121	0.009	13.44	0.0000
	$n = 27$	$R^2 = 0.7758$	$\hat{\sigma} = 22.577$	df $= 25$

the standard assumptions of least squares theory hold. The result of fitting Y/X and $1/X$ leads to estimates of β_0' and β_1' which can be used for the original model.

The equation for the transformed variables is $Y/X = 0.121 + 3.803/X$. In terms of the original variables, we have $\hat{Y} = 3.803 + 0.121X$. The results are summarized in Table 7.11. By comparing Tables 7.10 and 7.11 we see the reduction in standard errors that is accomplished by working with transformed variables. The variance of the estimate of the slope is reduced by 33%.

7.7 WEIGHTED LEAST SQUARES

Linear regression models with heteroscedastic errors can also be fitted by a method called the *weighted least squares* (WLS), where parameter estimates are obtained by minimizing a weighted sum of squares of residuals where the weights are inversely proportional to the variance of the errors. This is in contrast to ordinary least squares (OLS), where the parameter estimates are obtained by minimizing equally weighted sum of squares of residuals. In the preceding example, the WLS estimates are obtained by minimizing

$$\sum \frac{1}{x_i^2}(y_i - \beta_0 - \beta_1 x_i)^2 \qquad (7.15)$$

as opposed to minimizing

$$\sum (y_i - \beta_0 - \beta_1 x_i)^2. \qquad (7.16)$$

It can be shown that WLS is equivalent to performing OLS on the transformed variables Y/X and $1/X$. We leave this as an exercise for the reader. Weighted least squares as an estimation method is discussed in more detail in Chapter 8.

7.8 LOGARITHMIC TRANSFORMATION OF DATA

The logarithmic transformation is one of the most widely used transformations in regression analysis. Instead of working directly with the data, the statistical analysis is carried out on the logarithms of the data. This transformation is particularly useful when the variable analyzed has a large standard deviation compared to its mean. Working with the data on a log scale often has the effect of dampening variability and reducing asymmetry. This transformation is also effective in removing heteroscedasticity. We illustrate this point by using the industrial data given in Table 7.9, where heteroscedasticity has already been detected. Besides illustrating the use of log (logarithmic) transformation to remove heteroscedasticity, we also show in this example that for a given body of data there may exist several adequate descriptions (models).

Instead of fitting the model given in (7.10), we now fit the model

$$\ln y_i = \beta_0 + \beta_1 x_i + \varepsilon_i \qquad (7.17)$$

(i.e., instead of regressing Y on X, we regress $\ln Y$ on X). The corresponding scatter plot is given in Figure 7.16. The results of fitting (7.17) are given in Table 7.12. The coefficients are significant, and the value of R^2 (0.77) is comparable to that obtained from fitting the model given in (7.10).

The plot of the residuals against X is shown in Figure 7.17. The plot is quite revealing. Heteroscedasticity has been removed, but the plot shows distinct nonlinearity. The residuals display a quadratic effect, suggesting that a more appropriate model for the data may be

$$\ln y_i = \beta_0 + \beta_1 x_i + \beta_2 x_i^2 + \varepsilon_i. \qquad (7.18)$$

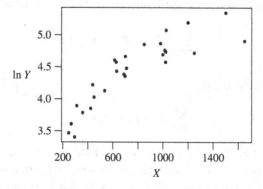

Figure 7.16 Scatter plot of $\ln Y$ versus X.

LOGARITHMIC TRANSFORMATION OF DATA

Table 7.12 Estimated Regression Coefficients When ln Y Is Regressed on X

Variable	Coefficient	s.e.	t-Test	p-Value
Constant	3.5150	0.1110	31.65	0.0000
X	0.0012	0.0001	9.15	0.0000
	n = 27	$R^2 = 0.77$	$\hat{\sigma} = 0.252$	df = 25

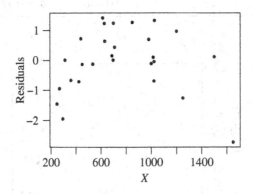

Figure 7.17 Plot of the standardized residuals against X when ln Y is regressed on X.

Equation (7.18) is a multiple regression model because it has two predictor variables, X and X^2. As discussed in Chapter 5, residual plots can also be used in the detection of model deficiencies in multiple regression. To show the effectiveness of residual plots in detecting model deficiencies and their ability to suggest possible corrections, we present the results of fitting model (7.18) in Table 7.13. Plots of the standardized residuals against the fitted values and against each of the predictor variables X and X^2 are presented in Figures 7.18–7.20, respectively.[5]

Residuals from the model containing a quadratic term appear satisfactory. There is no appearance of heteroscedasticity or nonlinearity in the residuals. We now have two equally acceptable models for the same data. The model given in Table 7.13 may be slightly preferred because of the higher value of R^2. The model given in Table 7.11 is, however, easier to interpret since it is based on the original variables.

---

[5] Recall from our discussion in Chapter 5 that in simple regression the plots of residuals against fitted values and against the predictor variable X_1 are identical; hence, one needs to examine only one of the two plots but not both. In multiple regression the plot of residuals against the fitted values is distinct from the plots of residuals against each of the predictors.

Table 7.13 Estimated Regression Coefficients When ln Y Is Regressed on X and X^2

Variable	Coefficient	s.e.	t-Test	p-Value
Constant	2.8516	0.1566	18.2	0.0000
X	$3.11267E-3$	0.0004	7.80	0.0000
X^2	$-1.10226E-6$	$0.220E-6$	-4.93	0.0000
	$n = 27$	$R^2 = 0.886$	$\hat{\sigma} = 0.1817$	df $= 24$

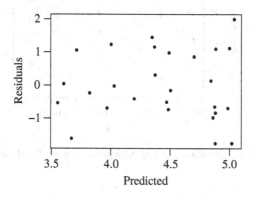

Figure 7.18 Plot of standardized residuals against the fitted values when ln Y is regressed on X and X^2.

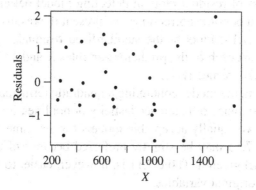

Figure 7.19 Plot of standardized residuals against X when ln Y is regressed on X and X^2.

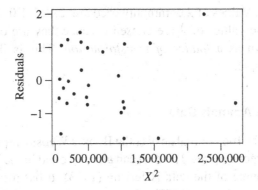

Figure 7.20 Plot of standardized residuals against X^2 when $\ln Y$ is regressed on X and X^2.

7.9 POWER TRANSFORMATION

In the previous sections we used several types of transformations (such as the reciprocal transformation, $1/Y$, the square root transformation, $\sqrt{Y}$, and the logarithmic transformation, $\ln Y$). These transformations have been chosen based on theoretical or empirical evidence to obtain linearity of the model, to achieve normality, and/or to stabilize the error variance. These transformations can be thought of as a general case of power transformation. In power transformation, we raise the response variable Y and/or some of the predictor variables to a power. For example, instead of using Y we use Y^λ, where λ is an exponent to be chosen by the data analyst based on either theoretical or empirical evidence. When $\lambda = -1$ we obtain the reciprocal transformation, $\lambda = 0.5$ gives the square root transformation, and when $\lambda = 0$ we obtain the logarithmic transformation.[6] Values of $\lambda = 1$ imply no transformation is needed.

If λ cannot be determined by theoretical considerations, the data can be used to determine the appropriate value of λ. This can be done using numerical methods. In practice, several values of λ are tried and the best

---

[6] Note that when $\lambda = 0$, $Y^\lambda = 1$ for all values of Y. To avoid this problem the transformation $(Y^\lambda - 1)/\lambda$ is used. It can be shown that as λ approaches zero, $(Y^\lambda - 1)/\lambda$ approaches $\ln Y$. This transformation is known as the Box–Cox power transformation. For more details, see Carroll and Ruppert (1988).

value is chosen. Values of λ commonly tried are: 2, 1.5, 1.0, 0.5, 0, −0.5, −1, −1.5, −2. These values of λ are chosen because they are easy to interpret. They are known as a *ladder of transformation*. This is illustrated in the following example.

Example: The Animals Data

The data set in the file Animals.csv at the Book's Website represents a sample taken from a larger data set. The data can also be found in the package MASS. The original source of the data is Jerison (1973). It has also been analyzed by Rousseeuw and Leroy (1987). The average brain weight (in grams), Y, and the average body weight (in kilograms), X, are measured for 28 animals. One purpose of the data is to determine whether a larger brain is required to govern a heavier body. Another purpose is to see whether the ratio of the brain weight to the body weight can be used as a measure of intelligence. The scatter plot of the data (Figure 7.21) does not show an obvious relationship. This is mainly due to the presence of very large animals (e.g., two elephants and three dinosaurs). Let us apply the power transformation to both Y and X. The scatter plots of Y^λ versus X^λ for several values of λ in the ladder of transformation are given in Figure 7.22. It can be seen that the values of $\lambda = 0$ (corresponding to the log transformation) are the most appropriate values. For $\lambda = 0$, the graph looks linear but the three dinosaurs do not conform to the linear pattern suggested by the other points. The graph suggests that either the brain weight of the dinosaurs is underestimated and/or their body weight is overestimated.

Note that in this example we transformed both the response and the predictor variables and that we used the same value of the power for both

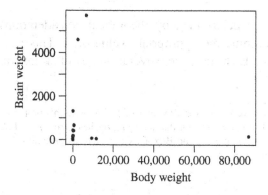

Figure 7.21 Animals data: scatter plots of brain weight versus body weight.

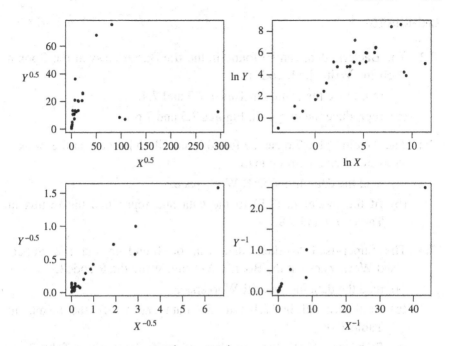

Figure 7.22 Scatter plots of Y^λ versus X^λ for various values of λ.

variables. In other applications, it may be more appropriate to raise each value to a different power and/or to transform only one variable. For further details on data transformation the reader is referred to Carroll and Ruppert (1988) and Atkinson (1985).

7.10 SUMMARY

After fitting a linear model one should examine the residuals for any evidence of heteroscedasticity. Heteroscedasticity is revealed if the residuals tend to increase or decrease with the values of the predictor variable, and is conveniently examined from a plot of the residuals. If heteroscedasticity is present, account should be taken of this in fitting the model. If no account is taken of the unequal error variance, the resulting least squares estimates will not have the maximum precision (smallest variances). Heteroscedasticity can be removed by working with transformed variables. Parameter estimates from the transformed model are then substituted for the appropriate parameters in the original model. The residuals from the appropriately transformed model should show no evidence of heteroscedasticity.

250 TRANSFORMATION OF VARIABLES

EXERCISES

7.1 The Bacteria data can be found in the file Bacteria.csv at the Book's Website. Write the R code to
  (a) reproduce the results in Tables 7.3 and 7.4.
  (b) reproduce the graphs in Figures 7.5 and 7.6

7.2 The data in Table 7.6 can be found in the file Injury.csv at the Book's Website. Write the R code to
  (a) read the data into your R Workspace.
  (b) fit the model in (7.8) to the data and reproduce the results in Tables 7.7 and 7.8.

7.3 The Supervised Workers data can be found in the file Supervised.Workers.csv at the Book's Website. Write the R code to
  (a) read the data into your R Workspace.
  (b) fit the model in (7.10) to the data and verify the results in Table 7.10.
  (c) fit the model in (7.13) to the data and verify the results in Table 7.11.

7.4 Refer again to the data in Exercise 7.3, write the R code to
  (a) fit the model in (7.17) to the data.
  (b) reproduce the results in Table 7.12.
  (c) reproduce the graph in Figure 7.17.

7.5 Refer again to the data in Exercise 7.3, write the R code to
  (a) fit the model in (7.18) to the data and reproduce the results in Table 7.13.
  (b) reproduce the graph in Figures 7.18–7.20.

7.6 Two variables, Y and X, are believed to be strongly nonlinearly related. A power transformation Y^λ was thought to make the relationship between Y^λ and X linear for some value of λ. Table 7.14 gives the value of the correlation coefficient between Y^λ and X for some values of λ.
  (a) What is the correlation coefficient between Y and X? Explain.

Table 7.14 Correlation Coefficient Between Y^λ and X for Some Values of λ

λ	1	0.5	0.001	−0.001	−0.5	−1	−2
Corr.	−0.777	−0.852	−0.930	0.930	0.985	0.999	0.943

(b) Observing the trend in Table 7.14, what is the best (and easy to explain or interpret) value of λ? Explain.

(c) Using your choice of λ in (b), write the equation that relates Y to X.

7.7 Two variables, Y and X, are believed to be strongly nonlinearly related. A power transformation Y^λ and X^λ was thought to make the relationship between Y^λ and X^λ linear for some value of λ. Table 7.15 gives the value of the correlation coefficient between Y^λ and X^λ for some values of λ.

Table 7.15 Correlation Coefficient Between Y^λ and X for Some Values of λ

λ	−1	−0.5	−0.001	0.001	0.5	1
Corr.	−0.524	−0.680	−0.992	0.992	0.698	0.543

(a) What is the correlation coefficient between Y and X? Explain.

(b) Observing the trend in Table 7.15, what is the best (and easy to explain or interpret) value of λ? Explain.

(c) Using your choice of λ in (b), write the equation that relates Y to X.

7.8 Download the Animals dataset, discussed in Section 7.9, either from the package MASS or from the Books Website, and store it in an object named df, then

(a) construct the corresponding scatter plot of the two variables in df.

(b) Use the function identify(df, labels = rownames(df)) to identify the unusual observations on the scatter plot by clicking near the points. Press the Esc key when done.

7.9 We have written a function in the file ladder.R. Download this file in your Working directory, then

(a) Source the file using source("ladder.R). Now, you have a function named ladder() in your Workspace.

(b) Call this function using ladder(df, lambda = seq(−2, 2, 0.5), tr.opt = 3). This function takes three arguments. The first is a two-column dataset df, the second is a sequence of lambda values set by default for transformation, and the third is tr.opt (1 means transform only df[, 1], 2 means transform only df[, 2], and 3 means transform both variables in df). The function will plot the data for each value of lambda. From the graphs you can visually determine which value of lambda makes the scatter of points more linear than all others. Experiment with this function using a dataset of your choice.

7.10 Magazine Advertising: In a study of revenue from advertising, data were collected for 41 magazines in 1986. The variables observed are number of pages of advertising and advertising revenue. The names of the magazines are listed. The data can be found in the file Magazine.Advertising.csv in the Book's Website.

(a) Fit a linear regression equation relating advertising revenue to advertising pages. Verify that the fit is poor.

(b) Choose an appropriate transformation of the data and fit the model to the transformed data. Evaluate the fit.

(c) You should not be surprised by the presence of a large number of outliers because the magazines are highly heterogeneous and it is unrealistic to expect a single relationship to connect all of them. Delete the outliers and obtain an acceptable regression equation that relates advertising revenue to advertising pages.

7.11 Wind Chill Factor: Table 7.16 gives the effective temperatures (W), which are due to the wind chill effect, for various values of the actual temperatures (T) in still air and wind speed (V). The zero-wind condition is taken as the rate of chilling when one is walking through still air [an apparent wind of 4 miles per hour (mph)]. The National Weather Service originally published the data; we have compiled it from a publication of the Museum of Science of Boston. The temperatures are measured in degrees Fahrenheit and the wind speed in mph.

(a) The data in Table 7.16 are not given in a format suitable for direct application of regression programs. You may need to construct another table containing three columns, one column for each of the variables W, T, and V. This table can be found in the file Wind.Chill.Factor.csv at the Book's Website. Read the data into your R Workspace.

(b) Fit a linear relationship between W, T, and V. The pattern of residuals should indicate the inadequacy of the linear model.

(c) After adjusting W for the effect of T (e.g., keeping T fixed), examine the relationship between W and V. Does the relationship between W and V appear linear?

Table 7.16 Wind Chill Factor (°F) for Various Values of Wind speed, V, in Miles/Hour, and Temperature (°F)

	\multicolumn{11}{c}{Actual Air Temperature (T)}
V	50	40	30	20	10	0	−10	−20	−30	−40	−50	−60
5	48	36	27	17	5	−5	−15	−25	−35	−46	−56	−66
10	40	29	18	5	−8	−20	−30	−43	−55	−68	−80	−93
15	35	23	10	−5	−18	−29	−42	−55	−70	−83	−97	−112
20	32	18	4	−10	−23	−34	−50	−64	−79	−94	−108	−121
25	30	15	−1	−15	−28	−38	−55	−72	−88	−105	−118	−130
30	28	13	−5	−18	−33	−44	−60	−76	−92	−109	−124	−134
35	27	11	−6	−20	−35	−48	−65	−80	−96	−113	−130	−137
40	26	10	−7	−21	−37	−52	−68	−83	−100	−117	−135	−140
45	25	9	−8	−22	−39	−54	−70	−86	−103	−120	−139	−143
50	25	8	−9	−23	−40	−55	−72	−88	−105	−123	−142	−145

(d) After adjusting W for the effect of V, examine the relationship between W and T. Does the relationship appear linear?

(e) Fit the model

$$W = \beta_0 + \beta_1 T + \beta_2 V + \beta_3 \sqrt{V} + \varepsilon. \quad (7.19)$$

Does the fit of this model appear adequate? The W numbers were produced by the National Weather Service according to the formula (except for rounding errors)

$$W = 0.0817(3.71\sqrt{V} + 5.81 - 0.25V)(T - 91.4) + 91.4. \quad (7.20)$$

Does the formula above give an accurate numerical description of W?

(f) Can you suggest a model better than those in (7.19) and (7.20)?

7.12 Refer to the Presidential Election Data in the file Election.csv at the Book's Website, where the response variable V is the proportion of votes obtained by a presidential candidate in the United States. Since the response is a proportion, it has a value between 0 and 1. The transformation $Y = \log[V/(1 - V)]$ takes the variable V with values between 0 and 1 to a variable Y with values between $-\infty$ to $+\infty$. It is therefore more reasonable to expect that Y satisfies the normality assumption than does V. Consider then fitting the model in (6.15) but replacing V by Y.

(a) For each of the two models, examine the appropriate residual plots discussed in Chapter 5 to determine which model satisfies the standard assumptions more than the other, the original variable V or the transformed variable Y.

(b) What does the fitted model above imply about the form of the model relating the original variables V in terms of the predictor variables? That is, find the form of the function

$$V = f(\beta_0 + \beta_1 I + \beta_2 D + \beta_3 W + \beta_4 (G \cdot I)$$
$$+ \beta_5 P + \beta_6 N + \varepsilon). \tag{7.21}$$

[Hint: This is a nonlinear function referred to as the *logistic function*, which is discussed in Chapter 13.]

7.13 Repeat Exercise 7.12 but when fitting the model in (6.16) but replacing V by Y and compare the results of the two exercises.

7.14 Oil Production Data: The data in Table 7.17 are the annual world crude oil production in millions of barrels for the period 1880–1988. The data are taken from Moore and McCabe (1993, p. 147). The data can be found in the file Oil.Production.csv at the Book's Website.

Table 7.17 Annual World Crude Oil Production in Millions of Barrels (1880–1988)

Year	OIL	Year	OIL	Year	OIL
1880	30	1940	2,150	1972	18,584
1890	77	1945	2,595	1974	20,389
1900	149	1950	3,803	1976	20,188
1905	215	1955	5,626	1978	21,922
1910	328	1960	7,674	1980	21,722
1915	432	1962	8,882	1982	19,411
1920	689	1964	10,310	1984	19,837
1925	1,069	1966	12,016	1986	20,246
1930	1,412	1968	14,104	1988	21,338
1935	1,655	1970	16,690		

(a) Construct a scatter plot of the oil production variable (OIL) versus Year and observe that the scatter of points on the graph is not linear. In order to fit a linear model to these data, OIL must be transformed.

(b) Construct a scatter plot of log(OIL) versus Year. The scatter of points now follows a straight line from 1880 to 1973. Political turmoil in the oil-producing regions of the Middle East affected patterns of oil production after 1973.

(c) Fit a linear regression of log(OIL) on Year. Assess the goodness of fit of the model.

(d) Construct the index plot of the standardized residuals. This graph shows clearly that one of the standard assumptions is violated. Which one?

Table 7.18 Average Price Per Megabyte in Dollars from 1988 to 1998

Year	Price	Year	Price
1988	11.54	1994	0.705
1989	9.30	1995	0.333
1990	6.86	1996	0.179
1991	5.23	1997	0.101
1992	3.00	1998	0.068
1993	1.46		

Source: Kindly provided by Jim Porter, Disk/Trends in Wired April 1998.

7.15 One of the remarkable technological developments in computer industry has been the ability to store information densely on hard disk. The cost of storage has steadily declined. Table 7.18 shows the average price per megabyte in dollars from 1988 to 1998. The data can be found in the file Megabite.csv at the Book's Website.

(a) Does a linear time trend describe the data? Define a new variable t by coding 1988 as 1, 1989 as 2, and so forth.

(b) Fit the model $P_t = P_0 e^{\beta t}$, where P_t is the price in period t. Does this model describe the data?

(c) Introduce an indicator variable which takes the value 0 for the years 1988–1991, and 1 for the remaining years. Fit a model to connecting $\log(P_t)$ with time t, the indicator variable, and the variable created by taking the product of time and the indicator variable. Interpret the coefficients of the fitted model.

CHAPTER 8

WEIGHTED LEAST SQUARES

8.1 INTRODUCTION

So far in our discussion of regression analysis it has been assumed that the underlying regression model is of the form

$$y_i = \beta_0 + \beta_1 x_{i1} + \cdots + \beta_p x_{ip} + \varepsilon_i, \qquad (8.1)$$

where the ε_i's are random errors that are independent and identically distributed (iid) with mean zero and variance σ^2. Various residual plots have been used to check these assumptions (Chapter 5). If the residuals are not consistent with the assumptions, the equation form may be inadequate, additional variables may be required, or some of the observations in the data may be outliers.

There has been one exception to this line of analysis. In the example based on the Supervisor Data of Section 7.5, it is argued that the underlying model does not have residuals that are iid. In particular, the residuals do not have constant variance. For these data, a transformation was applied to correct the situation so that better estimates of the original model parameters could be obtained [better than the ordinary least squares (OLS) method].

Regression Analysis By Example Using R, Sixth Edition. Ali S. Hadi and Samprit Chatterjee
© 2024 John Wiley & Sons, Inc. Published 2024 by John Wiley & Sons, Inc.
Companion website: www.wiley.com/go/hadi/regression_analysis_6e

In this chapter and in Chapter 9 we investigate situations where the underlying process implies that the errors are not iid. The present chapter deals with the *heteroscedasticity* problem, where the residuals do not have the same variance, and Chapter 9 treats the *autocorrelation* problem, where the residuals are not independent.

In Chapter 7 heteroscedasticity was handled by transforming the variables to stabilize the variance. The *weighted least squares* (WLS) method is equivalent to performing OLS on the transformed variables. The WLS method is presented here both as a way of dealing with heteroscedastic errors and as an estimation method in its own right. For example, WLS performs better than OLS in fitting *dose-response curves* (Section 8.5) and *logistic models* (Section 8.5 and Chapter 13).

In this chapter the assumption of equal variance is relaxed. Thus, the ε_i's are assumed to be independently distributed with mean zero and $\text{Var}(\varepsilon_i) = \sigma_i^2$. In this case, we use the WLS method to estimate the regression coefficients in (8.1). The WLS estimates of $\beta_0, \beta_1, \ldots, \beta_p$ are obtained by minimizing

$$\sum_{i=1}^{n} w_i (y_i - \beta_0 - \beta_1 x_{i1} - \cdots - \beta_p x_{ip})^2,$$

where w_i are weights inversely proportional to the variances of the residuals (i.e., $w_i = 1/\sigma_i^2$). Note that any observation with a small weight will be severely discounted by WLS in determining the values of $\beta_0, \beta_1, \ldots, \beta_p$. In the extreme case where $w_i = 0$, the effect of WLS is to exclude the ith observation from the estimation process.

Our approach to WLS uses a combination of prior knowledge about the process generating the data and evidence found in the residuals from an OLS fit to detect the heteroscedastic problem. If the weights are unknown, the usual solution prescribed is a two-stage procedure. In Stage 1, the OLS results are used to estimate the weights. In Stage 2, WLS is applied using the weights estimated in Stage 1. This is illustrated by examples in the rest of this chapter.

8.2 HETEROSCEDASTIC MODELS

Three different situations in which heteroscedasticity can arise will be distinguished. For the first two situations, estimation can be accomplished in one stage once the source of heteroscedasticity has been identified. The third type is more complex and requires the two-stage estimation procedure mentioned earlier. An example of the first situation is found in Chapter 7 and will be reviewed here. The second situation is described, but no data are analyzed. The third is illustrated with two examples.

8.2.1 Supervisors Data

In Section 7.5, data on the number of workers (X) in an industrial establishment and the number of supervisors (Y) were presented for 27 establishments. The regression model

$$y_i = \beta_0 + \beta_1 x_i + \varepsilon_i \tag{8.2}$$

was proposed. It was argued that the variance of ε_i depends on the size of the establishment as measured by x_i; that is, $\sigma_i^2 = k^2 x_i^2$, where k is a positive constant (see Section 7.5 for details). Empirical evidence for this type of heteroscedasticity is obtained by plotting the standardized residuals versus X. A pattern of points like the one in Figure 8.1 typifies the situation. The residuals tend to have a funnel-shaped distribution, either fanning out or closing in with the values of X. If corrective action is not taken and OLS is applied to the raw data, the resulting estimated coefficients will lack precision in a theoretical sense. In addition, for the type of heteroscedasticity present in these data, the estimated standard errors of the regression coefficients are often understated, giving a false sense of precision. The problem is resolved by using a version of weighted least squares, as described in Chapter 7.

This approach to heteroscedasticity may also be considered in multiple regression models. In (8.1) the variance of the residuals may be affected by only one of the predictor variables. (The case where the variance is a function of more than one predictor variable is discussed later.) Empirical evidence

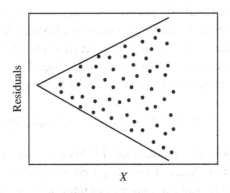

Figure 8.1 Example of heteroscedastic residuals.

is available from the plots of the standardized residuals versus the suspected variables. For example, if the model is given as (8.1) and it is discovered that the plot of the standardized residuals versus X_2 produces a pattern similar to that shown in Figure 8.1, then one could assume that $\text{Var}(\varepsilon_i)$ is proportional to x_{i2}^2, that is, $\text{Var}(\varepsilon_i) = k^2 x_{i2}^2$, where $k > 0$. The estimates of the parameters are determined by minimizing

$$\sum_{i=1}^{n} \frac{1}{x_{i2}^2}(y_i - \beta_0 - \beta_1 x_{i1} - \cdots - \beta_p x_{ip})^2.$$

If the software being used has a special weighted least squares procedure, we make the weighting variable equal to $1/x_{i2}^2$. On the other hand, if the software is only capable of performing OLS, we transform the data as described in Chapter 7. In other words, we divide both sides of (8.1) by x_{i2} to obtain

$$\frac{y_i}{x_{i2}} = \beta_0 \frac{1}{x_{i2}} + \beta_1 \frac{x_{i1}}{x_{i2}} + \cdots + \beta_p \frac{x_{ip}}{x_{i2}} + \frac{\varepsilon_i}{x_{i2}}.$$

The OLS estimate of the coefficient of the variable $1/X_2$ is the WLS estimate of β_0. The coefficient of the variable X_j/X_2 is an estimate of β_j for all $j \neq 2$. The constant term in this fitting is an estimate of β_2. Refer to Chapter 7 for a detailed discussion of this method applied to simple regression.

8.2.2 College Expense Data

A second type of heteroscedasticity occurs in large-scale surveys where the observations are averages of individual sampling units taken over well-defined groups or clusters. Typically, the average and number of sampling units are reported for each cluster. In some cases, measures of variability such as a standard deviation or range are also reported.

For example, consider a survey of undergraduate college students that is intended to estimate total annual college-related expenses and relate those expenses to characteristics of the institution attended. A list of variables chosen to explain expenses is shown in Table 8.1. Regression analysis with the model

$$Y = \beta_0 + \beta_1 X_1 + \beta_2 X_2 + \cdots + \beta_6 X_6 + \varepsilon \qquad (8.3)$$

may be used to study the relationship. In this example, a cluster is equated with a school and an individual sampling unit is a student. Data are collected by selecting a set of schools at random and interviewing a prescribed number of randomly selected students at each school. The response variable, Y,

Table 8.1 Variables in Cost of Education Survey

Name	Description
Y	Total annual expense (above tuition)
X_1	Size of city or town where school is located
X_2	Distance to nearest urban center
X_3	Type of school (public or private)
X_4	Size of student body
X_5	Proportion of entering freshman who graduate
X_6	Distance from home

in (8.3) is the average expenditure at the ith school. The predictor variables are characteristics of the school. The numerical values of these variables would be determined from the official statistics published for the school.

The precision of average expenditure is directly proportional to the square root of the sample size on which the average is based. That is, the standard deviation of $\bar{y}_i$ is $\sigma/\sqrt{n_i}$, where n_i represents the number of students interviewed at the ith institution and σ is the standard deviation for annual expense for the population of students. Then the standard deviation of ε_i in the model (8.1) is $\sigma_i = \sigma/\sqrt{n_i}$. Estimation of the regression coefficients is carried out using WLS with weights $w_i = 1/\sigma_i^2$. Since $\sigma_i^2 = \sigma^2/n_i$, the regression coefficients are obtained by minimizing the weighted sum of squared residuals,

$$S = \sum_{i=1}^{n} n_i \left(y_i - \beta_0 - \sum_{j=1}^{6} \beta_j x_{ij} \right)^2. \qquad (8.4)$$

Note that the procedure implicitly recognizes that observations from institutions where a large number of students were interviewed as more reliable and should have more weight in determining the regression coefficients than observations from institutions where only a few students were interviewed. The differential precision associated with different observations may be taken as a justification for the weighting scheme.

The estimated coefficients and summary statistics may be computed using a special WLS computer program or by transforming the data and using OLS on the transformed data. Multiplying both sides of (8.1) by $\sqrt{n_i}$, we obtain the new model

$$y_i\sqrt{n_i} = \beta_0\sqrt{n_i} + \beta_1 x_{i1}\sqrt{n_i} + \cdots + \beta_6 x_{i6}\sqrt{n_i} + \varepsilon_i\sqrt{n_i}. \qquad (8.5)$$

The error terms in (8.5), $\varepsilon_i \sqrt{n_i}$, now satisfy the necessary assumption of constant variance. Regression of $y_i \sqrt{n_i}$ against the seven new variables consisting of $\sqrt{n_i}$, and the six transformed predictor variables, $x_{ji} \sqrt{n_i}$ using OLS, will produce the desired estimates of the regression coefficients and their standard errors. Note that the regression model in (8.5) has seven predictor variables, a new variable $\sqrt{n_i}$, and the six original predictor variables multiplied by $\sqrt{n_i}$. Note also that there is no constant term in (8.5) because the intercept of the original model, β_0, is now the coefficient of $\sqrt{n_i}$. Thus the regression with the transformed variables must be carried out with the constant term constrained to be zero, that is, we fit a no-intercept model. More details on this point are given in the numerical example in Section 8.4.

8.3 TWO-STAGE ESTIMATION

In the two preceding problems heteroscedasticity was expected at the outset. In the first problem the nature of the process under investigation suggests residual variances that increase with the size of the predictor variable. In the second case, the method of data collection indicates heteroscedasticity. In both cases, homogeneity of variance is accomplished by a transformation. The transformation is constructed directly from information in the raw data. In the problem described in this section, there is also some prior indication that the variances are not equal. But here the exact structure of heteroscedasticity is determined empirically. As a result, estimation of the regression parameters requires two stages.

Detection of heteroscedasticity in multiple regression is not a simple matter. If present it is often discovered as a result of some good intuition on the part of the analyst on how observations may be grouped or clustered. For multiple regression models, the plots of the standardized residuals versus the fitted values and versus each predictor variable can serve as a first step. If the magnitude of the residuals appears to vary systematically with $\hat{y}_i$ or with x_{ij}, heteroscedasticity is suggested. The plot, however, does not necessarily indicate why the variances differ (see the following example).

One direct method for investigating the presence of nonconstant variance is available when there are replicated measurements on the response variable corresponding to a set of fixed values of the predictor variables. For example, in the case of one predictor variable, we may have measurements $y_{11}, y_{21}, \ldots, y_{n_1 1}$ at x_1; $y_{12}, y_{22}, \ldots y_{n_2 2}$ at x_2; and so on, up to $y_{1k}, y_{2k}, \ldots, y_{n_k k}$ at x_k. Taking $k = 5$ for illustrative purposes, a plot of the data appears as

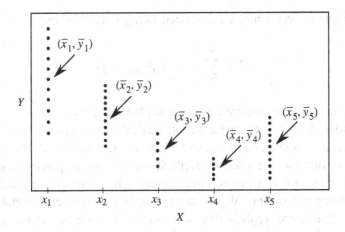

Figure 8.2 Nonconstant variance with replicated observations.

Figure 8.2.[1] With this wealth of data, it is not necessary to make restrictive assumptions regarding the nature of heteroscedasticity. It is clear from the graph that the nonconstancy of variance does not follow a simple systematic pattern such as $\text{Var}(\varepsilon_i) = k^2 x_i^2$. The variability first decreases as x increases up to x_4, then jumps again at x_5. The regression model could be stated as

$$y_{ij} = \beta_0 + \beta_j x_{ij} + \varepsilon_{ij}, \quad i = 1, 2, \ldots, n_j; \quad j = 1, 2, 3, 4, \qquad (8.6)$$

where $\text{Var}(\varepsilon_{ij}) = \sigma_j^2$.

The observed residual for the ith observation in the jth cluster or group is $e_{ij} = y_{ij} - \hat{y}_{ij}$. Adding and subtracting the mean of the response variable in for the jth cluster, $\bar{y}_j$, we obtain

$$e_{ij} = (y_{ij} - \bar{y}_j) + (\bar{y}_j - \hat{y}_{ij}), \qquad (8.7)$$

which shows that the residual is made up of two parts, the difference between y_{ij} and $\bar{y}_j$ and the difference between $\bar{y}_j$ and the point on the regression line, $\hat{y}_{ij}$. The first part is referred to as *pure error*. The second part measures lack of fit. An assessment of heteroscedasticity is based on the pure error.[2]

---

[1] The R code for producing this and all other graphs in this book can be found at the Book's Website at http://www.aucegypt.edu/faculty/hadi/RABE6.
[2] The notion of pure error can also be used to obtain a test for lack of fit [see, e.g., Draper and Smith (1998)].

The weights for WLS may be estimated as $w_{ij} = 1/s_j^2$, where

$$s_j^2 = \sum_{i=1}^{n_j} (y_{ij} - \bar{y}_j)^2 / (n_j - 1),$$

is the variance of the response variable for the *j*th group.

When the data are collected in a controlled laboratory setting, the researcher can choose to replicate the observations at any values of the predictor variables. But the presence of replications on the response variable for a given value of X is rather uncommon when data are collected in a nonexperimental setting. When there is only one predictor variable, it is possible that some replications will occur. If there are many predictor variables, it is virtually impossible to imagine coming upon two observations with identical values on all predictor values. However, it may be possible to form pseudoreplications by clustering responses where the predictor values are approximately identical. The reader is referred to Daniel and Wood (1980), where these methods are discussed in considerable detail. A more plausible way to investigate heteroscedasticity in multiple regression is by clustering observations according to prior, natural, and meaningful associations. As an example, we analyze data on state education expenditures. These data were used in Chapter 6.

8.4 EDUCATION EXPENDITURE DATA

The Education Expenditure data were used in Section 6.7 and it was suggested there that these data be looked at across time (the data are available for 1965, 1970, and 1975) to check on the stability of the coefficients. Here we use these data to demonstrate methods of dealing with heteroscedasticity in multiple regression and to analyze the effects of regional characteristics on the regression relationships. For the present analysis we shall work only with the 1975 data. The data can be found in the file Edu.Expend.csv in the Book's Website.[3] Note that the file Edu.Expend.csv contains the data for all three years combined. Use the following R code to read the data and to extract the data for 1975:

url="http://www.aucegypt.edu/faculty/hadi/RABE6/Data6/Edu.Expend.csv"

---
[3] http://www.aucegypt.edu/faculty/hadi/RABE6

```
download.file(url, "Edu.Expend.csv")
df <- read.csv("Edu.Expend.csv", header = TRUE)
d75=dplyr::filter(df, df$Year == 1975);
head(d75,2) # Display the first 2 lines of the dataset
```

The above code assumes that you have already installed the package dplyr. The 1975 data is now in the object d75.

The objective is to get the best representation of the relationship between expenditure on education and the other variables using data for all 50 states. The data are grouped in a natural way, by geographic region. Our assumption is that, although the relationship is structurally the same in each region, the coefficients and residual variances may differ from region to region. The different variances constitute a case of heteroscedasticity that can be treated directly in the analysis.

The variable names and definitions appear in Table 8.2. The model is

$$Y = \beta_0 + \beta_1 X_1 + \beta_2 X_2 + \beta_3 X_3 + \varepsilon. \tag{8.8}$$

States may be grouped into geographic regions based on the presumption that there exists a sense of regional homogeneity. The four broad geographic regions: (a) Northeast, (b) North Central, (c) South, and (d) West, are used to define the groups. It should be noted that data could be analyzed using indicator variables to look for special effects associated with the regions or to formulate tests for the equality of regressions across regions. However, our objective here is to develop one relationship that can serve as the best representation for all regions and all states. This goal is accomplished by taking regional differences into account through an extension of the method of weighted least squares.

It is assumed that there is a unique residual variance associated with each of the four regions. The variances are denoted as $(c_1\sigma)^2, (c_2\sigma)^2, (c_3\sigma)^2$,

Table 8.2 State Expenditures on Education, Variable List

Variable	Description
Y	Per capita expenditure on education projected for 1975
X_1	Per capita income in 1973
X_2	Number of residents per thousand under 18 years of age in 1974
X_3	Number of residents per thousand living in urban areas in 1970
Region	(a) Northeast, (b) North Central, (c) South, and (d) West

and $(c_4\sigma)^2$, where σ is the common part and the c_j's are unique to the regions. According to the principle of weighted least squares, the regression coefficients should be determined by minimizing

$$S_w = S_1 + S_2 + S_3 + S_4,$$

where

$$S_j = \sum_{i=1}^{n_j} \frac{1}{c_j^2}(y_i - \beta_0 - \beta_1 x_{i1} - \beta_2 x_{i2} - \beta_3 x_{i3})^2; \quad j = 1, 2, 3, 4. \qquad (8.9)$$

Each of S_1 through S_4 corresponds to a region, and the sum is taken over only those states that are in the region. The factors $1/c_j^2$ are the weights that determine how much influence each observation has in estimating the regression coefficients. The weighting scheme is intuitively justified by arguing that observations that are most erratic (large error variance) should have little influence in determining the coefficients.

The WLS estimates can also be justified by a second argument. The object is to transform the data so that the parameters of the model are unaffected, but the residual variance in the transformed model is constant. The prescribed transformation is to divide each observation by the appropriate c_j, resulting in a regression of Y/c_j on $1/c_j$, X_1/c_j, X_2/c_j, and X_3/c_j.[4] Then the error term, in concept, is also divided by c_j, the resulting residuals have a common variance, σ^2, and the estimated coefficients have all the standard least squares properties.

The values of the c_j's are unknown and must be estimated in the same sense that σ^2 and the β's must be estimated. We propose a two-stage estimation procedure. In the first stage perform a regression using the raw data as prescribed in the model of (8.8). Use the empirical residuals grouped by

---

[4] If we denote a variable with a double subscript, i and j, with j representing region and i representing observation within region, then each variable for an observation in region j is divided by c_j. Note that β_0 is the coefficient attached to the transformed variable $1/c_j$. The transformed model is

$$\frac{y_{ij}}{c_j} = \beta_0 \frac{1}{c_j} + \beta_1 \frac{x_{1ij}}{c_j} + \beta_2 \frac{x_{2ij}}{c_j} + \beta_3 \frac{x_{3ij}}{c_j} + \varepsilon'_{ij}$$

and the variance of ε'_{ij} is σ^2. Notice that the same regression coefficients appear in the transformed model as in the original model. The transformed model is also a no-intercept model.

region to compute an estimate of regional residual variance. The residual variance σ_j^2 for the region j is estimated by

$$\hat{\sigma}_j^2 = (n_j - 1)^{-1} \sum e_i^2, \quad j = 1, 2, 3, 4, \tag{8.10}$$

where the sum is taken over the n_j residuals corresponding to the n_j states in the region j and n_j is the number of states in the region j. Note here that the sum of all 50 residuals is 0, but the sum of the residuals for each region is not necessarily 0. We divide the sum in (8.10) by $n_j - 1$ because we lose one degree of freedom for each region.

In the second stage, c_j^2 in (8.9) is estimated by

$$\hat{c}_j^2 = \frac{\hat{\sigma}_j^2}{n^{-1} \sum_{i=1}^{n} e_i^2}. \tag{8.11}$$

The regression results for Stage 1 (OLS) using data from all 50 states for the year 1975 are given in Table 8.3. Two residual plots are prepared to check on specification. The standardized residuals are plotted versus the fitted values (Figure 8.3) and versus a categorical variable designating region (Figure 8.4). The purpose of Figure 8.3 is to look for patterns in the size and variation of the residuals as a function of the fitted values. The observed scatter of points has a funnel shape, indicating heteroscedasticity. The spread of the residuals in Figure 8.4 is different for the different regions, which also indicates that the variances are not equal. The scatter plots of standardized residual versus each of the predictor variables (Figures 8.5–8.7) indicate that the residual variance increases with the values of X_1.

Looking at the standardized residuals and the influence measures in this example is very revealing. The reader can verify that observation 49 (Alaska) is an outlier with a standardized residual value of 3.28. The standardized

Table 8.3 Regression Results: State Expenditures on Education for the Year 1975 ($n = 50$)

Variable	Coefficient	s.e.	t-Test	p-Value
Constant	−556.568	123.200	−4.52	0.0000
X_1	0.072	0.012	6.24	0.0000
X_2	1.552	0.315	4.93	0.0000
X_3	−0.004	0.051	−0.08	0.9342
$n = 50$	$R^2 = 0.591$	$R_a^2 = 0.565$	$\hat{\sigma} = 40.47$	df = 46

268 WEIGHTED LEAST SQUARES

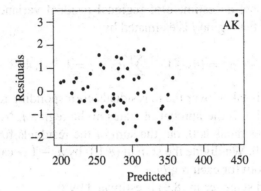

Figure 8.3 Plot of standardized residuals versus fitted values.

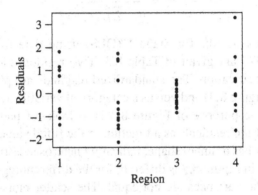

Figure 8.4 Plot of standardized residuals versus regions.

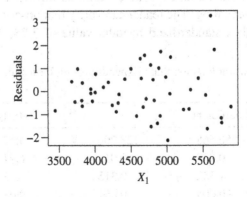

Figure 8.5 Plot of standardized residuals versus each of the predictor variable X_1.

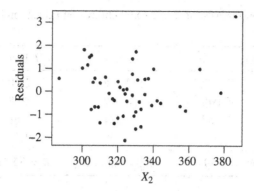

Figure 8.6 Plot of standardized residuals versus each of the predictor variable X_2.

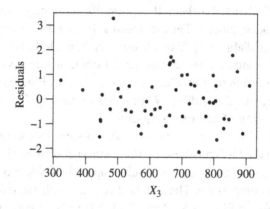

Figure 8.7 Plot of standardized residuals versus each of the predictor variable X_3.

residual for this observation can actually be seen to be separated from the rest of the residuals in Figure 8.3. Observations 44 (Utah) and 49 (Alaska) are high-leverage points with leverage values of 0.29 and 0.44, respectively. On examining the influence measures we find only one influential point 49, with a Cook's distance value of 2.13 and a DFITS value of 3.30 (see Section 5.9). Utah is a high-leverage point without being influential. Alaska, on the other hand, has high leverage and is also influential. Compared to other states, Alaska represents a very special situation: a state with a very small population and a boom in revenue from oil. The year is 1975! Alaska's education budget is therefore not strictly comparable with those of the other states. Consequently, this observation (Alaska) is excluded from the remainder of the analysis. It represents a special situation that has considerable influence on the regression results, thereby distorting the overall picture.

Table 8.4 Regression Results: State Expenditures on Education in 1975 ($n = 49$), Alaska Omitted

Variable	Coefficient	s.e.	t-Test	p-Value
Constant	−277.577	132.400	−2.10	0.0417
X_1	0.048	0.012	3.98	0.0003
X_2	0.887	0.331	2.68	0.0103
X_3	0.067	0.049	1.35	0.1826
$n = 49$	$R^2 = 0.497$	$R_a^2 = 0.463$	$\hat{\sigma} = 35.81$	df = 45

The data for Alaska may have an undue influence on determining the regression coefficients. To check this possibility, the regression was recomputed with Alaska excluded. The estimated values of the coefficients changed significantly [see Table 8.4]. This observation is excluded for the remainder of the analysis because it represents a special situation that has too much influence on the regression results. Plots similar to those of Figures 8.3 and 8.4 are presented as Figures 8.8 and 8.9. With Alaska removed, Figures 8.8 and 8.9 still show indication of heteroscedasticity.

To proceed with the analysis we must obtain the weights. They are computed from the OLS residuals by the method described above and appear in Table 8.5. The WLS regression results appear in Table 8.6 along with the OLS results for comparison. The standardized residuals from the transformed model are plotted in Figures 8.10 and 8.11. There is no pattern in the plot of the standardized residuals versus the fitted values (Figure 8.10). Also, from Figure 8.11, it appears that the spread of residuals by geographic region has

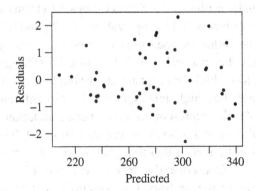

Figure 8.8 Plot of the standardized residuals versus fitted values (excluding Alaska).

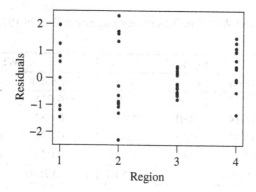

Figure 8.9 Plot of the standardized residuals versus region (excluding Alaska).

Table 8.5 Weights $\hat{c}_j$ for Weighted Least Squares

Region j	n_j	$\hat{\sigma}_j^2$	$\hat{c}_j$
Northeast	9	1632.50	1.177
North Central	12	2658.52	1.503
South	16	266.06	0.475
West	12	1036.83	0.938

evened out compared to Figures 8.4 and 8.9. The WLS solution is preferred to the OLS solution. Referring to Table 8.6, we see that the WLS solution does not fit the historical data as well as the OLS solution when considering $\hat{\sigma}$ or R^2 as indicators of goodness of fit.[5] This result is expected since one of the important properties of OLS is that it provides a solution with minimum $\hat{\sigma}$ or, equivalently, maximum R^2. Our choice of the WLS solution is based on the pattern of the residuals. The difference in the scatter of the standardized residuals when plotted against Region (compare Figures 8.9 and 8.11) shows that WLS has succeeded in taking account of heteroscedasticity.

It is not possible to make a precise test of significance because exact distribution theory for the two-stage procedure used to obtain the WLS

---

[5] Note that for comparative purposes, $\hat{\sigma}$ for the WLS solution is computed as the square root of

$$\hat{\sigma}^2 = \frac{1}{45} \sum_{i=1}^{n} (y_i - \hat{y}_i)^2,$$

and $\hat{y}_i = -316.024 + 0.062 x_{i1} + 0.874 x_{i2} + 0.029 x_{i3}$, are the fitted values computed in terms of the WLS estimated coefficients and the weights, c_j; weights play no further role in the computation of $\hat{\sigma}$.

Table 8.6 OLS and WLS Coefficients for Education Data in 1975 ($n = 49$), Alaska Omitted

	OLS			WLS		
Variable	Coefficient	s.e.	t	Coefficient	s.e.	t
Constant	−277.577	132.40	−2.10	−316.024	77.42	−4.08
X_1	0.048	0.01	3.98	0.062	0.01	8.00
X_2	0.887	0.33	2.68	0.874	0.20	4.41
X_3	0.067	0.05	1.35	0.029	0.03	0.85
$R^2 = 0.497$			$\hat{\sigma} = 35.81$	$R^2 = 0.7605$		$\hat{\sigma} = 36.52$

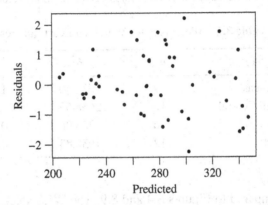

Figure 8.10 Standardized residuals versus fitted values for WLS solution.

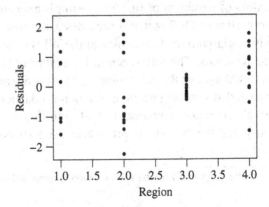

Figure 8.11 Standardized residuals by geographic region for WLS solution.

solution has not been worked out. If the weights were known in advance rather than as estimates from data, then the statistical tests based on the WLS procedure would be exact. Of course, it is difficult to imagine a situation similar to the one being discussed where the weights would be known in advance. Nevertheless, based on the empirical analysis above, there is a clear suggestion that weighting is required. In addition, since less than 50% of the variation in Y has been explained ($R^2 = 0.477$), the search for other factors must continue. It is suggested that the reader carry out an analysis of these data by introducing indicator variables for the four geographical regions. In any model with four categories, as has been pointed out in Chapter 6, only three indicator variables are needed. Heteroscedasticity can often be eliminated by the introduction of indicator variables corresponding to different subgroups in the data.

8.5 FITTING A DOSE–RESPONSE RELATIONSHIP CURVE

An important area for the application of weighted least squares analysis is the fitting of a linear regression line when the response variable Y is a proportion (values between zero and one). Consider the following situation: An experimenter can administer a stimulus at different levels. Subjects are assigned at random to different levels of the stimulus and for each subject a binary response is noted. From this set of observations, a relationship between the stimulus and the proportion responding to the stimulus is constructed. A very common example is in the field of pharmacology, in bioassay, where the levels of stimulus may represent different doses of a drug or poison, and the binary response is death or survival. Another example is the study of consumer behavior where the stimulus is the discount offered and the binary response is the purchase or nonpurchase of some merchandise.

Suppose that a pesticide is tried at k different levels. At the jth level of dosage x_j, let r_j be the number of insects dying out of a total n_j exposed ($j = 1, 2, \ldots, k$). We want to estimate the relationship between dose and the proportion dying. The sample proportion $p_j = r_j/n_j$ is a binomial random variable, with mean value π_j and variance $\pi_j(1 - \pi_j)/n_j$, where π_j is the population probability of death for a subject receiving dose x_j. The relationship between π and X is based on the notion that

$$\pi = f(X), \qquad (8.12)$$

where the function $f(\cdot)$ is increasing (or at least not decreasing) with X and is bounded between 0 and 1. The function should satisfy these properties because (a) π being a probability is bounded between 0 and 1, and (b) if the pesticide is toxic, higher doses should decrease the chances of survival (or increase the chances for death) for a subject. These considerations effectively rule out the linear model

$$\pi_j = \alpha + \beta x_j + \varepsilon_j, \tag{8.13}$$

because π_j would be unbounded.

Stimulus–response relationships are generally nonlinear. A nonlinear function which has been found to represent accurately the relationship between dose x_j and the proportion dying is

$$\pi_j = \frac{e^{\beta_0 + \beta_1 x_j}}{1 + e^{\beta_0 + \beta_1 x_j}}. \tag{8.14}$$

The relationship (8.14) is called the *logistic response function* and has the shape shown in Figure 8.12. It is seen that the logistic function is bounded between 0 and 1, and is monotonic. Physical considerations based on concepts of threshold values provide a heuristic justification for the use of (8.14) to represent a stimulus–response relationship (Cox, 1989).

The setup described above differs considerably from those of our other examples. In the present situation the experimenter has the control of dosages or stimuli and can use replication to estimate the variability of response at each dose level. This is a designed, experimental study, unlike the others, which were observational or nonexperimental.

The objectives for this type of analysis are not only to determine the nature of dose–response relationship but also to estimate the dosages which induce

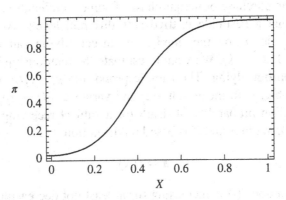

Figure 8.12 Logistic response function.

specified levels of response. Of particular interest is the dosage that produces a response in 50% of the population (median dose).

The logistic model (sometimes called *logit model*) has been used extensively in biological and epidemiological work. For analyzing proportions from binary response data, it is a very appealing model and easy to fit.

An alternative model in which the response function is represented by the cumulative distribution function of the normal probability distribution is also used. The cumulative curve of the normal distribution has a shape similar to that of the logistic function. This model is called the *probit model*, and for details we refer the reader to Finney (1964).

Besides medicine and pharmacology, the logistic model has been used in risk analysis, learning theory, the study of consumer behavior (choice models), and market promotion studies.

Since the response function in (8.14) is nonlinear, we can work with transformed variables. The transformation is chosen to make the response function linear. However, the transformed variables will have nonconstant variance. Then, we must use the weighted least squares methods for fitting the transformed data.

A whole chapter (Chapter 13) is devoted to the discussion of logistic regression models, for we believe that they have important and varied practical applications. General questions regarding the suitability and fitting of logistic models are considered there.

EXERCISES

8.1 Use the Educational Expenditure data for 1975, write the R code to reproduce the data in Table 8.3 and the graphs in Figures 8.3 and 8.4.

8.2 Repeat the analysis in Section 8.4 using the 1965 Education Expenditure Data in the file Edu.Expend.csv at the Book's Website.

8.3 Repeat the analysis in Section 8.4 using the 1970 Education Expenditure Data in the file Edu.Expend.csv at the Book's Website.

8.4 Use the 1975 Education Expenditure data found in the file Edu.Expend.csv at the Book's Website, compute the leverage values, the standardized residuals, Cook's distance, and DFITS for the regression model relating Y to the three predictor variables X_1, X_2, and X_3. Draw an appropriate graph for each of these measures. From the graph verify that Alaska and Utah are high-leverage points, but only Alaska is an influential point.

8.5 Refer again to the 1975 Educational Expenditure data in Exercise 8.4, fit a linear regression model relating Y to the three predictor variables X_1, X_2, and X_3 plus indicator variables for the region. Compare the results of the fitted model with the WLS results obtained in Section 8.4. Test for the equality of regressions across regions.

8.6 Repeat Exercise 8.5 for the 1965 data in the file Edu.Expend.csv at the Book's Website.

8.7 Use (8.11) to reproduce the results in Table 8.5.

8.8 Reproduce the OLD and WLS results in Table 8.6.

CHAPTER 9

THE PROBLEM OF CORRELATED ERRORS

9.1 INTRODUCTION: AUTOCORRELATION

One of the standard assumptions in the regression model is that the error terms ε_i and ε_j, associated with the ith and jth observations, are uncorrelated. Correlation in the error terms suggests that there is additional information in the data that has not been exploited in the current model. When the observations have a *natural* sequential order, the correlation is referred to as *autocorrelation*.

Autocorrelation may occur for several reasons. Adjacent residuals tend to be similar in both temporal and spatial dimensions. Successive residuals in economic time series tend to be positively correlated. Large positive errors are followed by other positive errors, and large negative errors are followed by other negative errors. Observations sampled from adjacent experimental plots or areas tend to have residuals that are correlated since they are affected by similar external conditions.

Regression Analysis By Example Using R, Sixth Edition. Ali S. Hadi and Samprit Chatterjee
© 2024 John Wiley & Sons, Inc. Published 2024 by John Wiley & Sons, Inc.
Companion website: www.wiley.com/go/hadi/regression_analysis_6e

The symptoms of autocorrelation may also appear as the result of a variable having been omitted from the right-hand side of the regression equation. If successive values of the omitted variable are correlated, the errors from the estimated model will appear to be correlated. When the variable is added to the equation, the apparent problem of autocorrelation disappears. The presence of autocorrelation has several effects on the analysis. These are summarized as follows:

1. Least squares estimates of the regression coefficients are unbiased but are not efficient in the sense that they no longer have minimum variance.

2. The estimate of σ^2 and the standard errors of the regression coefficients may be seriously understated; that is, from the data the estimated standard errors would be much smaller than they actually are, giving a spurious impression of accuracy.

3. The confidence intervals and the various tests of significance commonly employed would no longer be strictly valid.

The presence of autocorrelation can be a problem of serious concern for the preceding reasons and should not be ignored.

We distinguish between two types of autocorrelation and describe methods for dealing with each. The first type is only autocorrelation in appearance. It is due to the omission of a variable that should be in the model. Once this variable is uncovered, the autocorrelation problem is resolved. The second type of autocorrelation may be referred to as pure autocorrelation. The methods of correcting for pure autocorrelation involve a transformation of the data. Formal derivations of the methods can be found in Johnston (1984) and Kmenta (1986).

9.2 CONSUMER EXPENDITURE AND MONEY STOCK

Table 9.1 gives quarterly data from 1952 to 1956 on consumer expenditure (Y) and the stock of money (X), both measured in billions of current dollars for the United States. The data can be found in the file CEMS.csv at the Book's Website.[1]

A simplified version of the quantity theory of money suggests a model given by

$$y_t = \beta_0 + \beta_1 x_t + \varepsilon_t, \qquad (9.1)$$

---
[1] http://www.aucegypt.edu/faculty/hadi/RABE6.

Table 9.1 Consumer Expenditure and Money Stock

Year	Quarter	Consumer Expenditure	Money Stock	Year	Quarter	Consumer Expenditure	Money Stock
1952	1	214.6	159.3	1954	3	238.7	173.9
	2	217.7	161.2		4	243.2	176.1
	3	219.6	162.8	1955	1	249.4	178.0
	4	227.2	164.6		2	254.3	179.1
1953	1	230.9	165.9		3	260.9	180.2
	2	233.3	167.9		4	263.3	181.2
	3	234.1	168.3	1956	1	265.6	181.6
	4	232.3	169.7		2	268.2	182.5
1954	1	233.7	170.5		3	270.4	183.3
	2	236.5	171.6		4	275.6	184.3

Source: Adapted Friedman and Meiselman (1963, p. 266).

Table 9.2 Results When Consumer Expenditure Is Regressed on Money Stock, X

Variable	Coefficient	s.e.	t-Test	p-value
Constant	−154.72	19.850	−7.79	0.0000
X	2.30	0.115	20.10	0.0000
$n = 20$	$R^2 = 0.957$	$R_a^2 = 0.955$	$\hat{\sigma} = 3.983$	df = 18

where β_0 and β_1 are constants, ε_t the error term. Economists are interested in estimating β_1 and its standard error; β_1 is called the *multiplier* and has crucial importance as an instrument in fiscal and monetary policy. Since the observations are ordered in time, it is reasonable to expect that autocorrelation may be present. A summary of the regression results is given in Table 9.2. These results can be obtained using the following R code:

```
url="http://www.aucegypt.edu/faculty/hadi/RABE6/Data6/CEMS.csv"
download.file(url, "CEMS.csv")
df <- read.csv("CEMS.csv",header=TRUE)
head(df,2) # Display the first 2 lines of the dataset
reg=lm(Expenditure Stock, data=df) # Fits a linear model to the data
summary(reg) # Display the regression results
```

The regression coefficients are significant; the standard error of the slope coefficient is 0.115. For a unit change in the money supply the 95%

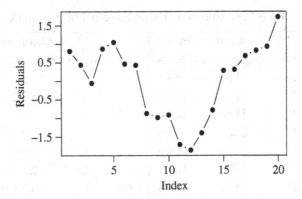

Figure 9.1 Index plot of the standardized residuals.

confidence interval for the change in the aggregate consumer expenditure would be $2.30 \pm 2.10 \times 0.115 = (2.06, 2.54)$. The value of R^2 indicates that roughly 96% of the variation in the consumer expenditure can be accounted for by the variation in money stock. The analysis would be complete if the basic regression assumptions were valid. To check on the model assumption, we examine the residuals. If there are indications that autocorrelation is present, the model should be reestimated after eliminating the autocorrelation.

For time series data a useful plot for analysis is the index plot (plot of the standardized residuals versus time). The graph is given in Figure 9.1.[2] The pattern of residuals is revealing and is characteristic of situations where the errors are correlated. Residuals of the same sign occur in clusters or bunches. The characteristic pattern would be that several successive residuals are positive, the next several are negative, and so on. From Figure 9.1 we see that the first seven residuals are positive, the next seven negative, and the last six positive. This pattern suggests that the error terms in the model are correlated and some additional analysis is required.

This visual impression can be formally confirmed by counting the number of runs in a plot of the signs of the residuals, the residuals taken in the order of the observations. These types of plots are called *sequence plots*. In our present example the sequence plot of the signs of the residuals is

$$+ + + + + + + - - - - - - - + + + + + +$$

---

[2] The R code for producing this and all other graphs in this book can be found at the Book's Website at http://www.aucegypt.edu/faculty/hadi/RABE6.

and it indicates three runs. With n_1 residuals positive and n_2 residuals negative, under the hypothesis of randomness the expected number of runs μ and its variance σ^2 would be

$$\mu = \frac{2n_1 n_2}{n_1 + n_2} + 1,$$

$$\sigma^2 = \frac{2n_1 n_2 (2n_1 n_2 - n_1 - n_2)}{(n_1 + n_2)^2 (n_1 + n_2 - 1)}.$$

In our case $n_1 = 13, n_2 = 7$, giving the expected number of runs to be 10.1 and a standard deviation of 1.97. The observed number of runs is three. The deviation of 5.1 from the expected number of runs is more than twice the standard deviation, indicating a significant departure from randomness. This formal *runs test* procedure merely confirms the conclusion arrived at visually that there is a pattern in the residuals.

Many computer packages now have the runs test as an available option. This approximate runs test for confirmation can therefore be easily executed. The runs test as we have described it should not, however, be used for small values of n_1 and n_2 (less than 10). For small values of n_1 and n_2 one needs exact tables of probability to judge significance. For more details on the runs test, the reader should refer to a book on nonparametric statistics such as Lehmann (1975), Conover (1980), Gibbons (1993), and Hollander and Wollfe (1999). Besides the graphical analysis, which can be confirmed by the runs test, autocorrelated errors can also be detected by the Durbin–Watson statistic (Durbin and Watson, 1950).

9.3 DURBIN–WATSON STATISTIC

The Durbin–Watson statistic is the basis of a popular test of autocorrelation in regression analysis. The test is based on the assumption that successive errors are correlated, namely,

$$\varepsilon_t = \rho \varepsilon_{t-1} + \omega_t, \qquad |\rho| < 1, \qquad (9.2)$$

where ρ is the correlation coefficient between ε_t and ε_{t-1}, and ω_t is normally independently distributed with zero mean and constant variance. In this case, the errors are said to have *first-order autoregressive structure* or *first-order autocorrelation*. In most situations the error ε_t may have a much more complex correlation structure. The first-order dependency structure, given in (9.2), is taken as a simple approximation to the actual error structure.

The Durbin–Watson statistic is defined as

$$d = \frac{\sum_{t=2}^{n}(e_t - e_{t-1})^2}{\sum_{t=1}^{n} e_t^2},$$

where e_i is the ith ordinary least squares (OLS) residual. The statistics d is used for testing the null hypothesis $H_0 : \rho = 0$ against an alternative $H_1 : \rho > 0$. Note that when $\rho = 0$ in (9.2), the ε's are uncorrelated.

Since ρ is unknown, we estimate the parameter ρ by $\hat{\rho}$, where

$$\hat{\rho} = \frac{\sum_{t=2}^{n} e_t e_{t-1}}{\sum_{t=1}^{n} e_t^2}. \tag{9.3}$$

An approximate relationship between d and $\hat{\rho}$ is $d \doteq 2(1 - \hat{\rho})$, ($\doteq$ means approximately equal to) showing that d has a range of 0–4. Since $\hat{\rho}$ is an estimate of ρ, it is clear that d is close to 2 when $\rho = 0$ and near to zero when $\rho = 1$. The closer the sample value of d to 2, the firmer the evidence that there is no autocorrelation present in the error. Evidence of autocorrelation is indicated by the deviation of d from 2. The formal test for positive autocorrelation operates as follows: Calculate the sample statistic d. Then, if

1. $d < d_L$, reject H_0.
2. $d > d_U$, do not reject H_0.
3. $d_L < d < d_U$, the test is inconclusive.

The values of (d_L, d_U) for different percentage points have been tabulated by Durbin and Watson (1951). But since we now use R, we no longer need these tables because we can compute the p-value of this test using the function DurbinWatsonTest in the package DescTools or the function durbinWatsonTest in the package car. The first three lines of the following R code will install the package DescTools, if needed, compute the regression, compute the Durbin Watson Test using DurbinWatsonTest:

if(!require("DescTools"))
install.packages("DescTools")
reg <- lm(Expenditure ~ Stock, data = df)
DurbinWatsonTest(reg$residuals ~ 1, alternative = "greater")
   Durbin–Watson test data: reg$residuals ~ 1
DW = 0.32821, p-value = 2.809e-07
alternative hypothesis: true autocorrelation is greater than 0.

Alternatively, the first three lines of the following R code will install the package car, if needed, compute the regression, compute the Durbin Watson Test using DurbinWatsonTest:

```
if(!require("car")) install.packages("car")
durbinWatsonTest(lm(Expenditure ~ Stock, data=df), alternative=
    "positive")
lag Autocorrelation D-W Statistic p-value
1    0.7506122    0.3282113    0
Alternative hypothesis: rho > 0
```

As can be seen from the above output, the value of d is 0.328 and the p-value is practically to 0, hence H_0 is rejected, showing that a positive autocorrelation is present. This essentially reconfirms our earlier conclusion, which was arrived at by looking at the index plot of the residuals.

The above test was carried out for positive autocorrelation. Tests for negative autocorrelation are seldom performed. If, however, a test is desired, then you need to replace alternative = "greater" by alternative = "less" or alternative = "positive" by alternative = "negative". Similarly, if a two-sided test is desired, you need to set the alternative = two.sided in the two functions above.

As pointed out earlier, the presence of correlated errors distorts estimates of standard errors, confidence intervals, and statistical tests, and therefore we should reestimate the equation. When autocorrelated errors are indicated, two approaches may be followed. These are (a) work with transformed variables, or (b) introduce additional variables that have time-ordered effects. We illustrate the first approach with the Money Stock data. The second approach is illustrated in Section 9.6.

9.4 REMOVAL OF AUTOCORRELATION BY TRANSFORMATION

When the residual plots and Durbin–Watson statistic indicate the presence of correlated errors, the estimated regression equation should be refitted taking the autocorrelation into account. One method for adjusting the model is the use of a transformation that involves the unknown autocorrelation parameter, ρ. The introduction of ρ causes the model to be nonlinear. The direct application of least squares is not possible. However, there are a number of procedures that may be used to circumvent the nonlinearity (Johnston, 1984). We use the method due to Cochrane and Orcutt (1949).

THE PROBLEM OF CORRELATED ERRORS

From model (9.1), ε_t and ε_{t-1} can be expressed as

$$\varepsilon_t = y_t - \beta_0 - \beta_1 x_t,$$
$$\varepsilon_{t-1} = y_{t-1} - \beta_0 - \beta_1 x_{t-1}.$$

Substituting these in (9.2), we obtain

$$y_t - \beta_0 - \beta_1 x_t = \rho(y_{t-1} - \beta_0 - \beta_1 x_{t-1}) + \omega_t.$$

Rearranging terms in the above equation, we get

$$\begin{aligned} y_t - \rho y_{t-1} &= \beta_0(1-\rho) + \beta_1(x_t - \rho x_{t-1}) + \omega_t, \\ y_t^* &= \beta_0^* + \beta_1^* \, x_t^* + \omega_t, \end{aligned} \qquad (9.4)$$

where

$$y_t^* = y_t - \rho y_{t-1},$$
$$x_t^* = x_t - \rho x_{t-1},$$
$$\beta_0^* = \beta_0(1-\rho),$$
$$\beta_1^* = \beta_1.$$

Since the ω's are uncorrelated, (9.4) represents a linear model with uncorrelated errors. This suggests that we run an ordinary least squares regression using y_t^* as a response variable and x_t^* as a predictor. The estimates of the parameters in the original equations are

$$\hat{\beta}_0 = \frac{\hat{\beta}_0^*}{1-\hat{\rho}} \quad \text{and} \quad \hat{\beta}_1 = \hat{\beta}_1^*. \qquad (9.5)$$

Therefore, when the errors in model (9.1) have an autoregressive structure as given in (9.2), we can transform both sides of the equation and obtain transformed variables which satisfy the assumption of uncorrelated errors.

The value of ρ is unknown and has to be estimated from the data. Cochrane and Orcutt (1949) have proposed an iterative procedure. The procedure operates as follows:

1. Compute the OLS estimates of β_0 and β_1 by fitting model (9.1) to the data.
2. Calculate the residuals and, from the residuals, estimate ρ using (9.3).

3. Fit the equation given in (9.4) using the variables $y_t - \hat{\rho} y_{t-1}$ and $x_t - \hat{\rho} x_{t-1}$ as response and predictor variables, respectively, and obtain $\hat{\beta}_0$ and $\hat{\beta}_1$ using (9.5).

4. Examine the residuals of the newly fitted equation. If the new residuals continue to show autocorrelation, repeat the entire procedure using the estimates $\hat{\beta}_0$ and $\hat{\beta}_1$ as estimates of β_0 and β_1 instead of the original least squares estimates. On the other hand, if the new residuals show no autocorrelation, the procedure is terminated and the fitted equation for the original data is

$$\hat{y}_t = \hat{\beta}_0 + \hat{\beta}_1 x_t.$$

As a practical rule we suggest that if the first application of the Cochrane–Orcutt procedure does not yield non-autocorrelated residuals, one should look for alternative methods of removing autocorrelation. We apply the Cochrane–Orcutt procedure to the data given in Table 9.1 using the following R code:

```
reg = lm(Expenditure ~ Stock, data = df) # Equation (9.1)
e = reg$residuals; n = length(e)
rho = sum(e[2:n]*e[1:(n - 1)])/sum(e^2) # Equation (9.3)
print(rho)
0.7506122
y = df$Expenditure; x = df$Stock
y = y[2:n] –rho*y[1:(n-1)]
x = x[2:n] – rho*x[1:(n – 1)]
reg1 = lm(y ~ x)
DurbinWatsonTest(reg1$residuals ~ 1)
Durbin–Watson test
data: reg1$residuals ~ 1
DW = 1.426, p-value = 0.09667
alternative hypothesis: true autocorrelation is greater than 0
reg1$coefficients # Display regression coefficients
(Intercept)      x
-53.695919    2.643443
```

It is interesting to see that the estimate of ρ, computed by the function durbinWatsonTest() earlier, agrees with the one-step application of the first

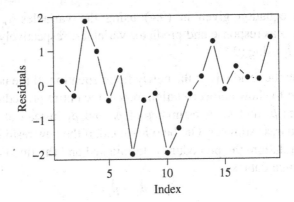

Figure 9.2 Index plot of standardized residuals after one iteration of the Cochrane–Orcutt method.

application of the Cochrane–Orcutt procedure. The d value for the original data is 0.328, which is highly significant. The value of $\hat{\rho}$ is 0.751. On fitting the regression equation to the variables $(y_t - \hat{\rho} y_{t-1})$ and $(x_t - \hat{\rho} x_{t-1})$, we have a d value of 1.426, which is not significant at the 5% level (the p-value is 0.09667). Consequently, $H_0 : \rho = 0$ is not rejected.[3] From the above output, the fitted equation is

$$\hat{y}_t^* = -53.70 + 2.64 x_t^*,$$

which, using (9.4), the fitted equation in terms of the original variables is

$$\hat{y}_t = -215.31 + 2.64 x_t.$$

The estimated standard error for the slope is 0.307, as opposed to the least squares estimate of the original equation, which was $y_t = -154.7 + 2.3 x_t$ with a standard error for the slope of 0.115. The newly estimated standard error is larger by a factor of almost 3. The residual plots for the fitted equation of the transformed variables are shown in Figure 9.2. The residual plots show less clustering of the adjacent residuals by sign, and the Cochrane–Orcutt procedure has worked to our advantage.

---

[3] The significance level of the test is not exact because $\hat{\rho}$ was used in the estimation process. The d value of 1.426 may be viewed as an index of autocorrelation that indicates an improvement from the previous value of 0.328.

9.5 ITERATIVE ESTIMATION WITH AUTOCORRELATED ERRORS

One advantage of the Cochrane–Orcutt procedure is that estimates of the parameters are obtained using standard least squares computations. Although two stages are required, the procedure is relatively simple. A more direct approach is to try to estimate values of ρ, β_0, and β_1 simultaneously. The model is formulated as before requiring the construction of transformed variables $y_t - \rho y_{t-1}$ and $x_t - \rho x_{t-1}$. Parameter estimates are obtained by minimizing the sum of squared errors, which is given as

$$S(\beta_0, \beta_1, \rho) = \sum_{t=2}^{n} [y_t - \rho y_{t-1} - \beta_0(1 - \rho) - \beta_1(x_t - \rho x_{t-1})]^2.$$

If the value of ρ were known, β_0 and β_1 would be easily obtained by regressing $y_t - \rho y_{t-1}$ on $x_t - \rho x_{t-1}$. Final estimates are obtained by searching through many values of ρ until a combination of ρ, β_0, and β_1 is found that minimizes $S(\rho, \beta_0, \beta_1)$. The search could be accomplished using a standard regression computer program, but the process can be much more efficient with an automated search procedure. This method is due to Hildreth and Lu (1960). For a discussion of the estimation procedure and properties of the estimates obtained, see Kmenta (1986).

Once the minimizing values, say $\tilde{\rho}$, $\tilde{\beta}_0$, and $\tilde{\beta}_1$, have been obtained, the standard error for the estimate of β_1 can be approximated using a version of (3.25) of Chapter 3. The formula is used as though $y_t - \rho y_{t-1}$ were regressed on $x_t - \rho x_{t-1}$ with ρ known; that is, the estimated standard error of $\tilde{\beta}_1$ is

$$\text{s.e.}(\tilde{\beta}_1) = \frac{\hat{\sigma}}{\sqrt{\sum [x_t - \tilde{\rho} x_{t-1} - \bar{x}(1 - \tilde{\rho})]^2}},$$

where $\hat{\sigma}$ is the square root of $S(\tilde{\rho}, \tilde{\beta}_0, \tilde{\beta}_1)/(n - 2)$. When adequate computing facilities are available such that the iterative computations are easy to accomplish, then the latter method is recommended. However, it is not expected that the estimates and standard errors for the iterative method and the two-stage Cochrane–Orcutt method would be appreciably different. The estimates from the three methods, OLS, Cochrane–Orcutt, and iterative for the data of Table 9.1, are given in Table 9.3 for comparison.

Table 9.3 Comparison of Regression Estimates

Method	$\hat{\rho}$	$\hat{\beta}_0$	$\hat{\beta}_1$	s.e.$(\hat{\beta}_1)$
OLS		−154.720	2.300	0.115
Cochrane–Orcutt	0.874	−324.440	2.758	0.444
Iterative	0.824	−235.509	2.753	0.436

9.6 AUTOCORRELATION AND MISSING VARIABLES

The characteristics of the regression residuals that suggest autocorrelation may also be indicative of other aspects of faulty model specification. In the preceding example, the index plot of residuals and the statistical test based on the Durbin–Watson statistic were used to conclude that the residuals are autocorrelated. Autocorrelation is only one of a number of possible explanations for the clustered type of residual plot or low Durbin–Watson value.

In general, a plot of residuals versus any one of the list of potential predictor variables may uncover additional information that can be used to further explain variation in the response variable. When an index plot of residuals shows a pattern of the type described in the preceding example, it is reasonable to suspect that it may be due to the omission of variables that change over time. Certainly, when the residuals appear in clusters alternating above and below the mean value line of zero, when the estimated autocorrelation coefficient is large and the Durbin–Watson statistic is significant, it would appear that the presence of autocorrelation is overwhelmingly supported. We shall see that this conclusion may be incorrect. The observed symptoms would be better interpreted initially as a general indication of some form of model misspecification.

All possible correction procedures should be considered. In fact, it is always better to explore fully the possibility of some additional predictor variables before yielding to an autoregressive model for the error structure. It is more satisfying and probably more useful to be able to understand the source of apparent autocorrelation in terms of an additional variable. The marginal effect of that variable can then be estimated and used in an informative way. The transformations that correct for pure autocorrelation may be viewed as an action of last resort.

9.7 ANALYSIS OF HOUSING STARTS

As an example of a situation where autocorrelation appears artificially because of the omission of another predictor variable, consider the following project undertaken by a midwestern construction industry association. The association wants to have a better understanding of the relationship between housing starts and population growth. They are interested in being able to forecast construction activity. Their approach is to develop annual data on regional housing starts and try to relate these data to potential home buyers in the region. Realizing that it is almost impossible to measure the number of potential house buyers accurately, the researchers settled for the size of the 22- to 44-year-old population group in the region as a variable that reflects the size of potential home buyers. With some diligent work they were able to bring together 25 years of historical data for the region. The data can be found in the file Housing.Starts.csv in the Book's Website. Their goal was to get a simple regression relationship between housing starts and population,

$$H_t = \beta_0 + \beta_1 P_t + \varepsilon_t. \tag{9.6}$$

Then using methods that they developed for projecting population changes, they would be able to estimate corresponding changes in the requirements for new houses. The construction association was aware that the relationship between population and housing starts could be very complex. It is even reasonable to suggest that housing affects population growth (by migration) instead of the other way around. Although the proposed model is undoubtedly naive, it serves a useful purpose as a starting point for their analysis.

Analysis

The regression results from fitting model (9.5) to the 25 years of data are given in Table 9.4. The proportion of variation in H accounted for by the

Table 9.4 Regression on Housing Starts (H) Versus Population (P)

Variable	Coefficient	s.e.	t-Test	p-value
Constant	−0.0609	0.0104	−5.85	0.0000
P	0.0714	0.0042	16.90	0.0000
$n = 25$	$R^2 = 0.925$	$d = 0.621$	$\hat{\sigma} = 0.0041$	df = 23

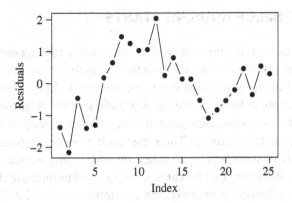

Figure 9.3 Index plot of standardized residuals from the regression of H_t on P_t for the housing starts data.

variability in P is $R^2 = 0.925$. We also see that an increase in population of 1 million leads to an increase in housing starts of about 71,000. The Durbin–Watson statistic and the index plot of the residuals (Figure 9.3) suggest strong autocorrelation. However, it is fairly simple to conjecture about other variables that may further explain housing starts and could be responsible for the appearance of autocorrelation. These variables include the unemployment rate, social trends in marriage and family formation, government programs in housing, and the availability of construction and mortgage funds. The first choice was an index that measures the availability of mortgage money for the region. Adding that variable to the equation the model becomes

$$H_t = \beta_0 + \beta_1 P_t + \beta_2 D_t + \varepsilon_t. \tag{9.7}$$

The introduction of the additional variable has the effect of removing autocorrelation. From Table 9.5 we see that the Durbin–Watson statistic has

Table 9.5 Results of the Regression of Housing Starts (H) on Population (P) and Index (D)

Variable	Coefficient	s.e.	t-Test	p-value
Constant	−0.0104	0.0103	−1.01	0.3220
P	0.0347	0.0064	5.39	0.0000
D	0.7605	0.1216	6.25	0.0000
$n = 25$	$R^2 = 0.973$	$d = 1.85$	$\hat{\sigma} = 0.0025$	df $= 22$

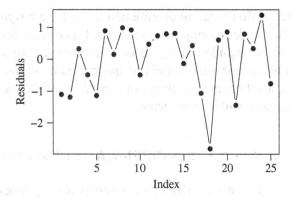

Figure 9.4 Index plot of the standardized residuals from the regression of H_t on P_t and D_t for the housing starts data.

the new value 1.852, well into the acceptable region. The index plot of the residuals (Figure 9.4) is also improved. The regression coefficients and their corresponding t-values show that there is a significant population effect but that it was overstated by a factor of more than 2 in the first equation. In a certain sense, the effect of changes in the availability of mortgage money for a fixed level of population is more important than a similar change in population.

If each variable in the regression equation is replaced by the standardized version of the variable (the variables transformed so as to have mean 0, and unit variance), the resulting regression equation is $\tilde{H}_t = 0.4668\tilde{P}_t + 0.5413\tilde{D}_t$, where $\tilde{H}$ denotes the standardized value of H, $\tilde{H} = (H - \bar{H})/s_H$. A unit increase in the standardized value of $\tilde{P}_t$ is worth an additional 0.4668 to the standardized value of H_t; that is, if the population increases by standard deviation, then H_t increases by 0.4668 standard deviation. Similarly, if D_t increases by 1 standard deviation, H_t increases by 0.5413 standard deviation. Therefore, in terms of the standardized variables, the mortgage index is more important (has a larger effect) than population size.

The example on housing starts illustrates two important points. First, a large value of R^2 does not imply that the data have been fitted and explained well. Any pair of variables that show trends over time are usually highly correlated. A large value of R^2 does not necessarily confirm that the relationship between the two variables has been adequately characterized. Second, the Durbin–Watson statistic as well as the residual plots may indicate the presence of autocorrelation among the errors when, in fact, the errors are independent but the omission of a variable or variables has given rise to the observed situation. Even though the Durbin–Watson statistic was designed to detect first-order autocorrelation it can have a significant value when some other model assumptions are violated such as misspecification of the variables to be included in the model. In general, a significant value of the Durbin–Watson statistic should be interpreted as an indication that a problem exists, and both the possibility of a missing variable or the presence of autocorrelation should be considered.

9.8 LIMITATIONS OF THE DURBIN–WATSON STATISTIC

In the previous examples on Expenditure versus Money Stock and Housing Starts versus Population Size the residuals from the initial regression equations indicated model misspecifications associated with time dependence. In both cases the Durbin–Watson statistic was small enough to conclude that positive autocorrelation was present. The index plot of residuals further confirmed the presence of a time-dependent error term. In each of the two problems the presence of autocorrelation was dealt with differently. In one case (Housing Starts) an additional variable was uncovered that had been responsible for the appearance of autocorrelation, and in the other case (Money Stock) the Cochrane–Orcutt method was used to deal with what was perceived as pure autocorrelation. It should be noted that the time dependence observed in the residuals in both cases is a first-order type of dependence. Both the Durbin–Watson statistic and the pattern of residuals indicate dependence between residuals in adjacent time periods. If the pattern of time dependence is other than first order, the plot of residuals will still be informative. However, the Durbin–Watson statistic is not designed to measure higher-order time dependence and may not yield much valuable information.

As an example we consider the efforts of a company that produces and markets ski equipment in the United States to obtain a simple aggregate relationship of quarterly sales to some leading economic indicator. The indicator chosen is personal disposable income, PDI, in billions of current dollars. The initial model is

$$S_t = \beta_0 + \beta_1 \text{PDI}_t + \varepsilon_t, \tag{9.8}$$

where S_t is ski sales in period t in millions of dollars and PDI_t is the personal disposable income for the same period. Data for 10 years (40 quarters) can be found in file Ski.Sales.csv at the Book's Website. The regression output is in Table 9.6 and the index plot of residuals is given in Figure 9.5.

Table 9.6 Ski Sales Versus PDI

Variable	Coefficient	s.e.	t-Test	p-value
Constant	12.3921	2.539	4.88	0.0000
PDI	0.1979	0.016	12.40	0.0000
$n = 40$	$R^2 = 0.801$	$d = 1.968$	$\hat{\sigma} = 3.019$	df = 38

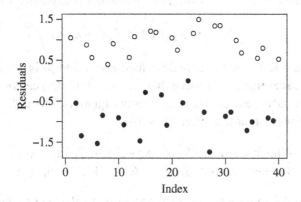

Figure 9.5 Index plot of the standardized residuals. Quarters 1 and 4 are indicated by an open circle and Quarters 2 and 3 are indicated by a solid circle.

At first glance the results in Table 9.6 are encouraging. The proportion of variation in sales accounted for by PDI is 0.80. The marginal contribution of an additional dollar unit of PDI to sales is between $165,420 and $230,380 ($\hat{\beta}_1 = 0.1979$) with a confidence coefficient of 95%. In addition, the Durbin–Watson statistic is 1.968, indicating no first-order autocorrelation.

It should be expected that PDI would explain a large proportion of the variation in sales since both variables are increasing over time. Therefore, although the R^2 value of 0.80 is good, it should not be taken as a final evaluation of the model. Also, the Durbin–Watson value is in the acceptable range, but it is clear from Figure 9.5 that there is some sort of time dependence of the residuals. We notice that residuals from the first and fourth quarters are positive, while residuals from the second and third quarters are negative for all the years. Since skiing activities are affected by weather conditions, we suspect that a seasonal effect has been overlooked. The pattern of residuals suggests that there are two seasons that have some bearing on ski sales: the second and third quarters, which correspond to the warm weather season, and the fourth and first quarters, which correspond to the winter season, when skiing is in full progress. This seasonal effect can be simply characterized by defining an indicator (dummy) variable that takes the value 1 for each winter quarter and is set equal to zero for each summer quarter (see Chapter 6). The expanded data set in the file Ski.Sales.expanded.csv at the Book's Website.

9.9 INDICATOR VARIABLES TO REMOVE SEASONALITY

Using the additional seasonal variable, the model is expanded to be

$$S_t = \beta_0 + \beta_1 \text{PDI}_t + \beta_2 Z_t + \varepsilon_t, \tag{9.9}$$

where Z_t is the zero-one variable described above and β_2 is a parameter that measures the seasonal effect. Note that the model in (9.9) can be represented by the two models (one for the cold weather quarters where $Z_t = 1$) and the other for the warm quarters where $Z_t = 0$):

Winter season: $S_t = (\beta_0 + \beta_2) + \beta_1 \text{PDI}_t + \varepsilon_t,$
Summer season: $S_t = \quad \beta_0 \quad + \beta_1 \text{PDI}_t + \varepsilon_t.$

Thus, the model represents the assumption that sales can be approximated by a linear function of PDI, in one line for the winter season and one for the summer season. The lines are parallel; that is, the marginal effect of changes in PDI is the same in both seasons. The level of sales, as reflected by the intercept, is different in each season (Figure 9.6).

The regression results are summarized in Table 9.7 and the index plot of the standardized residuals is shown in Figure 9.7. We see that all indications of the seasonal pattern have been removed. Furthermore, the precision of

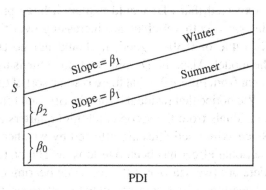

Figure 9.6 Model for ski sales and PDI adjusted for season.

Table 9.7 Ski Sales Versus PDI and Seasonal Variables

Variable	Coefficient	s.e.	t-Test	p-value
Constant	9.5402	0.9748	9.79	0.3220
PDI	0.1987	0.0060	32.90	0.0000
Z	5.4643	0.3597	15.20	0.0000
$n = 40$	$R^2 = 0.972$	$d = 1.772$	$\hat{\sigma} = 1.137$	df = 37

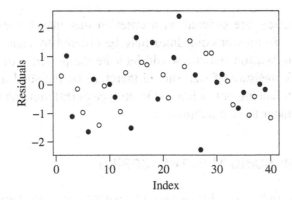

Figure 9.7 Index plot of the standardized residuals with seasonal variables (quarters indicated). Quarters 1 and 4 are indicated by an open circle and Quarters 2 and 3 are indicated by a solid circle.

the estimated marginal effect of PDI increased. The confidence interval is now ($186,520, $210,880). Also, the seasonal effect has been quantified and we can say that for a fixed level of PDI the winter season brings between $4,734,109 and $6,194,491 over the summer season (with 95% confidence).

The Ski Sales data illustrate two important points concerning autocorrelation. First, the Durbin–Watson statistic is only sensitive to correlated errors when the correlation occurs between adjacent observations (first-order autocorrelation). In the Ski Sales data the first-order correlation is −0.001. The second-, fourth-, sixth-, and eighth-order correlations are −0.81, 0.76, −0.71, and 0.73, respectively. The Durbin–Watson test does not show significance in this case. There are other tests that may be used for the detection of higher-order autocorrelations [see Box and Pierce (1970)]. But in all cases, the graph of residuals will show the presence of time dependence in the error term when it exists.

Second, when autocorrelation is indicated the model should be refitted. Often the autocorrelation appears because a time-dependent variable is missing from the model. The inclusion of the omitted variable often removes the observed autocorrelation. Sometimes, however, no such variable is present. Then one has to make a differencing type of transformation on the original variables to remove the autocorrelation.

If the observations are not ordered in time, the Durbin–Watson statistic is not strictly relevant. The statistic may still, however, be a useful diagnostic tool. If the data are ordered by an extraneous criterion, for example, an alphabetic listing, the value of the Durbin–Watson statistic should be near 2.0. Small values are suspicious, and the data should be scrutinized very carefully.

Many data sets are ordered on a criterion that may be relevant to the study. A list of cities or companies may be ordered by size. A low value of the Durbin–Watson statistic would indicate the presence of a significant size effect. A measure of size should therefore be included as a predictor variable. Differencing or Cochrane–Orcutt type of differencing would not be appropriate under these conditions.

9.10 REGRESSING TWO TIME SERIES

The data sets analyzed in this chapter all have the common characteristic that they are time series data (i.e., the observations arise in successive periods of time). This is quite unlike the data sets studied in previous chapters (a notable exception being the bacteria data in Chapter 7), where all the observations are generated at the same point in time. The observations in these examples were contemporaneous and gave rise to *cross-sectional data*. When the observations are generated simultaneously (and relate to a single time period), we have cross-sectional data. The contrast between time series and cross-sectional data can be seen by comparing the Ski Sales data discussed in this chapter (data arising sequentially in time), and the supervisor performance data in Section 4.3, where all the data were collected in an attitude survey and relate to one historical point in time.

Regression analysis of one time series on another is performed extensively in economics, business, public health, and other social sciences. There are some special features in time series data that are not present in cross-sectional data. We draw attention to these features and suggest possible techniques for handling them.

The concept of autocorrelation is not relevant in cross-sectional data. The ordering of the observations is often arbitrary. Consequently, the correlation of adjacent residuals is an artifact of the organization of the data. For time series data, however, autocorrelation is often a significant factor. The presence of autocorrelation shows that there are hidden structures in the data (often time related) which have not been detected. In addition, most time series data exhibit seasonality, and an investigator should look for seasonal patterns. A regular time pattern in the residuals (as in the Ski Sales data) will often indicate the presence of seasonality. For quarterly or monthly data, introduction of indicator variables, as has been pointed out, is a satisfactory solution. For quarterly data, four indicator variables would be needed but only three used in the analysis (see the discussion in Chapter 4). For monthly data, we will need 12 indicator variables but use only 11, to avoid problems of collinearity (this is discussed in Chapter 6). Not all of the indicator variables will be significant and some of them may well be deleted in the final stages of the analysis.

In attempting to find a relationship between y_t and $x_{1t}, x_{2t}, \cdots, x_{pt}$ one may expand the set of predictor variables by including lagged values of the predictor variables. A model such as

$$y_t = \beta_0 + \beta_1 x_{1t} + \beta_2 x_{1t-1} + \beta_3 x_{2t} + \varepsilon_t \qquad (9.10)$$

is meaningful in an analysis of time series data but not with cross-sectional data. The model given above implies that the value of Y in a given period is affected not only by the values of X_1 and X_2 of that period but also by the value of X_1 in the preceding period (i.e., there is a lingering effect of X_1 on Y for one period). Variables lagged by more than one period are also possibilities and could be included in the set of predictor variables.

Time series data are also likely to contain trends. Data in which time trends are likely to occur are often analyzed by including variables that are direct functions of time (t). Variables such as t and t^2 are included in the list of predictor variables. They are used to account for possible linear or quadratic trend. Simple first differencing ($y_t - y_{t-1}$), or more complex lagging of the type ($y_t - ay_{t-1}$) as in the Cochrane–Orcutt procedure, are also possibilities. For a fuller discussion, the reader should consult a book on time series analysis such as Shumway (1988) and Hamilton (1994).

To summarize, when performing regression analysis with time series data the analyst should be watchful for autocorrelation and seasonal effects, which are often present in the data. The possibility of using lagged predictor variables should also be explored.

EXERCISES

9.1 Use the function ols_regress in the package olsrr to fit the model in (9.1) to the Consumer Expenditure and Money Stock data, then
  (a) verify the results in Table 9.2 and the graph in Figure 9.1.
  (b) verify that the 95% confidence interval for β_1 is (2.06, 2.54).

9.2 Write the R code to reproduce the results in Table 9.3.

9.3 Fit model (9.6) to the data in the file Housing.Starts.csv at the Book's Website and
  (a) reproduce the results in Table 9.4 and the graph in Figure 9.3.
  (b) compute the Durbin–Watson statistic d. What conclusion regarding the presence of autocorrelation would you draw from d?
  (c) compare the number of runs to their expected value and standard deviation when fitting model (9.6) to the data in the file Housing.Starts.csv at the Book's Website. What conclusion regarding the presence of autocorrelation would you draw from this comparison?

9.4 Fit model (9.7) to the data in the file Housing.Starts.csv at the Book's Website and reproduce the results in Table 9.5 and the graph in Figure 9.4.

9.5 Fit model (9.8) to the data in the file Ski.Sales.csv at the Book's Website and
  (a) reproduce the results in Table 9.6 and the graph in Figure 9.5.
  (b) compute the Durbin–Watson statistic d. What conclusion regarding the presence of autocorrelation would you draw from d?

9.6 Fit model (9.9) to the data in the file Ski.Sales.csv at the Book's Website and reproduce the results in Table 9.7 and the graph in Figure 9.7.

9.7 Oil Production Data: Refer to the oil production data in Table 7.17. The index plot of the residuals obtained after fitting a linear regression of log(OIL) on Year shows a clear cyclical pattern.
  (a) Compute the Durbin–Watson statistic d. What conclusion regarding the presence of autocorrelation would you draw from d?
  (b) Compare the number of runs to their expected value and standard deviation. What conclusion regarding the presence of autocorrelation would you draw from this comparison?

9.8 Refer to the Presidential Election Data in the file Election.csv at the Book's Website. Since the data come over time (for 1916–1996 election years), one might suspect the presence of the autocorrelation problem when fitting the model in (6.15) to the data.

(a) Do you agree? Explain.

(b) Would adding a time trend (e.g., year) as an additional predictor variable improve or exacerbate the autocorrelation? Explain.

9.9 Dow Jones Industrial Average (DJIA): The data in the file DJIA.csv in the Book's Website contain the values of the daily DJIA for all the trading days in 1996. The data can be found at the Book's Website. DJIA is a very popular financial index and is meant to reflect the level of stock prices in the New York Stock Exchange. The Index is composed of 30 stocks. The variable Day denotes the trading day of the year. There were 262 trading days in 1996, and as such the variable Day goes from 1 to 262.

(a) Fit a linear regression model connecting DJIA with Day using all 262 trading days in 1996. Is the linear trend model adequate? Examine the residuals for time dependencies.

(b) Regress $DJIA_{(t)}$ against $DJIA_{(t-1)}$, that is, regress DJIA against its own value lagged by one period. Is this an adequate model? Are there any evidences of autocorrelation in the residuals?

(c) The variability (volatility) of the daily DJIA is large, and to accommodate this phenomenon the analysis is carried out on the logarithm of DJIA. Repeat the above exercises using log(DJIA) instead of DJIA. Are your conclusions similar? Do you notice any differences?

9.10 Refer again to the DJIA data in Exercise 9.9.

(a) Use the form of the model you found adequate in Exercise 9.9 and refit the model but using only the trading days in the first six months of 1996 (the first 130 days in the file DJIA.csv at the Book's Website). Compute the residual mean square.

(b) Use the above model to predict the daily DJIA for the first 15 trading days in July 1996. Compare your predictions with the actual values of the DJIA by computing the *prediction errors*, which is the difference between the actual values of the DJIA for the first 15 days of July, 1996 and their corresponding values predicted by the model.

(c) Compute the average of the squared prediction errors and compare with the residual mean square.

(d) Repeat the above exercise but using the model to predict the daily DJIA for the second half of the year (132 days).

(e) Explain the results you obtained above in the light of the scatter plot the DJIA versus Day.

9.11 Continuing with modeling the DJIA data in Exercises 9.9 and 9.10. A simplified version of the so-called *random walk model* of stock prices states that the best prediction of the stock index at Day t is the value of the index at Day $t-1$. In regression model terms it would mean that for the models fitted in Exercises 9.9 and 9.10 the constant term is 0, and the slope is 1.

(a) Carry out the appropriate statistical tests of significance. (Test the values of the coefficients individually and then simultaneously.) Which test is the appropriate one: the individual or the simultaneous?

(b) The random walk theory implies that the first differences of the index (the difference between successive values) should be independently normally distributed with zero mean and constant variance. Examine the first differences of DJIA and log(DJIA) to see if this hypothesis holds.

(c) DJIA is widely available. Collect the latest values available to see if the findings for 1996 hold for the latest period.

CHAPTER 10

ANALYSIS OF COLLINEAR DATA

10.1 INTRODUCTION

Interpretation of the multiple regression equation depends implicitly on the assumption that the predictor variables are not strongly interrelated. It is usual to interpret a regression coefficient as measuring the change in the response variable when the corresponding predictor variable is increased by one unit and all other predictor variables are held constant. This interpretation may not be valid if there are strong linear relationships among the predictor variables. It is always conceptually possible to increase the value of one variable in an estimated regression equation while holding the others constant. However, there may be no information about the result of such a manipulation in the estimation data. Moreover, it may be impossible to change one variable while holding all others constant in the process being studied. When these conditions exist, simple interpretation of the regression coefficient as a marginal effect is lost.

When there is a complete absence of linear relationship among the predictor variables, they are said to be *orthogonal*. In most regression applications the predictor variables are not orthogonal. Usually, the lack of orthogonality

---

Regression Analysis By Example Using R, Sixth Edition. Ali S. Hadi and Samprit Chatterjee
© 2024 John Wiley & Sons, Inc. Published 2024 by John Wiley & Sons, Inc.
Companion website: www.wiley.com/go/hadi/regression_analysis_6e

302 ANALYSIS OF COLLINEAR DATA

is not serious enough to affect the analysis. However, in some situations the predictor variables are so strongly interrelated that the regression results are ambiguous. Typically, it is impossible to estimate the unique effects of individual variables in the regression equation. The estimated values of the coefficients are very sensitive to slight changes in the data and to the addition or deletion of variables in the equation. The regression coefficients have large sampling errors, which affect both inference and forecasting that is based on the regression model.

The condition of severe nonorthogonality (that is, the existence of strong linear relationships among the predictor variables) is also referred to as the problem of collinear data, *collinearity*, or *multicollinearity*. The problem can be difficult to detect. It is not a specification error that may be uncovered by exploring regression residual. In fact, collinearity is not a modeling error. It is a condition of deficient data. In any event, it is important to know when collinearity is present and to be aware of its possible consequences. It is recommended that one should be very cautious about any and all substantive conclusions based on a regression analysis in the presence of collinearity.

This chapter focuses on three questions:

1. How does collinearity affect statistical inference and forecasting?

2. How can collinearity be detected?

3. What can be done to resolve the difficulties associated with collinearity?

When analyzing data, these questions cannot be answered separately. If collinearity is a potential problem, the three issues must be treated simultaneously by necessity.

The discussion begins with two examples. They have been chosen to demonstrate the effects of collinearity on inference and forecasting, respectively. A treatment of methods for detecting collinearity follows and the chapter concludes with a presentation of methods for resolving problems of collinearity. The obvious prescription to collect better data is considered, but the discussion is mostly directed at improving interpretation of the existing data. Alternatives to the ordinary least squares (OLS) estimation method that perform efficiently in the presence of collinearity are considered in Chapter 11.

10.2 EFFECTS OF COLLINEARITY ON INFERENCE

This first example demonstrates the ambiguity that may result when attempting to identify important predictor variables from among a linearly dependent collection of predictor variables. The context of the example is borrowed from research on equal opportunity in public education as reported by Coleman et al. (1966), Mosteller and Moynihan (1972), and others.

In conjunction with the Civil Rights Act of 1964, the Congress of the United States ordered a survey "concerning the lack of availability of equal educational opportunities for individuals by reason of race, color, religion or national origin in public educational institutions...." Data were collected from a cross section of school districts throughout the country. In addition to reporting summary statistics on variables such as level of student achievement and school facilities, regression analysis was used to try to establish factors that are the most important determinants of achievement. The data for this example consist of measurements taken in 1965 for 70 schools selected at random. The data consist of variables that measure student achievements (ACHV), faculty credentials (FAM), the influence of their peer group in the school (PEER), and school facilities (SCHOOL). The objective is to evaluate the effect of school inputs on achievement.

Assume that an acceptable index has been developed to measure those aspects of the school environment that would be expected to affect achievement. The index includes evaluations of the physical plant, teaching materials, special programs, training and motivation of the faculty, and so on. Achievement can be measured by using an index constructed from standardized test scores. There are also other variables that may affect the relationship between school inputs and achievement. Students' performances may be affected by their home environments and the influence of their peer group in the school. These variables must be accounted for in the analysis before the effect of school inputs can be evaluated. We assume that indexes have been constructed for these variables that are satisfactory for our purposes. The data can be found in the file EEO.csv at the Book's Website[1] and can be read into your R Workspace using the following commands:

url="http://www.aucegypt.edu/faculty/hadi/RABE6/Data6/EEO.csv"
download.file(url, "EEO.csv")
df <- read.csv("EEO.csv", header = TRUE)
head(df,2) # Display the first 2 lines of the dataset

Adjustment for the two basic variables (achievement and school) can be accomplished by using the regression model

$$\text{ACHV} = \beta_0 + \beta_1 \text{FAM} + \beta_2 \text{PEER} + \beta_3 \text{SCHOOL} + \varepsilon. \tag{10.1}$$

---

[1] http://www.aucegypt.edu/faculty/hadi/RABE6

The contribution of the school variable can be tested using the t-value for β_3. Recall that the t-value for β_3 tests whether SCHOOL is necessary in the equation when FAM and PEER are already included. Effectively, the model above is being compared to

$$\text{ACHV} - \beta_1\text{FAM} - \beta_2\text{PEER} = \beta_0 + \beta_3\text{SCHOOL} + \varepsilon, \quad (10.2)$$

that is, the contribution of the school variable is being evaluated after adjustment for FAM and PEER. Another view of the adjustment notion is obtained by noting that the left-hand side of (10.2) is an adjusted achievement index where adjustment is accomplished by subtracting the linear contributions of FAM and PEER. The equation is in the form of a regression of the adjusted achievement score on the SCHOOL variable. This representation is used only for the sake of interpretation. The estimated β's are obtained from the original model given in (10.1). The regression results are summarized in Table 10.1 and a plot of the residuals against the predicted values of ACHV appears as Figure 10.1.[2]

Checking first the residual plot we see that there are no glaring indications of misspecification. The point located in the lower left of the graph has a

Table 10.1 EEO Data: Regression Results

ANOVA Table

Source	Sum of Squares	df	Mean Square	F-Test
Regression	73.506	3	24.502	5.72
Residuals	282.873	66	4.286	

Coefficients Table

Variable	Coefficient	s.e.	t-Test	p-Value
Constant	−0.070	0.251	−0.28	0.7810
FAM	1.101	1.411	0.78	0.4378
PEER	2.322	1.481	1.57	0.1218
SCHOOL	−2.281	2.220	−1.03	0.3080
$n = 70$	$R^2 = 0.206$	$R_a^2 = 0.170$	$\hat{\sigma} = 2.07$	df = 66

---

[2] The R code for producing this and all other graphs in this book can be found at the Book's Website.

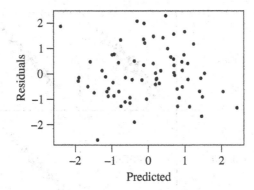

Figure 10.1 Standardized residuals against fitted values of ACHV.

residual value that is about 2.5 standard deviations from the mean of zero and should possibly be looked at more closely. However, when it is deleted from the sample, the regression results show almost no change. Therefore, the observation has been retained in the analysis.

From Table 10.1 we see that about 21% of the variation in achievement score is accounted for by the three predictors jointly ($R^2 = 0.206$). The F-value is 5.72 based on 3 and 66 degrees of freedom and is significant at better than the 0.01 level. Therefore, even though the total explained variation is estimated at only 20%, it is accepted that FAM, PEER, and SCHOOL are valid predictor variables. However, the individual t-values are all small. In total, the summary statistics say that the three predictors taken together are important but the t-values indicate that none of the variables individually are significant. It follows that any one predictor may be deleted from the model provided the other two are retained.

These results are typical of a situation where extreme collinearity is present. The predictor variables are so highly correlated that each one may serve as a proxy for the others in the regression equation without affecting the total explanatory power. The low t-values confirm that any one of the predictor variables may be dropped from the equation. Hence the regression analysis has failed to provide any information for evaluating the importance of school inputs on achievement. The culprit is clearly collinearity. The pairwise correlation coefficients of the three predictor variables and the corresponding scatter plots (Figure 10.2) all show strong linear relationships among all pairs of predictor variables. All pairwise correlation coefficients are high. In all scatter plots, all the observations lie close to a straight line.

Collinearity in this instance could have been expected. It is the nature of these three variables that each is determined by and helps to determine the

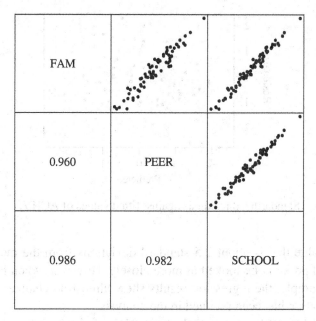

Figure 10.2 Pairwise scatter plots of the three predictor variables FAM, PEER, and SCHOOL and the corresponding pairwise correlation coefficients.

others. It is not unreasonable to conclude that there are not three variables but in fact only one. Unfortunately, that conclusion does not help to answer the original question about the effects of school facilities on achievement. There remain two possibilities. First, collinearity may be present because the sample data are deficient, but can be improved with additional observations. Second, collinearity may be present because the interrelationships among the variables are an inherent characteristic of the process under investigation. Both situations are discussed in the following paragraphs.

In the first case the sample should have been selected to ensure that the correlations between the predictor variables were not large. For example, in the scatter plot of FAM versus SCHOOL (the graph in the top right corner in Figure 10.2), there are no schools in the sample with values in the upper-left or lower-right regions of the graph. Hence there is no information in the sample on achievement when the value of FAM is high and SCHOOL is low, or FAM is low and SCHOOL is high. But it is only with data collected under these two conditions that the individual effects of FAM and SCHOOL on ACHV can be determined. For example, assume that there were some observations in the upper-left quadrant of the graph. Then it would at least be possible to compare average ACHV for low and high values of SCHOOL when FAM is held constant.

Since there are three predictor variables in the model, there are eight distinct combinations of data that should be included in the sample. Using + to represent a value above the average and − to represent a value below the average, the eight possibilities are represented in Table 10.2.

The large correlations that were found in the analysis suggest that only combinations 1 and 8 are represented in the data. If the sample turned out this way by chance, the prescription for resolving the collinearity problem is to collect additional data on some of the other combinations. For example, data based on combinations 1 and 2 alone could be used to evaluate the effect of SCHOOL on ACHV holding FAM and PEER at a constant level, both above average. If these were the only combinations represented in the data, the analysis would consist of the simple regression of ACHV against SCHOOL. The results would give only a partial answer, namely, an evaluation of the school–achievement relationship when FAM and PEER are both above average.

The prescription for additional data as a way to resolve collinearity is not a panacea. It is often not possible to collect more data because of constraints on budgets, time, and staff. It is always better to be aware of impending data deficiencies beforehand. Whenever possible, the data should be collected according to design. Unfortunately, prior design is not always feasible. In surveys, or observational studies such as the one being discussed, the values of the predictor variables are usually not known until the sampling unit is selected for the sample and some costly and time-consuming measurements are developed. Following this procedure, it is fairly difficult to ensure that a balanced sample will be obtained.

Table 10.2 Data Combinations for Three Predictor Variables

Combination	FAM	PEER	SCHOOL
1	+	+	+
2	+	+	−
3	+	−	+
4	−	+	+
5	+	−	−
6	−	+	−
7	−	−	+
8	−	−	−

The second reason that collinearity may appear is because the relationships among the variables are an inherent characteristic of the process being sampled. If FAM, PEER, and SCHOOL exist in the population only as data combinations 1 and 8 of Table 10.2, it is not possible to estimate the individual effects of these variables on achievement. The only recourse for continued analysis of these effects would be to search for underlying causes that may explain the interrelationships of the predictor variables. Through this process, one may discover other variables that are more basic determinants affecting equal opportunity in education and achievement.

10.3 EFFECTS OF COLLINEARITY ON FORECASTING

We shall examine the effects of collinearity in forecasting when the forecasts are based on a multiple regression equation. A historical data set with observations indexed by time is used to estimate the regression coefficients. Forecasts of the response variable are produced by using future values of the predictor variables in the estimated regression equation. The future values of the predictor variables must be known or forecasted from other data and models. We shall not treat the uncertainty in the forecasted predictor variables. In our discussion it is assumed that the future values of the predictor variables are given.

We have chosen an example based on aggregate data concerning import activity in the French economy. The data have been analyzed by Malinvaud (1968). Our discussion follows his presentation. The variables are imports (IMPORT), domestic production (DOPROD), stock formation (STOCK), and domestic consumption (CONSUM), all measured in billions of French francs for the years 1949–1966. The data can be found in the file Import.csv at the Book's Website. The model being considered is

$$\text{IMPORT} = \beta_0 + \beta_1 \text{DOPROD} + \beta_2 \text{STOCK} + \beta_3 \text{CONSUM} + \varepsilon. \quad (10.3)$$

The regression results are presented in Table 10.3. The index plot of residuals (Figure 10.3) shows a distinctive pattern, suggesting that the model is not well specified. Even though collinearity appears to be present ($R^2 = 0.973$ and all t-values are small), it should not be pursued further in this model. Collinearity should only be attacked after the model specification is satisfactory. The difficulty with the model is that the European Common Market began operations in 1960, causing changes in import–export relationships. Since our objective in this chapter is to study the effects of collinearity,

EFFECTS OF COLLINEARITY ON FORECASTING   309

Table 10.3 Import Data (1949–1966): Regression Results

ANOVA Table

Source	Sum of Squares	df	Mean Square	F-Test
Regression	2576.92	3	858.974	168.449
Residuals	71.39	14	5.099	

Coefficients Table

Variable	Coefficient	s.e.	t-Test	p-Value
Constant	−19.725	4.125	−4.78	0.000
DOPROD	0.032	0.187	0.17	0.866
STOCK	0.414	0.322	1.29	0.220
CONSUM	0.243	0.285	0.85	0.409

$n = 18$ $R^2 = 0.973$ $R_a^2 = 0.967$ $\hat{\sigma} = 2.258$ df = 14

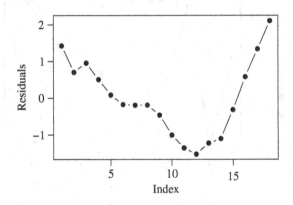

Figure 10.3 Import data (1949–1966): index plot of the standardized residuals.

we shall not complicate the model by attempting to capture the behavior after 1959. We shall assume that it is now 1960 and look only at the 11 years 1949–1959. The regression results for those data are summarized in Table 10.4. The residual plot is now satisfactory (Figure 10.4).

The value of $R^2 = 0.99$ is high. However, the coefficient of DOPROD is negative and not statistically significant, which is contrary to prior expectation. We believe that if STOCK and CONSUM were held constant, an increase in DOPROD would cause an increase in IMPORT, probably for raw materials or manufacturing equipment. Collinearity is a possibility here

Table 10.4 Import Data (1949–1959): Regression Results

ANOVA Table				
Source	Sum of Squares	df	Mean Square	F-Test
Regression	204.776	3	68.259	285.61
Residuals	1.673	7	0.239	

Coefficients Table				
Variable	Coefficient	s.e.	t-Test	p-Value
Constant	−10.128	1.212	−8.36	0.0000
DOPROD	−0.051	0.070	−0.73	0.4883
STOCK	0.587	0.095	6.20	0.0004
CONSUM	0.287	0.102	2.81	0.0263
$n = 11$	$R^2 = 0.992$	$R_a^2 = 0.988$	$\hat{\sigma} = 0.4889$	df = 7

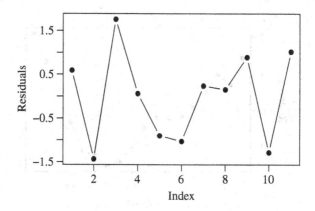

Figure 10.4 Import data (1949–1959): index plot of the standardized residuals.

and in fact this is the case. The simple correlation between CONSUM and DOPROD is 0.997. Upon further investigation it turns out that CONSUM has been about two-thirds of DOPROD throughout the 11-year period. The estimated relationship between the two quantities is

$$\text{CONSUM} = 6.259 + 0.686\,\text{DOPROD}.$$

Even in the presence of such severe collinearity the regression equation may produce some good forecasts. From Table 10.4, the forecasting equation is

$$\text{IMPORT} = -10.13 - 0.051\,\text{DOPROD} + 0.587\,\text{STOCK} + 0.287\,\text{CONSUM}.$$

Recall that the fit to the historical data is very good and the residual variation appears to be purely random. To forecast we must be confident that the character and strength of the overall relationship will hold into future periods. This matter of confidence is a problem in all forecasting models whether or not collinearity is present. For the purpose of this example we assume that the overall relationship does hold into future periods.[3] Implicit in this assumption is the relationship between DOPROD and CONSUM. The forecast will be accurate as long as the future values of DOPROD, STOCK, and CONSUM have the relationship that CONSUM is approximately equal to 0.67× DOPROD.

For example, let us forecast the change in IMPORT next year corresponding to an increase in DOPROD of 10 units while holding STOCK and CONSUM at their current levels. The resulting forecast is

$$\text{IMPORT}_{1960} = \text{IMPORT}_{1959} - 0.051(10),$$

which means that IMPORT will decrease by 0.51 units. However, if the relationship between DOPROD and CONSUM is kept intact, CONSUM will increase by $10(2/3) = 6.67$ units and the forecasted result is

$$\text{IMPORT}_{1960} = \text{IMPORT}_{1959} - 0.51 + 0.287 \times 6.67 = \text{IMPORT}_{1959} + 1.4.$$

The case where DOPROD increases alone corresponds to a change in the basic structure of the data that were used to estimate the model parameters and cannot be expected to produce meaningful forecasts.

In summary, the two examples demonstrate that multicollinear data can seriously limit the use of regression analysis for inference and forecasting. Extreme care is required when attempting to interpret regression results when collinearity is suspected. In Section 10.4 we discuss methods for detecting extreme collinearity among predictor variables.

10.4 DETECTION OF COLLINEARITY

In the preceding examples some of the ideas for detecting collinearity were already introduced. In this section we review those ideas and introduce additional criteria that indicate collinearity.

---

[3] For the purpose of convenient exposition we ignore the difficulties that arise because of our previous finding that the formation of the European Common Market has altered the relationship since 1960. But we are impelled to advise the reader that changes in structure make forecasting a very delicate endeavor even when the historical fit is excellent.

10.4.1 Simple Signs of Collinearity

Collinearity is associated with unstable estimated regression coefficients. This situation results from the presence of strong linear relationships among the predictor variables. It is not a problem of misspecification. Therefore, the empirical investigation of problems that result from a collinear data set should begin only after the model has been satisfactorily specified. However, there may be some indications of collinearity that are encountered during the process of adding, deleting, and transforming variables or data points in search of the good model. Indications of collinearity that appear as instability in the estimated coefficients are as follows:

- Large changes in the estimated coefficients when a variable is added or deleted.
- Large changes in the estimated coefficients when a data point is altered or dropped.

Once the residual plots indicate that the model has been satisfactorily specified, collinearity may be present if:

- The algebraic signs of the estimated coefficients do not conform to prior expectations; and/or
- Coefficients of variables that are expected to be important have large standard errors (small t-values).

For the IMPORT data discussed previously, the coefficient of DOPROD was negative and not significant. Both results are contrary to prior expectations. The effects of dropping or adding a variable can be seen in Table 10.5. There we see that the presence or absence of certain variables has a large effect on the other coefficients. For the EEO data (in the file EEO.csv at the Book's Website) the algebraic signs are all correct, but their standard errors are so large that none of the coefficients are statistically significant. It was expected that they would all be important.

The presence of collinearity is also indicated by the size of the correlation coefficients that exist among the predictor variables. A large correlation between a pair of predictor variables indicates a strong linear relationship between those two variables. The correlations for the EEO data (Figure 10.2) are large for all pairs of predictor variables. For the IMPORT data, the correlation coefficient between DOPROD and CONSUM is 0.997.

Table 10.5 Import Data (1949–1959): Regression Coefficients for All Possible Regressions

		Variable		
Regression	Constant	DOPROD	STOCK	CONSUM
1	−6.558	0.146	−	−
2	19.611	−	0.691	−
3	−8.013	−	−	0.214
4	−8.440	0.145	0.622	−
5	−8.884	−0.109	−	0.372
6	−9.743	−	0.596	0.212
7	−10.128	−0.051	0.587	0.287

The source of collinearity may be more subtle than a simple relationship between two variables. A linear relation can involve many of the predictor variables. It may not be possible to detect such a relationship with a simple correlation coefficient. As an example, we shall look at an analysis of the effects of advertising expenditures (A_t), promotion expenditures (P_t), and sales expense (E_t) on the aggregate sales of a firm in year t. The data represent a period of 23 years during which the firm was operating under fairly stable conditions. The data can be found in the file Advertising.csv at the Book's Website.

The proposed regression model is

$$S_t = \beta_0 + \beta_1 A_t + \beta_2 P_t + \beta_3 E_t + \beta_4 A_{t-1} + \beta_5 P_{t-1} + \varepsilon_t, \quad (10.4)$$

where A_{t-1} and P_{t-1} are the lagged one-year variables. The regression results are given in Table 10.6. The plot of residuals versus fitted values and the index plot of residuals (Figures 10.5 and 10.6), as well as other plots of the residuals versus the predictor variables (not shown), do not suggest any problems of misspecification. Furthermore, the correlation coefficients between the predictor variables are small (Table 10.7). However, if we do a little experimentation to check the stability of the coefficients by dropping the contemporaneous advertising variable A from the model, many things change. The coefficient of P_t drops from 8.37 to 3.70; the coefficients of lagged advertising A_{t-1} and lagged promotions P_{t-1} change signs. But the coefficient of sales expense is stable and R^2 does not change much (see Exercise 10.5).

Table 10.6 Regression Results for the Advertising Data

ANOVA Table

Source	Sum of Squares	df	Mean Square	F-Test
Regression	307.572	5	61.514	35.3
Residuals	27.879	16	1.742	

Coefficients Table

Variable	Coefficient	s.e.	t-Test	p-Value
Constant	−14.194	18.715	−0.76	0.4592
A_t	5.361	4.028	1.33	0.2019
P_t	8.372	3.586	2.33	0.0329
E_t	22.521	2.142	10.51	0.0000
A_{t-1}	3.855	3.578	1.08	0.2973
P_{t-1}	4.125	3.895	1.06	0.3053

$n = 22$ $\quad R^2 = 0.917 \quad R_a^2 = 0.891 \quad \hat{\sigma} = 1.320 \quad$ df $= 16$

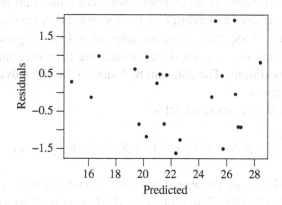

Figure 10.5 Standardized residuals versus fitted values of sales.

The evidence suggests that there is some type of relationship involving the contemporaneous and lagged values of the advertising and promotions variables. The regression of A_t on P_t, A_{t-1}, and P_{t-1} returns an R^2 of 0.973. The equation takes the form

$$\hat{A}_t = 4.63 - 0.87 P_t - 0.86 A_{t-1} - 0.95 P_{t-1}. \tag{10.5}$$

Upon further investigation into the operations of the firm, it was discovered that close control was exercised over the expense budget during those 23 years

DETECTION OF COLLINEARITY 315

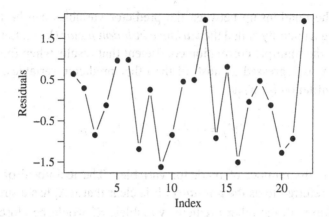

Figure 10.6 Index plot of the standardized residuals.

Table 10.7 Pairwise Correlation Coefficients for the Advertising Data

	A_t	P_t	E_t	A_{t-1}	P_{t-1}
A_t	1.000				
P_t	−0.357	1.000			
E_t	−0.129	0.063	1.000		
A_{t-1}	−0.140	−0.316	−0.166	1.000	
P_{t-1}	−0.496	−0.296	0.208	−0.358	1.000

of stability. In particular, there was an approximate rule imposed on the budget that the sum of A_t, A_{t-1}, P_t, and P_{t-1} was to be held to approximately five units over every two-year period. The relationship

$$A_t + P_t + A_{t-1} + P_{t-1} \doteq 5 \tag{10.6}$$

is the cause of the collinearity.

The above indicators of the presence of collinearity (e.g., the computational instability of the regression coefficients, the regression results do not conform to prior expectations, and large values of pairwise correlation coefficients) are not sufficient to see a complete picture of collinearity. Two methods for measuring collinearity are the *variance inflation factors* and the *condition indices*. These are explained below.

10.4.2 Variance Inflation Factors

A thorough investigation of collinearity will involve examining the value of R^2 that results from regressing each of the predictor variables against all the others. The relationship between the predictor variables can be judged by examining a quantity called the *variance inflation factor* (VIF). Let R_j^2 be the square of the multiple correlation coefficient that results when the predictor variable X_j is regressed against all the other predictor variables. Then the variance inflation for X_j is

$$\text{VIF}_j = \frac{1}{1 - R_j^2}, \quad j = 1,\ldots,p, \tag{10.7}$$

where p is the number of predictor variables. The reciprocal of the VIF_j, $1 - R_j^2$ is referred to as the *tolerance*. It is clear that if X_j has a strong linear relationship with the other predictor variables, R_j^2 would be close to 1, the tolerance would be close to 0, and VIF_j would be large. Values of VIFs greater than 10 is often taken as a signal that the data have collinearity problems.

In the absence of any linear relationship between the predictor variables (i.e., if the predictor variables are orthogonal), R_j^2 would be 0 and VIF_j would be 1. The deviation of the VIF_j value from 1 indicates departure from orthogonality and tendency toward collinearity.

The value of VIF_j also measures the amount by which the variance of the jth regression coefficient is increased due to the linear association of X_j with other predictor variables relative to the variance that would result if X_j were not related to them linearly. This explains the naming of this particular diagnostic.

As R_j^2 tends toward 1, indicating the presence of a linear relationship in the predictor variables, the VIF for $\hat{\beta}_j$ tends to infinity. It is suggested that a VIF in excess of 10 is an indication that collinearity may be causing problems in estimation.

The precision of an OLS estimated regression coefficient is measured by its variance, which is proportional to σ^2, the variance of the error term in the regression model. The constant of proportionality is the VIF. Thus, the VIFs may be used to obtain an expression for the expected squared distance of the OLS estimators from their true values. Denoting the square of the distance by D^2, it can be shown that, on average,

$$D^2 = \sigma^2 \sum_{j=1}^{p} \text{VIF}_j.$$

This distance is another measure of precision of the least squares estimators. The smaller the distance, the more accurate are the estimates. If the predictor variables were orthogonal, the VIFs would all be 1 and D^2 would be $p\sigma^2$. It follows that the ratio

$$\frac{\sigma^2 \sum_{i=1}^{p} \text{VIF}_i}{p\sigma^2} = \frac{\sum_{i=1}^{p} \text{VIF}_i}{p} = \overline{\text{VIF}},$$

which shows that the average of the VIFs measures the squared error in the OLS estimators relative to the size of that error if the data were orthogonal. Hence, $\overline{\text{VIF}}$ may also be used as an index of collinearity.

Most computer packages now furnish values of VIF_j routinely. Some have built-in messages when high values of VIF_j are observed. In any regression analysis the values of VIF_j should always be examined to avoid the pitfalls resulting from fitting a regression model to collinear data by least squares.

In each of the three examples (EEO, Import, and Advertising) we have seen evidence of collinearity. The VIF_j's and their average values for these data sets are given in Table 10.8. For the EEO data the values of VIF_j range from 30.2 to 83.2, showing that all three variables are strongly intercorrelated and that dropping one of the variables will not eliminate collinearity. The average value of VIF of 50.3 indicates that the squared error in the OLS estimators is 50 times as large as it would be if the predictor variables were orthogonal.

For the Import data, the squared error in the OLS estimators is 313 times as large as it would be if the predictor variables were orthogonal. However, the VIF_j's indicate that domestic production and consumption are strongly correlated but are not correlated with the STOCK variable. A regression equation containing either CONSUM or DOPROD along with STOCK will eliminate collinearity.

For the Advertising data, VIF_E (for the variable E) is 1.1, indicating that this variable is not correlated with the remaining predictor variables.

Table 10.8 Variance Inflation Factors for Three Data Sets

| EEO || Import 1949–1966 || Advertising ||
Variable	VIF	Variable	VIF	Variable	VIF
FAM	37.6	DOPROD	469.7	A_t	36.9
PEER	30.2	STOCK	1.0	P_t	33.5
SCHOOL	83.2	CONSUM	469.4	E_t	1.1
				A_{t-1}	25.9
				P_{t-1}	43.5
Average	50.3	Average	313.4	Average	28.2
Median	37.6	Median	469.4	Median	33.5

The VIF_j's for the other four variables are large, ranging from 26.6 to 44.1. This indicates that there is a strong linear relationship among the four variables, a fact that we have already noted. Here the prescription might be to regress sales S_t against E_t and three of the remaining four variables $(A_t, P_t, A_{t-1}, S_{t-1})$ and examine the resulting VIF_j's to see if collinearity has been eliminated.

We should note, however, that deleting some predictor variables is not always the best way to reduce collinearity and sometimes it does not work at all. For example, we have seen in Hamilton's data (see Sections 5.5.2 and 5.5.3), that the two predictor variables are mildly collinear but the response variable Y depends on them collectively but not individually. So, when we delete one of the predictor variables, one cannot predict Y using only one of the predictor variables. Better ways of dealing with collinearity are presented in Chapter 11.

10.4.3 The Condition Indices

Another way to detect collinearity in the data is to examine the condition indices of the *correlation matrix* of the predictor variables. The pairwise correlation coefficients of a set of p variables, $X_1, \ldots, X_p$, can be neatly displayed in a square matrix, which is known as the *correlation matrix*. For example, the correlation matrix of the three predictor variables in the Import data for the years 1949–1959 is

$$\begin{array}{c} \\ \text{DOPROD} \\ \text{STOCK} \\ \text{CONSUM} \end{array} \begin{pmatrix} \text{DOPROD} & \text{STOCK} & \text{CONSUM} \\ 1.000 & 0.026 & 0.997 \\ 0.026 & 1.000 & 0.036 \\ 0.997 & 0.036 & 1.000 \end{pmatrix}. \quad (10.8)$$

Thus, for example, cor(DOPROD, CONSUM) = 0.997, which indicates that the two variables are highly correlated. The elements on the diagonal that runs from the upper-left corner to the lower-right corner of the matrix are known as the *diagonal elements*. Note that all the diagonal elements of the correlation matrix are equal to one.

Recall that a set of variables is said to be orthogonal if there exists no linear relationships among them. If the standardized predictor variables are orthogonal, their matrix of variances and covariances consists of 1 for the diagonal elements and 0 for the off-diagonal elements.

DETECTION OF COLLINEARITY

Every correlation matrix of p predictor variables has a set of p-values called *eigenvalues*.[4] These can be arranged in decreasing order and denoted by $\lambda_1 \geq \lambda_2 \geq \cdots \geq \lambda_p$. If any one of the λ's is exactly equal to zero, there is a perfect linear relationship among the original variables, which is an extreme case of collinearity. If one of the λ's is much smaller than the others (and near zero), collinearity is present.

An empirical criterion for the presence of collinearity is given by the sum of the reciprocals of the eigenvalues, that is,

$$\sum_{j=1}^{p} \frac{1}{\lambda_j}. \qquad (10.9)$$

If this sum is greater than, say, five times the number of predictor variables, collinearity is present.

Another measure of the overall collinearity of the variables can be obtained by computing the *condition indices* of the correlation matrix (see, e.g., Belsely, 1991). The jth condition index is defined by

$$\kappa_j = \sqrt{\frac{\lambda_1}{\lambda_j}}, \quad j = 1, 2, \ldots, p. \qquad (10.10)$$

The first condition index, κ_1, is always 1 but the remaining indices are larger than 1. The largest condition index

$$\kappa_p = \sqrt{\frac{\text{Maximum eigenvalue of the correlation matrix}}{\text{Minimum eigenvalue of the correlation matrix}}} = \sqrt{\frac{\lambda_1}{\lambda_p}}, \qquad (10.11)$$

is known as the *condition number* of the correlation matrix. If the condition number κ_p is small, then the predictor variables are not collinear. But a large condition number indicates evidence of strong collinearity. The harmful effects of collinearity in the data become strong when the values of the condition number exceed 15 (which means that λ_1 is more than 225 times λ_p). The cutoff value of 15 is not based on any theoretical considerations but arises from empirical observation. Corrective action should always be taken when the condition number of the correlation matrix exceeds 30.

---

[4] Readers not familiar with eigenvalues may benefit from reading Chapter 9 in the book, *Matrix Algebra as a Tool*, by Hadi (1996).

Table 10.9 Condition Indices for Three Data Sets

j	EEO λ_j	EEO κ_j	Import 1949–1966 λ_j	Import 1949–1966 κ_j	Advertising λ_j	Advertising κ_j
1	2.952	1	2.084	1	1.701	1
2	0.040	8.59	0.915	1.51	1.288	1.15
3	0.008	19.26	0.001	44.23	1.145	1.22
4					0.859	1.41
5					0.007	15.29

The eigenvalues and condition numbers for the three data sets EEO, Import, and Advertising are shown in Table 10.9. The condition numbers for the three data sets indicate that collinearity is present in all three data sets.

We end this section by noting that some data sets contain more than one subset of collinear variables. The number of these subsets is indicated by the number of large condition indices. In each of the three data sets, there is only one large condition index, which means that there is one set of collinearity in each of them.

If a data set was found to be collinear as indicated by a large condition number, the next question is: Which variables are involved in this collinearity? The answer to this question involves the *eigenvectors* of the correlation matrix. We have seen above that every correlation matrix of p predictor variables has a set of p eigenvalues $\lambda_1 \geq \lambda_2 \geq \cdots \geq \lambda_p$. For every eigenvalue λ_j, there exists an *eigenvector*, $V_j, j = 1, 2, \ldots, p$. The p eigenvectors $V_1, V_2, \ldots, V_p$ are pairwise orthogonal. They can be arranged in a $p \times p$ matrix as follows:

$$V = \begin{pmatrix} V_1 & V_2 & \cdots & V_p \\ \downarrow & \downarrow & \downarrow & \downarrow \end{pmatrix}$$

$$= \begin{pmatrix} v_{11} & v_{12} & \cdots & v_{1p} \\ v_{21} & v_{22} & \cdots & v_{2p} \\ \vdots & \vdots & \ddots & \vdots \\ v_{p1} & v_{p2} & \cdots & v_{pp} \end{pmatrix}. \quad (10.12)$$

For the Import data, the eigenvalues and the corresponding condition indices and eigenvectors of the correlation matrix are shown in Table 10.10. Since there is only one large condition index (κ_3), there is only one set of collinearity. The variables involved in this set can be determined by setting a linear function of the predictor variables to a constant. The constant is the smallest eigenvalue, λ_3, and the coefficients of this linear function are the

Table 10.10 Import Data (1949–1959): Eigenvalues and Corresponding Condition Indices and Eigenvectors of the Correlation Matrix of the Predictor Variables

λ_j	1.999	0.998	0.003
κ_j	1	1.42	27.26
V_j	V_1	V_2	V_3
DOPROD	0.706	−0.036	−0.707
STOCK	0.044	0.999	−0.007
CONSUM	0.707	−0.026	0.707

elements of the corresponding eigenvectors, V_3. That is, the variables are related by

$$v_{13}\tilde{X}_1 + v_{23}\tilde{X}_2 + v_{33}\tilde{X}_3 \doteq \lambda_3$$
$$-0.707\,\tilde{X}_1 - 0.007\,\tilde{X}_2 + 0.707\,\tilde{X}_3 \doteq 0.003, \qquad (10.13)$$

where $\tilde{X}_j$ is the *standardized*[5] version of the jth predictor variable. In (10.13), if we approximate λ_3 (0.003) and the coefficient of $\tilde{X}_2$ (−0.007) by zero, we obtain

$$-0.707\,\tilde{X}_1 + 0.707\,\tilde{X}_3 \doteq 0, \qquad (10.14)$$

or, equivalently

$$\tilde{X}_1 \doteq \tilde{X}_3, \qquad (10.15)$$

which represents the approximate relationship that exists between the standardized versions of CONSUM and DOPROD. This result is consistent with our previous finding based on the high simple correlation coefficient ($r = 0.997$) between the predictor variables CONSUM and DOPROD. (The reader can confirm this high value of r by examining the scatter plot of CONSUM versus DOPROD.) Since κ_3 is the only large condition index, the analysis tells us that the dependence structure among the predictor variables as reflected in the data is no more complex than the simple relationship between CONSUM and DOPROD as given in (10.15).

EXERCISES

10.1 Fit model (10.1) to the data in the file EEO.csv at the Book's Website and

---

[5] See Section 4.6.

ANALYSIS OF COLLINEAR DATA

    (a) reproduce the results in Table 10.1.

    (b) compute the p-value and the critical value at the 5% level of significance for the F-test in Table 10.1.

    (c) compute the critical value at the 5% level of significance for the t-values in Table 10.1.

    (d) reproduce the graph in Figure 10.1.

    (e) use the function identify(df, labels = 1:nrow(df)) to identify the observation with the smallest residual on the scatter plot in Figure 10.1 by clicking near the points. Press the Esc key when done.

    (f) reproduce the scatter plot matrix in Figure 10.2.

10.2 Read the data in the file Import.csv at the Book's Website into your R Workspace.

    (a) Compute the corresponding pairwise correlation matrix.

    (b) Reproduce the scatter plot matrix similar to that in Figure 10.2 for this dataset.

    (c) Fit the model in 10.3 to the data and reproduce the results in Table 10.3.

    (d) Reproduce the scatter plot matrix in Figure 10.2.

    (e) Reproduce the index plot of residuals in Figure 10.3.

10.3 Read the data in the file Import.csv at the Book's Website into your R Workspace in an object named df.

    (a) Use df1 <- df[df[, 1]<=1959,] to extract the data for the Year 1949–1959 and assign it to an object named df1.

    (b) Fit the model in 10.3 to the data in df1 using the function lm() and reproduce the results in Table 10.4. Store the results in an object named reg.

    (c) Reproduce the index plot of residuals in Figure 10.4.

    (d) Reproduce Table 10.5.

10.4 Consider again the data in Exercise 10.3.

    (a) Use the functions cor() and eigen() to compute the correlation matrix of the predictor variables only and the corresponding eigen values and eigen vectors (see Table 10.10).

    (b) Use the eigen values found above to compute the condition indices of the predictor variables.

(c) Compute the VIF and tolerance for all the predictor variables using the VIF(reg) in the package DescTools.

10.5 Use the Advertising data in the file Advertising.csv at the Book's Website:

10.6 Read the data in the file Advertising.csv at the Book's Website into your R Workspace in an object named df.
   (a) Fit the model in 10.4 to the data and reproduce the results in Table 10.6 and the graphs in Figures 10.5 and 10.6.
   (b) Construct the scatter plot of the residuals versus each of the predictor variables. Does any of these plots suggest any problems of model misspecification?
   (c) Verify the results in (10.5) and hence (10.6).
   (d) Fit the model in 10.4 to the data but after dropping the contemporaneous advertising variable A. Compare the obtained results with those in Table 10.6.

10.7 The eigenvalues of the correlation matrix of a set of predictor variables are 4.603, 1.175, 0.203, 0.015, 0.003, 0.001 and the corresponding eigenvectors are given in Table 10.11.

Table 10.11 Six Eigenvectors of the Correlation Matrix of the Predictors

	V_1	V_2	V_3	V_4	V_5	V_6
X_1	−0.46	0.06	−0.15	−0.79	0.34	−0.13
X_2	−0.46	0.05	−0.28	0.12	−0.15	0.82
X_3	−0.32	−0.60	0.73	−0.01	0.01	0.11
X_4	−0.20	0.80	0.56	0.08	0.02	0.02
X_5	−0.46	−0.05	−0.20	0.59	0.55	−0.31
X_6	−0.46	0.00	−0.13	0.05	−0.75	−0.45

   (a) How many sets of collinearity are there in this data set? Explain.
   (b) What are the variables involved in each set? Explain.

10.8 In the analysis of the Advertising data in Section 10.4 it is suggested that the regression of sales S_t against E_t and three of the remaining four variables $(A_t, P_t, A_{t-1}, S_{t-1})$ may resolve the collinearity problem. Run the four suggested regressions and, for each of them, examine the resulting VIF_j's to see if collinearity has been eliminated.

10.9 Gasoline Consumption: To study the factors that determine the gasoline consumption of cars, data were collected on 30 models of cars. Besides the gasoline consumption (Y), measured in miles per gallon for each car, 11 other measurements representing physical and mechanical characteristics are made. Definitions of variables are given in Table 10.12. The data can be found in the file Gasoline.Consumption.csv at the Book's Website. The source of the data is *Motor Trend* magazine for the year 1975. We wish to determine whether the data set is collinear.

Table 10.12 Variables for the Gasoline Consumption Data

Variable	Definition
Y	Miles/gallon
X_1	Displacement (cubic inches)
X_2	Horsepower (feet/pound)
X_3	Torque (feet/pound)
X_4	Compression ratio
X_5	Rear axle ratio
X_6	Carburetor (barrels)
X_7	Number of transmission speeds
X_8	Overall length (inches)
X_9	Width (inches)
X_{10}	Weight (pounds)
X_{11}	Type of transmission (1 = automatic; 0 = manual)

(a) Compute the correlation matrix of the predictor variables $X_1, \ldots, X_{11}$ and the corresponding pairwise scatter plots. Identify any evidence of collinearity.

(b) Compute the eigenvalues, eigenvectors, and the condition number of the correlation matrix. Is collinearity present in the data?

(c) Identify the variables involved in collinearity by examining the eigenvectors corresponding to small eigenvalues.

(d) Regress Y on the 11 predictor variables. Compute the VIF for each of the predictors. Which predictors are affected by the presence of collinearity?

10.10 Refer to the Presidential Election Data in the file Election.csv at the Book's Website. Consider fitting a model relating V to all the variables (including a time trend representing year of election) plus as many interaction terms involving two or three variables as you possibly can.

(a) What is the maximum number of terms (coefficients) in a linear regression model that you can fit to these data? [*Hint*: Consider the number of observations in the data.]

(b) Examine the predictor variables in the above model for the presence of collinearity. (Compute the correlation matrix, the condition number, and the VIFs.)

(c) Identify the subsets of variables involved in collinearity. Attempt to solve the collinearity problem by deleting some of the variables involved in collinearity.

(d) Fit a model relating V to the set of predictors you found to be free from collinearity.

10.11 Refer again to the Presidential Election Data in Exercise 10.10 and consider fitting the model in (6.16).

(a) Examine the predictor variables in this model for the presence of collinearity. (Compute the correlation matrix, the condition number, and the VIFs.)

(b) Identify the subsets of variables involved in collinearity. Attempt to solve the collinearity problem by deleting some of the variables involved in collinearity.

(c) Fit a model relating V to the set of predictors you found to be free from collinearity.

(d) Compare the results with those obtained in Exercise 10.10.

CHAPTER 11

WORKING WITH COLLINEAR DATA

11.1 INTRODUCTION

It was demonstrated in Chapter 10 that when collinearity is present in a set of predictor variables, the ordinary least squares estimates of the individual regression coefficients tend to be unstable and can lead to erroneous inferences. In this chapter, we present ways for dealing with collinearity when it is present in the data. We have seen in Section 10.4.2 that deleting some predictor variables as a way of eliminating or reducing collinearity is not always the best way to deal with the problem and sometimes it does not work at all. We consider here two alternative approaches for dealing with collinearity: (a) Imposing or searching for constraints on the regression parameters and (b) Two estimation methods (principal components regression and ridge regression) as alternative to the ordinary least squares method. We start first by presenting the *Principal Components* approach, which is needed for the explanation of the above methods.

Regression Analysis By Example Using R, Sixth Edition. Ali S. Hadi and Samprit Chatterjee
© 2024 John Wiley & Sons, Inc. Published 2024 by John Wiley & Sons, Inc.
Companion website: www.wiley.com/go/hadi/regression_analysis_6e

11.2 PRINCIPAL COMPONENTS

The principal components method is based on the fact that any set of p variables, $X_1, X_2, \ldots, X_p$, can be transformed to a set of p orthogonal variables. The new orthogonal variables are known as the *principal components* (PCs) and are denoted by $C_1, \ldots, C_p$. Each variable C_j is a linear function of the standardized[1] variables $\tilde{X}_1, \ldots, \tilde{X}_p$. That is,

$$C_j = v_{1j}\tilde{X}_1 + v_{2j}\tilde{X}_2 + \ldots + v_{pj}\tilde{X}_p, \quad j = 1, 2, \ldots, p. \quad (11.1)$$

The coefficients of the linear functions are chosen so that the variables $C_1, \ldots, C_p$ are orthogonal.[2] The coefficients of the jth Principal Component, C_j, are the elements of the jth eigenvector that corresponds to λ_j, the jth largest eigenvalue of the correlation matrix of the p variables.[3] The eigenvectors of the correlation matrix are

$$\mathbf{V} = (V_1 \ V_2 \ \ldots \ V_p) = \begin{pmatrix} v_{11} & v_{12} & \ldots & v_{1p} \\ v_{21} & v_{22} & \ldots & v_{2p} \\ \vdots & \vdots & \ddots & \vdots \\ v_{p1} & v_{p2} & \ldots & v_{pp} \end{pmatrix}. \quad (11.2)$$

It can be shown that the variance of the jth principal components is

$$\text{Var}(C_j) = \lambda_j; \quad j = 1, 2, \ldots, p. \quad (11.3)$$

The variance–covariance matrix of the PCs is of the form:

$$\begin{array}{c} \\ C_1 \\ C_2 \\ \vdots \\ C_p \end{array} \begin{array}{cccc} C_1 & C_2 & \ldots & C_p \end{array} \\ \begin{pmatrix} \lambda_1 & 0 & \ldots & 0 \\ 0 & \lambda_2 & \ldots & 0 \\ \vdots & \vdots & \ddots & \vdots \\ 0 & 0 & \ldots & \lambda_p \end{pmatrix}.$$

All the off-diagonal elements are zero because the PCs are orthogonal. The value on the jth diagonal element, λ_j, is the variance of C_j, the jth PC.

---

[1] See Section 4.6.
[2] A description of this technique employing matrix algebra is given in the Appendix to this chapter.
[3] Readers not familiar with eigenvalues may benefit from reading Chapter 9 of the book *Matrix Algebra as a Tool*, by Hadi (1996).

PRINCIPAL COMPONENTS 329

The PCs are arranged so that $\lambda_1 \geq \lambda_2 \geq \ldots \geq \lambda_p$, that is, the first PC has the largest variance and the last PC has the smallest variance. As mentioned in Section 10.4.2, if any one of the λ's is exactly equal to zero, there is a perfect linear relationship among the original variables, which is an extreme case of collinearity. If one of the λ's is much smaller than the others (and near zero), collinearity is present.

For example, the correlation matrix of the three predictor variables in the Import data for the years 1949–1959 is

$$\begin{array}{c} \\ \text{DOPROD} \\ \text{STOCK} \\ \text{CONSUM} \end{array} \begin{array}{ccc} \text{DOPROD} & \text{STOCK} & \text{CONSUM} \\ \begin{pmatrix} 1.000 & 0.026 & 0.997 \\ 0.026 & 1.000 & 0.036 \\ 0.997 & 0.036 & 1.000 \end{pmatrix} \end{array}. \qquad (11.4)$$

The eigenvalues of the correlation matrix in (11.4) are $\lambda_1 = 1.999$, $\lambda_2 = 0.998$, and $\lambda_3 = 0.003$. The corresponding eigenvectors are

$$V_1 = \begin{pmatrix} 0.706 \\ 0.044 \\ 0.707 \end{pmatrix}, \quad V_2 = \begin{pmatrix} -0.036 \\ 0.999 \\ -0.026 \end{pmatrix}, \quad V_3 = \begin{pmatrix} -0.707 \\ -0.007 \\ 0.707 \end{pmatrix}.$$

Thus, the PCs for the Import data for the years 1949–1959 are

$$\begin{aligned} C_1 &= 0.706\tilde{X}_1 + 0.044\tilde{X}_2 + 0.707\tilde{X}_3, \\ C_2 &= -0.036\tilde{X}_1 + 0.999\tilde{X}_2 - 0.026\tilde{X}_3, \\ C_3 &= -0.707\tilde{X}_1 - 0.007\tilde{X}_2 + 0.707\tilde{X}_3. \end{aligned} \qquad (11.5)$$

These PCs are given in Table 11.1. These are obtained, using R, by

pc = princomp(scale(df[1:11,3:5])) # Standardizing the relevant data matrix

pc$scores # These are the PCs

Note that the PCs are unique up to the direction. That is, if C_j is the jth PC, then $-C_j$ is also the jth PC. The variance–covariance matrix of the new variables PCs is

$$\begin{array}{c} \\ C_1 \\ C_2 \\ C_3 \end{array} \begin{array}{ccc} C_1 & C_2 & C_3 \\ \begin{pmatrix} 1.999 & 0 & 0 \\ 0 & 0.998 & 0 \\ 0 & 0 & 0.003 \end{pmatrix} \end{array}.$$

330 WORKING WITH COLLINEAR DATA

Table 11.1 The PCs for the Import Data (1949–1959)

Year	C_1	C_2	C_3
49	−2.126	0.639	−0.021
50	−1.619	0.556	−0.071
51	−1.115	−0.073	−0.022
52	−0.894	−0.082	0.011
53	−0.644	−1.307	0.073
54	−0.190	−0.659	0.027
55	0.360	−0.744	0.043
56	0.972	1.354	0.063
57	1.559	0.964	0.024
58	1.767	1.015	−0.045
59	1.931	−1.663	−0.081

The PCs lack simple interpretation since each is, in a sense, a mixture of the original variables. However, these new variables provide a unified approach for obtaining information about collinearity and serve as the basis of one of the alternative estimation techniques described later in this chapter.

One additional piece of information is available through this type of analysis. Since λ_j is the variance of the jth PC, if λ_j is approximately zero, the corresponding PC, C_j, is approximately equal to a constant. It follows that the equation defining the PC gives some idea about the type of relationship among the predictor variables that is causing collinearity. For example, in the Import data, $\lambda_3 = 0.003 \doteq 0$. Therefore, C_3 is approximately constant. The constant is the mean value of C_3 which is zero. The PCs all have means of zero since they are linear functions of the standardized variables and each standardized variable has a zero mean. Therefore

$$C_3 = -0.707\tilde{X}_1 - 0.007\tilde{X}_2 + 0.707\tilde{X}_3 \doteq 0.$$

Rearranging the terms yields

$$\tilde{X}_1 \doteq \tilde{X}_3, \quad (11.6)$$

where the coefficient of $\tilde{X}_2$ (−0.007) has been approximated as zero. The equation in (11.6) represents the approximate relationship that exists between the standardized versions of CONSUM and DOPROD. This result is consistent with our previous finding based on the high simple correlation coefficient ($r = 0.997$) between the predictor variables CONSUM and DOPROD. (The reader can confirm this high value of r by examining the scatter plot of CONSUM versus DOPROD.) Since λ_3 is the only small eigenvalue, the analysis of the PCs tells us that the dependence structure among the predictor variables as reflected in the data is no more complex than the simple relationship between CONSUM and DOPROD as given in (11.6).

For the advertising data, the eigenvalues of the correlation matrix of the five predictor variables and the corresponding eigenvectors are given in Table 11.2. The smallest eigenvalue is $\lambda_5 = 0.007$. The corresponding PC is

$$C_5 = 0.514\tilde{X}_1 + 0.489\tilde{X}_2 - 0.010\tilde{X}_3 + 0.428\tilde{X}_4 + 0.559\tilde{X}_5. \qquad (11.7)$$

Setting C_5 to zero and solving for $\tilde{X}_1$ leads to the approximate relationship,

$$\tilde{X}_1 \doteq -0.951\tilde{X}_2 - 0.833\tilde{X}_4 - 1.087\tilde{X}_5, \qquad (11.8)$$

where we have taken the coefficient of $\tilde{X}_3$ to be approximately zero. This equation reflects our earlier findings about the relationship between A_t, P_t, A_{t-1}, and P_{t-1}. Furthermore, since $\lambda_4 = 0.859$ and the other λ's are all large, we can be confident that the relationship involving A_t, P_t, A_{t-1}, and P_{t-1} in (11.8) is the only source of collinearity in the data.

Throughout this section, investigations concerning the presence of collinearity have been based on judging the magnitudes of various indicators, either a correlation coefficient or an eigenvalue. Although we speak in terms of large and small, there is no way to determine these threshold values.

Table 11.2 Advertising Data: The Eigenvalues and Corresponding Eigenvectors of the Correlation Matrix of the Predictor Variables

	Eigenvalues				
λ_j	1.701	1.288	1.145	0.859	0.007
	Eigenvectors				
V_j	V_1	V_2	V_3	V_4	V_5
$\tilde{A}_t$	0.532	−0.024	−0.668	0.074	−0.514
$\tilde{P}_t$	−0.232	0.825	0.158	−0.037	−0.489
$\tilde{E}_t$	−0.389	−0.022	−0.217	0.895	0.010
$\tilde{A}_{t-1}$	0.395	−0.260	0.692	0.338	−0.428
$\tilde{P}_{t-1}$	−0.596	−0.501	−0.057	−0.279	−0.559

The size is relative and is used to give an indication either that everything seems to be in order or that something is amiss. The only reasonable criterion for judging size is to decide whether the ambiguity resulting from the perceived collinearity is of material importance in the underlying problem.

We should also caution here that the data analyzed may contain one or a few observations that can have an undue influence on the various measures of collinearity (e.g., correlation coefficients, eigenvalues, or the condition number). These observations are called *collinearity-influential observations*. For more details the reader is referred to Hadi (1988).

11.3 COMPUTATIONS USING PRINCIPAL COMPONENTS

The computations required for this analysis involve something in addition to a standard least squares computer program. The raw data must be processed through a principal components subroutine that operates on the correlation matrix of the predictor variables in order to compute the eigenvalues, the eigenvectors, and the PCs. For example, for the Advertising data, the regression model stated in terms of the original variables is

$$S_t = \beta_0 + \beta_1 A_t + \beta_2 P_t + \beta_3 E_t + \beta_4 A_{t-1} + \beta_5 P_{t-1} + \varepsilon_t \quad (11.9)$$

and the regression model stated in terms of the standardized variables is

$$\tilde{Y} = \theta_1 \tilde{X}_1 + \theta_2 \tilde{X}_2 + \theta_3 \tilde{X}_3 + \theta_4 \tilde{X}_4 + \theta_5 \tilde{X}_5 + \varepsilon', \quad (11.10)$$

where $\tilde{Y}$ denotes the standardized version of the response variable and $\tilde{X}_j$ denotes the standardized version of the jth predictor variable. The regression coefficients in (11.10) are often referred to as the *beta coefficients*. They represent marginal effects of the predictor variables in standard deviation units. For example, θ_1 measures the change in standardized units of sales (S) corresponding to an increase of one standard deviation unit in advertising (A).

Let $\hat{\beta}_j$ be the least squares estimate of β_j when model (11.9) is fit to the data. Similarly, let $\hat{\theta}_j$ be the least squares estimate of θ_j obtained from fitting model (11.10). Then $\hat{\beta}_j$ and $\hat{\theta}_j$ are related by

$$\hat{\beta}_j = \frac{s_y}{s_j} \hat{\theta}_j, \quad j = 1, 2, 3, 4, 5,$$
$$\hat{\beta}_0 = \bar{y} - \sum_{j=1}^{5} \hat{\beta}_j \bar{x}_j, \quad (11.11)$$

where $\bar{y}$ is the mean of Y and s_y and s_j are standard deviations of the response and jth predictor variable, respectively.

From the eigenvectors in Table 11.2, the five PCs are

$$C_1 = 0.532\tilde{X}_1 - 0.232\tilde{X}_2 - 0.389\tilde{X}_3 + 0.395\tilde{X}_4 - 0.595\tilde{X}_5,$$
$$C_2 = -0.024\tilde{X}_1 + 0.825\tilde{X}_2 - 0.022\tilde{X}_3 - 0.260\tilde{X}_4 - 0.501\tilde{X}_5,$$
$$C_3 = -0.668\tilde{X}_1 + 0.158\tilde{X}_2 - 0.217\tilde{X}_3 + 0.692\tilde{X}_4 - 0.057\tilde{X}_5, \quad (11.12)$$
$$C_4 = 0.074\tilde{X}_1 - 0.037\tilde{X}_2 + 0.895\tilde{X}_3 + 0.338\tilde{X}_4 - 0.279\tilde{X}_5,$$
$$C_5 = -0.514\tilde{X}_1 - 0.489\tilde{X}_2 + 0.010\tilde{X}_3 - 0.428\tilde{X}_4 - 0.559\tilde{X}_5.$$

The coefficients in the equation defining C_1 are the components of the eigenvector corresponding to the largest eigenvalue of the correlation matrix of the predictor variables (see Table 11.2). Similarly, the coefficients defining C_2 through C_5 are components of the eigenvectors corresponding to the remaining eigenvalues in order by size. The variables $C_1, \ldots, C_5$ are the PCs associated with the standardized versions of the predictor variables, as described in Section 11.2.

In terms of the PCs, the model can then be written as

$$\tilde{Y} = \alpha_1 C_1 + \alpha_2 C_2 + \alpha_3 C_3 + \alpha_4 C_4 + \alpha_5 C_5 + \varepsilon'. \quad (11.13)$$

The estimates $\hat{\theta}_1, \ldots, \hat{\theta}_p$ can be computed in two equivalent ways. They can be obtained directly from a regression of the standardized variables as represented in (11.10). The results of this regression are given in Table 11.3. Alternatively, we can fit the model in (11.13) by the least squares regression of the standardized response variable on the five PCs and obtain the estimates $\hat{\alpha}_1, \ldots, \hat{\alpha}_5$. The results of this regression are shown in Table 11.4. The estimates $\hat{\theta}_1, \ldots, \hat{\theta}_p$ can be obtained from the estimates $\hat{\alpha}_1, \ldots, \hat{\alpha}_5$ as

$$\hat{\theta}_1 = 0.532\hat{\alpha}_1 - 0.024\hat{\alpha}_2 - 0.668\hat{\alpha}_3 + 0.074\hat{\alpha}_4 - 0.514\hat{\alpha}_5,$$
$$\hat{\theta}_2 = -0.232\hat{\alpha}_1 + 0.825\hat{\alpha}_2 + 0.158\hat{\alpha}_3 - 0.037\hat{\alpha}_4 - 0.489\hat{\alpha}_5,$$
$$\hat{\theta}_3 = -0.389\hat{\alpha}_1 - 0.022\hat{\alpha}_2 - 0.217\hat{\alpha}_3 + 0.895\hat{\alpha}_4 + 0.010\hat{\alpha}_5, \quad (11.14)$$
$$\hat{\theta}_4 = 0.395\hat{\alpha}_1 - 0.260\hat{\alpha}_2 + 0.692\hat{\alpha}_3 + 0.338\hat{\alpha}_4 - 0.428\hat{\alpha}_5,$$
$$\hat{\theta}_5 = -0.595\hat{\alpha}_1 - 0.501\hat{\alpha}_2 - 0.057\hat{\alpha}_3 - 0.279\hat{\alpha}_4 - 0.559\hat{\alpha}_5.$$

WORKING WITH COLLINEAR DATA

Table 11.3 Regression Results Obtained from Fitting the Model in (11.10)

Variable	Coefficient	s.e.	t-Test	p-Value
$\tilde{X}_1$	0.583	0.438	1.33	0.2019
$\tilde{X}_2$	0.973	0.417	2.33	0.0329
$\tilde{X}_3$	0.786	0.075	10.50	0.0000
$\tilde{X}_4$	0.395	0.367	1.08	0.2973
$\tilde{X}_5$	0.503	0.476	1.06	0.3053
$n = 22$	$R^2 = 0.917$	$R_a^2 = 0.891$	$\hat{\sigma} = 0.3303$	df = 16

Table 11.4 Regression Results Obtained from Fitting the Model in (11.13)

Variable	Coefficient	s.e.	t-Test	p-Value
C_1	−0.346	0.053	−6.55	0.0000
C_2	0.418	0.064	6.58	0.0000
C_3	−0.151	0.067	−2.25	0.0391
C_4	0.660	0.078	8.46	0.0000
C_5	−1.220	0.846	−1.44	0.1683
$n = 22$	$R^2 = 0.917$	$R_a^2 = 0.891$	$\hat{\sigma} = 0.3303$	df = 16

Thus, for example,

$$\hat{\theta}_1 = (0.532)(-0.3460) + (-0.024)(0.4179) + (-0.668)(-0.1513)$$
$$+ (0.074)(0.6599) + (-0.514)(-1.2203)$$
$$= 0.583,$$

which is the same as the coefficient of $\tilde{X}_1$ in Table 11.3.

Using the coefficients in (11.14), the standard error of $\hat{\theta}_1, \ldots, \hat{\theta}_5$ can be computed. For example the estimated variance of $\hat{\theta}_1$ is

$$\text{Var}(\hat{\theta}_1) = (0.532 \times \text{s.e.}(\hat{\alpha}_1))^2 + (-0.024 \times \text{s.e.}(\hat{\alpha}_2))^2 + (-0.668 \times \text{s.e.}(\hat{\alpha}_3))^2$$
$$+ (0.074 \times \text{s.e.}(\hat{\alpha}_4))^2 + (-0.514 \times \text{s.e.}(\hat{\alpha}_5))^2$$
$$= (0.532 \times 0.0529)^2 + (-0.024 \times 0.0635)^2 + (-0.668 \times 0.0674)^2$$
$$+ (0.074 \times 0.0780)^2 + (-0.514 \times 0.8456)^2 = 0.1918,$$

which means that the standard error of $\hat{\theta}_1$ is

$$\text{s.e.}(\hat{\theta}_1) = \sqrt{0.1918} = 0.438,$$

which is the same as the standard error of the coefficient of $\tilde{X}_1$ in Table 11.3.

It should be noted that the t-values for testing β_j and θ_j equal to zero are identical. The beta coefficient, θ_j is a scaled version of β_j. When constructing the t-values as either $\hat{\beta}_j/\text{s.e.}(\hat{\beta}_j)$ or $\hat{\theta}_j/\text{s.e.}(\hat{\theta}_j)$, the scale factor is canceled.

11.4 IMPOSING CONSTRAINTS

We have noted that collinearity is a condition associated with deficient data and not due to misspecification of the model. It is assumed that the form of the model has been carefully structured and that the residuals are acceptable before questions of collinearity are considered. Since it is usually not practical and often impossible to improve the data, we shall focus our attention on methods of better interpretation of the given data than would be available from a direct application of least squares. In this section, rather than trying to interpret individual regression coefficients, we shall attempt to identify and estimate informative linear functions of the regression coefficients. Alternative estimation methods for the individual coefficients are treated later in this chapter.

Before turning to the problem of searching the data for informative linear functions of the regression coefficients, one additional point concerning model specification should be discussed. A subtle step in specifying a relationship that can have a bearing on collinearity is acknowledging the presence of theoretical relationships among the regression coefficients. For example, in the model for the Import data,

$$\text{IMPORT} = \beta_0 + \beta_1 \text{DOPROD} + \beta_2 \text{STOCK} + \beta_3 \text{CONSUM} + \varepsilon, \quad (11.15)$$

one may argue that the marginal effects of DOPROD and CONSUM are equal. That is, on the basis of economic reasoning, and before looking at the data, it is decided the $\beta_1 = \beta_3$ or equivalently, $\beta_1 - \beta_3 = 0$. As described in Section 4.10.3, the model in (11.15) becomes

$$\text{IMPORT} = \beta_0 + \beta_1 \text{DOPROD} + \beta_2 \text{STOCK} + \beta_1 \text{CONSUM} + \varepsilon,$$
$$= \beta_0 + \beta_2 \text{STOCK} + \beta_1 (\text{DOPROD} + \text{CONSUM}) + \varepsilon.$$

WORKING WITH COLLINEAR DATA

Table 11.5 Regression Results of Import Data (1949–1959) with the Constraint $\beta_1 = \beta_3$

ANOVA Table

Source	Sum of Squares	df	Mean Square	F-Test
Regression	203.856	2	101.928	314
Residuals	2.593	8	0.324	

Coefficients Table

Variable	Coefficient	s.e.	t-Test	p-Value
Constant	−9.007	1.245	−7.23	0.0000
STOCK	0.612	0.109	5.60	0.0005
NEWVAR	0.086	0.004	24.30	0.0000
$n = 11$	$R^2 = 0.987$	$R_a^2 = 0.984$	$\hat{\sigma} = 0.5693$	df = 8

Thus, the common value of β_1 and β_3 is estimated by regressing IMPORT on STOCK and a new variable constructed as NEWVAR = DOPROD + CONSUM. The new variable has significance only as a technical manipulation to extract an estimate of the common value of β_1 and β_3. The results of the regression appear in Table 11.5. The correlation between the two predictor variables, STOCK and NEWVAR, is 0.03 and the eigenvalues are $\lambda_1 = 1.030$ and $\lambda_2 = 0.970$. There is no longer any indication of collinearity. The residual plots against time and the fitted values indicate that there are no other problems of specification (Figures 11.1 and 11.2,[4] respectively). The estimated model is

IMPORT = −9.007 + 0.086 DOPROD + 0.612 STOCK + 0.086 CONSUM.

Note that following the methods outlined in Section 4.10.3, it is also possible to test the constraint, $\beta_1 = \beta_3$, as a hypothesis. Even though the argument for $\beta_1 = \beta_3$ may have been imposed on the basis of existing theory, it is still interesting to evaluate the effect of the constraint on the explanatory power of the full model. The values of R^2 for the full and restricted models

---

[4] The R code for producing these and all other graphs in this book can be found at the Book's Website at http://www.aucegypt.edu/faculty/hadi/RABE6.

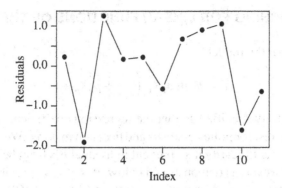

Figure 11.1 Index plot of the standardized residuals. Import data (1949–1959) with the constraint $\beta_1 = \beta_3$.

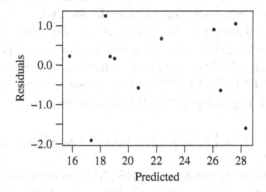

Figure 11.2 Standardized residuals against fitted values of Import data (1949–1959) with the constraint $\beta_1 = \beta_3$.

are 0.992 and 0.987, respectively. The F-ratio for testing $H_0 : (\beta_1 = \beta_3)$ is 3.36 with 1 and 8 degrees of freedom. Both results suggest that the constraint is consistent with the data.

The constraint that $\beta_1 = \beta_3$ is, of course, only one example of the many types of constraints that may be used when specifying a regression model. The general class of possibilities is found in the set of linear constraints described in Chapter 4. Constraints are usually justified on the basis of underlying theory. They may often resolve what appears to be a problem of collinearity. In addition, any particular constraint may be viewed as a testable hypothesis and judged by the methods described in Chapter 4.

11.5 SEARCHING FOR LINEAR FUNCTIONS OF THE β'S

We assume that the model

$$Y = \beta_0 + \beta_1 X_1 + \ldots + \beta_p X_p + \varepsilon$$

has been carefully specified so that the regression coefficients appearing are of primary interest for policy analysis and decision making. We have seen that the presence of collinearity may prevent individual β's from being accurately estimated. However, as demonstrated below, it is always possible to estimate some linear functions of the β's accurately (Silvey, 1969). The obvious questions are: Which linear functions can be estimated, and of those that can be estimated, which are of interest in the analysis? In this section we use the data to help identify those linear functions that can be accurately estimated and, at the same time, have some value in the analysis.

First we shall demonstrate in an indirect way that there are always linear functions of the β's that can be accurately estimated.[5] Consider once again the Import data. We have argued that there is a historical relationship between CONSUM and DOPROD that is approximated as CONSUM = (2/3) DOPROD. Replacing CONSUM in the original model,

$$\text{IMPORT} = \beta_0 + \left(\beta_1 + \frac{2}{3}\beta_3\right) \text{DOPROD} + \beta_2 \text{STOCK} + \varepsilon. \qquad (11.16)$$

Equivalently stated, by dropping CONSUM from the equation we are able to obtain accurate estimates of $\beta_1 + (2/3)\beta_3$ and β_2. Collinearity is no longer present. The correlation between DOPROD and STOCK is 0.026. The results are given in Table 11.6. R^2 is almost unchanged and the residual plots (not shown) are satisfactory. In this case we have used information in addition to the data to argue that the coefficient of DOPROD in the regression of IMPORT on DOPROD and STOCK is the linear function $\beta_1 + (2/3)\beta_3$. Also, we have demonstrated that this linear function can be estimated accurately even though collinearity is present in the data. Whether or not it is useful to know the value of $\beta_1 + (2/3)\beta_3$, of course, is another question. At least it is important to know that the estimate of the coefficient of DOPROD in this regression is not measuring the pure marginal effect of DOPROD, but includes part of the effect of CONSUM.

The above example demonstrates in an indirect way that there are always linear functions of the β's that can be accurately estimated. However, there

---

[5] Refer to the Appendix to this chapter for further treatment of this problem.

SEARCHING FOR LINEAR FUNCTIONS OF THE β'S

Table 11.6 Regression Results When Fitting Model (11.16) to the Import Data (1949–1959)

ANOVA Table

Source	Sum of Squares	df	Mean Square	F-Test
Regression	202.894	2	101.447	228
Residuals	3.555	8	0.444	

Coefficients Table

Variable	Coefficient	s.e.	t-Test	p-Value
Constant	−8.440	1.435	−5.88	0.0004
DOPROD	0.145	0.007	20.70	0.0000
STOCK	0.622	0.128	4.87	0.0012
$n = 11$	$R^2 = 0.983$	$R_a^2 = 0.978$	$\hat{\sigma} = 0.667$	df = 8

is a constructive approach for identifying the linear functions of the β's that can be accurately estimated. We shall use the advertising data introduced in Section 10.4 to demonstrate the method. The concepts are less intuitive than those found in the other sections of the chapter. We have attempted to keep things simple. A formal development of this problem is given in the Appendix to this chapter.

We begin with the linear transformation introduced in Sections 11.2 and 11.3 that takes the standardized predictor variables into a new orthogonal set of variables. For the Advertising data, the transformation that takes $X_1, \ldots, X_5$ into the new set of orthogonal variables $C_1, \ldots, C_5$ is given in (11.12). Note that the transformation involves the same weights that are used to define (11.12). The advantage of the transformed model is that the PCs are orthogonal. The precision of the estimated regression coefficients as measured by the variance of the $\hat{\alpha}$'s is easily evaluated. The estimated variance of $\hat{\alpha}_j$ is $\hat{\sigma}^2/\lambda_j$. It is inversely proportional to the ith eigenvalue. All but $\hat{\alpha}_5$ may be accurately estimated since only λ_5 is small. (Recall that $\lambda_1 = 1.701, \lambda_2 = 1.288, \lambda_3 = 1.145, \lambda_4 = 0.859$, and $\lambda_5 = 0.007$.)

Our interest in the $\hat{\alpha}$'s is only as a vehicle for analyzing the $\hat{\theta}$'s. From the representation in (11.14) it is a simple matter to compute and analyze the variances and, in turn, the standard errors of the $\hat{\theta}$'s. The variance of $\hat{\theta}_j$ is

$$\text{Var}(\hat{\theta}_j) = \sum_{i=1}^{p} v_{ij}^2 \text{Var}(\hat{\alpha}_i), \quad j = 1, \ldots, p, \quad (11.17)$$

where v_{ij} is the coefficient of $\hat{\alpha}_i$ in the jth equation in (11.14). Since the estimated variance of $\hat{\alpha}_i = \hat{\sigma}^2/\lambda_i$, where $\hat{\sigma}^2$ is the residual mean square, (11.17) becomes

$$\text{Var}(\hat{\theta}_j) = \hat{\sigma}^2 \sum_{i=1}^{p} \frac{v_{ij}^2}{\lambda_i}. \qquad (11.18)$$

For example, the estimated variance of $\hat{\theta}_1$ is

$$\hat{\sigma}^2 \left[\frac{(0.532)^2}{\lambda_1} + \frac{(-0.024)^2}{\lambda_2} + \frac{(-0.668)^2}{\lambda_3} + \frac{(0.074)^2}{\lambda_4} + \frac{(-0.514)^2}{\lambda_5} \right]. \qquad (11.19)$$

Recall that $\lambda_1 \geq \lambda_2 \geq \ldots \geq \lambda_5$ and only λ_5 is small ($\lambda_5 = 0.007$). Therefore, it is only the last term in the expression for the variance that is large and could destroy the precision of $\hat{\theta}_1$. Since expressions for the variances of the other $\hat{\theta}_j$'s are similar to (11.19), a requirement for small variance is equivalent to the requirement that the coefficient of $1/\lambda_5$ be small. Scanning the equations that define the transformation from $\{\hat{\alpha}_i\}$ to $\{\hat{\theta}_j\}$, we see that $\hat{\theta}_3$ is the most precise estimate since the coefficient of $1/\lambda_5$ in the variance expression for $\hat{\theta}_3$ is $(-0.01)^2 = 0.0001$.

Expanding this type of analysis, it may be possible to identify meaningful linear functions of the θ's that can be more accurately estimated than individual θ's. For example, we may be more interested in estimating $\theta_1 - \theta_2$ than θ_1 and θ_2 separately. In the sales model, $\theta_1 - \theta_2$ measures the increment to sales that corresponds to increasing the current year's advertising budget, X_1, by one unit and simultaneously reducing the current year's promotions budget, X_2, by one unit. In other words, $\theta_1 - \theta_2$ represents the effect of a shift in the use of resources in the current year. The estimate of $\theta_1 - \theta_2$ is $\hat{\theta}_1 - \hat{\theta}_2$. The variance of this estimate is obtained simply by subtracting the equation for $\hat{\theta}_2$ from that of $\hat{\theta}_1$ in (11.14) and using the resulting coefficients of the $\hat{\alpha}$'s as before. That is,

$$\hat{\theta}_1 - \hat{\theta}_2 = 0.764\hat{\alpha}_1 - 0.849\hat{\alpha}_2 - 0.826\hat{\alpha}_3 + 0.111\hat{\alpha}_4 - 0.025\hat{\alpha}_5,$$

from which we obtain the estimated variance of $(\hat{\theta}_1 - \hat{\theta}_2)$ as

$$(0.764)^2 \text{Var}(\hat{\alpha}_1) + (-0.849)^2 \text{Var}(\hat{\alpha}_2) + (-0.826)^2 \text{Var}(\hat{\alpha}_3)$$
$$+ (0.111)^2 \text{Var}(\hat{\alpha}_4) + (-0.025)^2 \text{Var}(\hat{\alpha}_1), \qquad (11.20)$$

or, equivalently, as

$$\hat{\sigma}^2 \left[\frac{(0.764)^2}{\lambda_1} + \frac{(-0.849)^2}{\lambda_2} + \frac{(-0.826)^2}{\lambda_3} + \frac{(0.111)^2}{\lambda_4} + \frac{(-0.025)^2}{\lambda_5} \right]. \tag{11.21}$$

The small coefficient of $1/\lambda_5$ makes it possible to estimate $\theta_1 - \theta_2$ accurately.

The estimate of $\theta_1 - \theta_2$ is $0.583 - 0.973 = -0.390$. The variance of $\hat{\theta}_1 - \hat{\theta}_2$ can be computed from (11.20) as 0.008. A 95% confidence interval for $\theta_1 - \theta_2$ is $-0.390 \pm 2.12\sqrt{0.008}$ or -0.58 to -0.20. That is, the effect of shifting one unit of expenditure from promotions to advertising in the current year is a loss of between 0.20 and 0.58 standardized sales unit.

There are other linear functions that may also be accurately estimated. Generalizing the above procedure, one can see that any linear function of the θ's that results in a small coefficient for $1/\lambda_5$ in the variance expression can be estimated with precision. Any function that produces a small coefficient for $1/\lambda_5$ in the variance expression is a possibility. For example, the equations in (11.14) suggest that all differences involving $\hat{\theta}_1, \hat{\theta}_2, \hat{\theta}_4$, and $\hat{\theta}_5$ can be considered. However, some of the differences are meaningful in the problem, whereas others are not. For example, the difference $(\theta_1 - \theta_2)$ is meaningful, as described previously. It represents a shift in current expenditures from promotions to advertising. The difference $\theta_1 - \theta_4$ is not particularly meaningful. It represents a shift from current advertising expenditure to a previous year's advertising expenditure. A shift of resources backward in time is impossible. Even though $\theta_1 - \theta_4$ could be accurately estimated, it is not of interest in the analysis of sales.

In general, when the weights in the equations in (11.14) are displayed and the corresponding values of the eigenvalues are known, it is always possible to scan the weights and identify those linear functions of the original regression coefficients that can be accurately estimated. Of those linear functions that can be accurately estimated, only some will be of interest for the problem being studied.

To summarize, where collinearity is indicated and it is not possible to supplement the data, it may still be possible to estimate some regression coefficients and some linear functions accurately. To investigate which coefficients and linear functions can be estimated, we recommend the analysis (transformation to principal components) that has just been described. This method of analysis will not overcome collinearity if it is present. There will still be regression coefficients and functions of regression coefficients that cannot be estimated. But the recommended analysis will indicate those functions that are estimable and indicate the structural dependencies that exist among the predictor variables.

11.6 BIASED ESTIMATION OF REGRESSION COEFFICIENTS

Two alternative estimation methods that provide a more informative analysis of the data than the OLS method when collinearity is present are considered. The estimators discussed here are biased but tend to have more precision (as measured by mean square error) than the OLS estimators [see Draper and Smith (1998), McCallum (1970), and Hoerl and Kennard (1970)]. These alternative methods do not reproduce the estimation data as well as the OLS method; the sum of squared residuals is not as small and, equivalently, the multiple correlation coefficient is not as large. However, the two alternatives have the potential to produce more precision in the estimated coefficients and smaller prediction errors when the predictions are generated using data other than those used for estimation.

Unfortunately, the criteria for deciding when these methods give better results than the OLS method depend on the true but unknown values of the model regression coefficients. That is, there is no completely objective way to decide when OLS should be replaced in favor of one of the alternatives. Nevertheless, when collinearity is suspected, the alternative methods of analysis are recommended. The resulting estimated regression coefficients may suggest a new interpretation of the data that, in turn, can lead to a better understanding of the process under study.

We consider here two specific alternatives to OLS: (a) principal components regression and (b) ridge regression. Principal components regression is introduced and discussed in Sections 11.7, 11.8, and 11.10. It will be demonstrated that the principal components estimation method can be interpreted in two ways; one interpretation relates to the nonorthogonality of the predictor variables and the other has to do with constraints on the regression coefficients (Section 11.9). Ridge regression also involves constraints on the coefficients. The ridge method is introduced and discussed in Sections 11.11, 11.12, and 11.13 and it is applied again in Chapter 12 to the problem of variable selection. Both methods, principal components and ridge regression, are examined using the French import data that were analyzed in Chapter 10.

11.7 PRINCIPAL COMPONENTS REGRESSION

The model under consideration is

$$\text{IMPORT} = \beta_0 + \beta_1 \text{DOPROD} + \beta_2 \text{STOCK} + \beta_3 \text{CONSUM} + \varepsilon. \quad (11.22)$$

The variables are defined in Section 10.3. The model in (11.22) stated in terms of standardized variables (see Section 4.6) is

$$\tilde{Y} = \theta_1 \tilde{X}_1 + \theta_2 \tilde{X}_2 + \theta_3 \tilde{X}_3 + \varepsilon', \quad (11.23)$$

where $\bar{y}$ and $\bar{x}_j$ are the means of Y and X_j, respectively; s_y and s_j are the standard deviations of Y and X_j, respectively; and $\tilde{Y} = (y_i - \bar{y})/s_y$ is the standardized version of the response variable and $\tilde{X}_j = (x_{ij} - \bar{x}_j)/s_j$ is the standardized version of the jth predictor variable. Many regression packages produce values for both the regular and standardized regression coefficients in (11.22) and (11.23). The estimated coefficients satisfy

$$\hat{\beta}_j = \frac{s_y}{s_j} \hat{\theta}_j, \qquad j = 1, 2, 3,$$
$$\hat{\beta}_0 = \bar{y} - \hat{\beta}_1 \bar{x}_1 - \hat{\beta}_2 \bar{x}_2 - \hat{\beta}_3 \bar{x}_3. \quad (11.24)$$

The principal components of the standardized predictor variables are [see (11.5)]

$$\begin{aligned} C_1 &= 0.706 \tilde{X}_1 + 0.044 \tilde{X}_2 + 0.707 \tilde{X}_3, \\ C_2 &= -0.036 \tilde{X}_1 + 0.999 \tilde{X}_2 - 0.026 \tilde{X}_3, \\ C_3 &= -0.707 \tilde{X}_1 - 0.007 \tilde{X}_2 + 0.707 \tilde{X}_3. \end{aligned} \quad (11.25)$$

These principal components are given in Table 11.1. The model in (11.23) may be written in terms of the principal components as

$$\tilde{Y} = \alpha_1 C_1 + \alpha_2 C_2 + \alpha_3 C_3 + \varepsilon'. \quad (11.26)$$

The equivalence of (11.23) and (11.26) follows since there is a unique relationship between the α's and θ's. In particular,

$$\begin{aligned} \alpha_1 &= 0.706 \theta_1 + 0.044 \theta_2 + 0.707 \theta_3, \\ \alpha_2 &= -0.036 \theta_1 + 0.999 \theta_2 - 0.026 \theta_3, \\ \alpha_3 &= -0.707 \theta_1 - 0.007 \theta_2 + 0.707 \theta_3. \end{aligned} \quad (11.27)$$

344 WORKING WITH COLLINEAR DATA

Table 11.7 Regression Results of Fitting Model (11.23) to the Import Data 1949–1959)

Variable	Coefficient	s.e.	t-Test	p-Value
$\tilde{X}_1$	−0.339	0.464	−0.73	0.4883
$\tilde{X}_2$	0.213	0.034	6.20	0.0004
$\tilde{X}_3$	1.303	0.464	2.81	0.0263
$n = 11$	$R^2 = 0.992$	$R_a^2 = 0.988$	$\hat{\sigma} = 0.1076$	df = 7

Conversely,

$$\theta_1 = 0.706\alpha_1 - 0.036\alpha_2 - 0.707\alpha_3,$$
$$\theta_2 = 0.044\alpha_1 + 0.999\alpha_2 - 0.007\alpha_3, \quad (11.28)$$
$$\theta_3 = 0.707\alpha_1 - 0.026\alpha_2 + 0.707\alpha_3.$$

These same relationships hold for the least squares estimates, the $\hat{\alpha}$'s and $\hat{\theta}$'s of the α's and θ's, respectively. Therefore, the $\hat{\alpha}$'s and $\hat{\theta}$'s may be obtained by the regression of $\tilde{Y}$ against the principal components C_1, C_2, and C_3, or against the original standardized variables. The regression results of fitting models (11.23) and (11.26) to the import data are shown in Tables 11.7 and 11.8. From Table 11.7, the estimates of θ_1, θ_2, and θ_3 are −0.339, 0.213, and 1.303, respectively. Similarly, from Table 11.8, the estimates of α_1, α_2, and α_3 are 0.690, 0.191, and 1.160, respectively. The results in one of these tables can be obtained from the other table using (11.27) and (11.28).

Although the equations in (11.23) and (11.26) are equivalent, the C's in (11.26) are orthogonal. Observe, however, that the regression relationship given in terms of the principal components in (11.25) is not easily interpreted. The predictor variables of that model are linear functions of the original predictor variables. The α's, unlike the θ's, do not have simple interpretations

Table 11.8 Regression Results of Fitting Model (11.26) to the Import Data (1949–1959)

Variable	Coefficient	s.e.	t-Test	p-Value
C_1	0.690	0.024	28.70	0.0000
C_2	0.191	0.034	5.62	0.0008
C_3	1.160	0.656	1.77	0.1204
$n = 11$	$R^2 = 0.992$	$R_a^2 = 0.988$	$\hat{\sigma} = 0.1076$	df = 7

as marginal effects of the original predictor variables. Therefore, we use principal components regression only as a means for analyzing the collinearity problem. The final estimation results are always restated in terms of the θ's for interpretation.

11.8 REDUCTION OF COLLINEARITY IN THE ESTIMATION DATA

It has been mentioned that the principal components regression has two interpretations. We shall first use the principal components technique to reduce collinearity in the estimation data. The reduction is accomplished by using less than the full set of principal components to explain the variation in the response variable. Note that when all three principal components are used, the OLS solution is reproduced exactly by applying (11.28).

The C's have sample variances $\lambda_1 = 1.999$, $\lambda_2 = 0.998$, and $\lambda_3 = 0.003$, respectively. Recall that the λ's are the eigenvalues of the correlation matrix of DOPROD, STOCK, and CONSUM. Since C_3 has variance equal to 0.003, the linear function defining C_3 is approximately equal to zero and is the source of collinearity in the data. We exclude C_3 and consider regressions of $\tilde{Y}$ against C_1 alone as well as against C_1 and C_2, that is, we consider the two possible regression models

$$\tilde{Y} = \alpha_1 C_1 + \varepsilon \tag{11.29}$$

and

$$\tilde{Y} = \alpha_1 C_1 + \alpha_2 C_2 + \varepsilon. \tag{11.30}$$

Both models lead to estimates for all three of the original coefficients, θ_1, θ_2, and θ_3. The estimates are biased since some information [C_3 in (11.30), C_2 and C_3 in (11.29)] has been excluded in both cases.

The estimated values of α_1 or α_1 and α_2 may be obtained by regressing $\tilde{Y}$ in turn against C_1 and then against C_1 and C_2. However, a simpler computational method is available that exploits the orthogonality of C_1 C_2 and C_3.[6] For example, the same estimated value of α_1 will be obtained from regression using (11.26), (11.29), or (11.30). Similarly, the value of α_2 may

---

[6] In any regression equation where the full set of potential predictor variables under consideration are orthogonal, the estimated values of regression coefficients are not altered when subsets of these variables are either introduced or deleted.

be obtained from (11.26) or (11.30). It also follows that if we have the OLS estimates of the θ's, estimates of the α's may be obtained from (11.27). Then principal components regression estimates of the θ's corresponding to (11.29) and (11.30) can be computed by referring back to (11.28) and setting the appropriate α's to zero. The following example clarifies the process.

Using $\alpha_1 = 0.690$ and $\alpha_2 = \alpha_3 = 0$ in (11.28) yields estimated θ's corresponding to regression on only the first principal component, that is,

$$\hat{\theta}_1 = 0.706 \times 0.690 = 0.487,$$
$$\hat{\theta}_2 = 0.044 \times 0.690 = 0.030, \quad (11.31)$$
$$\hat{\theta}_3 = 0.707 \times 0.690 = 0.487,$$

which yields

$$\tilde{Y} = 0.487\tilde{X}_1 + 0.030\tilde{X}_2 + 0.487\tilde{X}_3.$$

The estimates using the first two principal components, as in (11.30), are obtained in a similar fashion using $\alpha_1 = 0.690$, $\alpha_2 = 0.191$, and $\alpha_3 = 0$ in (11.28). The estimated of the regression coefficients, β_0, β_1, β_2, and β_3, of the original variables in (11.22), can be obtained by substituting $\hat{\theta}_1$, $\hat{\theta}_2$, and $\hat{\theta}_3$ in (11.24).

The estimates of the standardized and original regression coefficients using the three principal component models are shown in Table 11.9. It is evident that using different numbers of principal components gives substantially different results. It has already been argued that the OLS estimates are unsatisfactory. The negative coefficient of $\tilde{X}_1$ (DOPROD) is unexpected and cannot be sensibly interpreted. Furthermore, there is extensive collinearity which enters through the principal component, C_3. This variable has almost zero variance ($\lambda_3 = 0.003$) and is therefore approximately equal to zero. Of the two remaining principal components, it is fairly clear that the first one is associated with the combined effect of DOPROD and CONSUM. The second principal component is uniquely associated with STOCK. This conclusion is apparent in Table 11.9. The coefficients of DOPROD and CONSUM are completely determined from the regression of IMPORT on C_1 alone. These coefficients do not change when C_2 is used. The addition of C_2 causes the coefficient of STOCK to increase from 0.083 to 0.609. Also, R^2 increases from 0.952 to 0.988. Selecting the model based on the first two principal components, the resulting equation stated in original units is

$$\text{IMPORT} = -9.106 + 0.073\,\text{DOPROD}$$
$$+ 0.609\,\text{STOCK} + 0.106\,\text{CONSUM}. \quad (11.32)$$

CONSTRAINTS ON THE REGRESSION COEFFICIENTS 347

Table 11.9 Estimated Regression Coefficients for the Standardized and Original Variables Using Different Numbers of Principal Components for IMPORT Data (1949–1959)

	First PC Equation (11.29)		First and Second PCs Equation (11.30)		All PCs Equation (11.26)	
Variable	Stand.	Original	Stand.	Original	Stand.	Original
Constant	0	−7.735	0	−9.106	0	−10.130
DOPROD	0.487	0.074	0.480	0.073	−0.339	−0.051
STOCK	0.030	0.083	0.221	0.609	0.213	0.587
CONSUM	0.487	0.107	0.483	0.106	1.303	0.287
$\hat{\sigma}$		0.232		0.121		0.108
R^2		0.952		0.988		0.992

It provides a different and more plausible representation of the IMPORT relationship than was obtained from the OLS results. In addition, the analysis has led to an explicit quantification (in standardized variables) of the linear dependency in the predictor variables. We have $C_3 = 0$ or equivalently [from (11.25)]

$$-0.707\tilde{X}_1 - 0.007\tilde{X}_2 + 0.707\tilde{X}_3 \doteq 0.$$

The standardized values of DOPROD and CONSUM are essentially equal. This information can be useful qualitatively and quantitatively if (11.32) is used for forecasting or for analyzing policy decisions.

11.9 CONSTRAINTS ON THE REGRESSION COEFFICIENTS

There is a second interpretation of the results of the principal components regression equation. The interpretation is linked to the notion of imposing constraints on the θ's which was introduced in Chapter 10. The estimates in (11.30) were obtained by setting α_3 equal to zero in (11.28). From (11.27), $\alpha_3 = 0$ implies that

$$-0.707\theta_1 - 0.007\theta_2 + 0.707\theta_3 = 0 \qquad (11.33)$$

or $\theta_1 \doteq \theta_3$. In original units, (11.33) becomes

$$-6.60\beta_1 + 4.54\beta_3 = 0 \qquad (11.34)$$

or $\beta_1 = 0.69\beta_3$. Therefore, the estimates obtained by regression on C_1 and C_2 could have been obtained using OLS as in Chapter 10 with a linear constraint on the coefficients given by (11.34).

Recall that in Chapter 10 we conjectured that $\beta_1 = \beta_3$ as a prior constraint on the coefficients. It was argued that the constraint was the result of a qualitative judgment based on knowledge of the process under study. It was imposed without looking at the data. Now, using the data, we have found that principal components regression on C_1 and C_2 gives a result that is equivalent to imposing the constraint in (11.34). The result suggests that the marginal effect of domestic production on imports is about 69% of the marginal effect of domestic consumption on imports.

To summarize, the method of principal components regression provides both alternative estimates of the regression coefficients as well as other useful information about the underlying process that is generating the data. The structure of linear dependence among the predictor variables is made explicit. Principal components with small variances (eigenvalues) exhibit the linear relationships among the original variables that are the source of collinearity. Also elimination of collinearity by dropping one or more principal components from the regression is equivalent to imposing constraints on the regression coefficients. It provides a constructive way of identifying those constraints that are consistent with the proposed model and the information contained in the data.

11.10 PRINCIPAL COMPONENTS REGRESSION: A CAUTION

We have seen in Chapter 10 that principal components analysis is an effective tool for the detection of collinearity. In this chapter we have used the principal components as an alternative to the least squares method to obtain estimates of the regression coefficients in the presence of collinearity. The method has worked to our advantage in the Import data, where the first two of the three principal components have succeeded in capturing most of the variability in the response variable (see Table 11.9). This analysis is not guaranteed to work for all data sets. In fact, the principal components regression can fail in accounting for the variability in the response variable. To illustrate this point Hadi and Ling (1998) use a data set known as Hald's data and a constructed response variable U. The original data set can be found in Draper and Smith (1998, p. 348). The data can also be found in the file Hald.csv at the Book's

Table 11.10 Response Variable U and Set of Principal Components of Four Predictor Variables

U	C_1	C_2	C_3	C_4
0.955	1.467	1.903	−0.530	0.039
−0.746	2.136	0.238	−0.290	−0.030
−2.323	−1.130	0.184	−0.010	−0.094
−0.820	0.660	1.577	0.179	−0.033
0.471	−0.359	0.484	−0.740	0.019
−0.299	−0.967	0.170	0.086	−0.012
0.210	−0.931	−2.135	−0.173	0.008
0.558	2.232	−0.692	0.460	0.023
−1.119	0.352	−1.432	−0.032	−0.045
0.496	−1.663	1.828	0.851	0.020
0.781	1.641	−1.295	0.494	0.031
0.918	−1.693	−0.392	−0.020	0.037
0.918	−1.746	−0.438	−0.275	0.037

Website.[7] The data set has four predictor variables. The response variable U and the four PCs, $C_1, \ldots, C_4$, corresponding to the four predictor variables are given in Table 11.10. The variable U is already in a standardized form. The sample variances of the four PCs are $\lambda_1 = 2.2357$, $\lambda_2 = 1.5761$, $\lambda_3 = 0.1866$, and $\lambda_4 = 0.0016$. The condition number, $\kappa = \sqrt{\lambda_1/\lambda_4} = \sqrt{2.236/0.002} = 37$, is large, indicating the presence of collinearity in the original data.

The regression results obtained from fitting the model

$$U = \alpha_1 C_1 + \alpha_2 C_2 + \alpha_3 C_3 + \alpha_4 C_4 + \varepsilon \qquad (11.35)$$

to the data are shown in Table 11.11. The coefficient of the last PC, C_4, is highly significant and the other three coefficients are not significant. Now if we drop C_4, the PC with the smallest variance, we obtain the results in Table 11.12. As it is clear from a comparison of Tables 11.11 and 11.12, all four PCs capture almost all the variability in U, while the first three account for none of the variability in U. Therefore, one should be careful before dropping any of the PCs.

---

[7] http://www.aucegypt.edu/faculty/hadi/RABE6

WORKING WITH COLLINEAR DATA

Table 11.11 Regression Results Using All Four PCs of Hald's Data

Variable	Coefficient	s.e.	t-Test	p-Value
C_1	−0.002	0.001	−1.45	0.1842
C_2	−0.003	0.002	−1.77	0.1154
C_3	0.002	0.005	0.49	0.6409
C_4	24.761	0.049	502.12	0.0000
$n = 13$	$R^2 = 1.00$	$R_a^2 = 1.00$	$\hat{\sigma} = 0.0069$	df = 8

Table 11.12 Regression Results Using the First Three PCs of Hald's Data

Variable	Coefficient	s.e.	t-Test	p-Value
C_1	−0.001	0.223	−0.06	0.996
C_2	−0.000	0.266	−0.00	0.996
C_3	0.004	0.772	0.05	0.996
$n = 13$	$R^2 = 0.00$	$R_a^2 = -0.33$	$\hat{\sigma} = 1.155$	df = 9

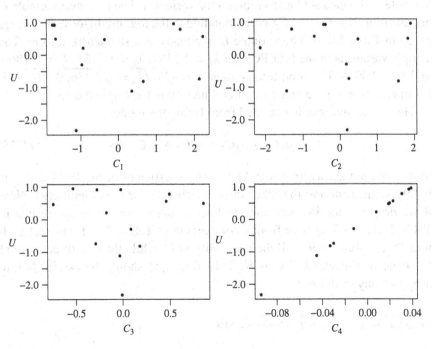

Figure 11.3 Scatter plots of U versus each of the PCs of Hald's data.

Another problem with the principal component regression is that the results can be unduly influenced by the presence of high-leverage points and outliers (see Chapter 5 for detailed discussion of outliers and influence). This is because the PCs are computed from the correlation matrix, which itself can be seriously affected by outliers in the data. A scatter plot of the response variable versus each of the PCs and the pairwise scatter plots of the PCs versus each other would point out outliers if they are present in the data. The scatter plot of U versus each of the PCs (Figure 11.3) show that there are no outliers in the data and U is related only to C_4, which is consistent with the results in Tables 11.11 and 11.12. The pairwise scatter plots of the PCs versus each other (not shown) also show no outliers in the data. For other possible pitfalls of principal components regression see Hadi and Ling (1998).

11.11 RIDGE REGRESSION

Ridge regression[8] provides another alternative estimation method that may be used to advantage when the predictor variables are highly collinear. There are a number of alternative ways to define and compute ridge estimates (see the Appendix to this chapter). We have chosen to present the method associated with the *ridge trace*. It is a graphical approach and may be viewed as an exploratory technique. Ridge analysis using the ridge trace represents a unified approach to problems of detection and estimation when collinearity is suspected. The estimators produced are biased but tend to have a smaller mean squared error than OLS estimators (Hoerl and Kennard, 1970).

Ridge estimates of the regression coefficients may be obtained by solving a slightly altered form of the normal equations (introduced in Chapter 4). Assume that the standardized form of the regression model is given as

$$\tilde{Y} = \theta_1 \tilde{X}_1 + \theta_2 \tilde{X}_2 + \ldots + \theta_p \tilde{X}_p + \varepsilon'. \quad (11.36)$$

The estimating equations for the ridge regression coefficients are

$$\begin{aligned}
(1+k)\theta_1 + r_{12}\theta_2 + \ldots + r_{1p}\theta_p &= r_{1y}, \\
r_{21}\theta_1 + (1+k)\theta_2 + \ldots + r_{2p}\theta_p &= r_{2y}, \\
\vdots \quad \vdots \quad \vdots \quad &\quad \vdots \\
r_{p1}\theta_1 + r_{p2}\theta_2 + \ldots + (1+k)\theta_p &= r_{py},
\end{aligned} \quad (11.37)$$

---

[8] Hoerl (1959) named the method *ridge regression* because of its similarity to ridge analysis used in his earlier work to study second-order response surfaces in many variables.

where r_{ij} is the correlation between the ith and jth predictor variables and r_{iy} is the correlation between the ith predictor variable and the response variable $\tilde{Y}$. The solution to (11.37), $\hat{\theta}_1, \ldots, \hat{\theta}_p$, is the set of estimated ridge regression coefficients. The ridge estimates may be viewed as resulting from a set of data that has been slightly altered. See the Appendix to this chapter for a formal treatment.

The essential parameter that distinguishes ridge regression from OLS is k. Note that when $k = 0$, the $\hat{\theta}$'s are the OLS estimates. The parameter k may be referred to as the bias parameter. As k increases from zero, bias of the estimates increases. On the other hand, the *total variance* (the sum of the variances of the estimated regression coefficients) is

$$\text{Total Variance}(k) = \sum_{j=1}^{p} \text{Var}(\hat{\theta}_j(k)) = \sigma^2 \sum_{j=1}^{p} \frac{\lambda_j}{(\lambda_j + k)^2}, \qquad (11.38)$$

which is a decreasing function of k. The formula in (11.38) shows the effect of the ridge parameter on the total variance of the ridge estimates of the regression coefficients. Substituting $k = 0$ in (11.38), we obtain

$$\text{Total Variance}(0) = \sigma^2 \sum_{j=1}^{p} \frac{1}{\lambda_j}, \qquad (11.39)$$

which shows the effect of small eigenvalue on the total variance of the OLS estimates of the regression coefficients.

As k continues to increase without bound, the regression estimates all tend toward zero.[9] The idea of ridge regression is to pick a value of k for which the reduction in total variance is not exceeded by the increase in bias.

It has been shown that there is a positive value of k for which the ridge estimates will be stable with respect to small changes in the estimation data (Hoerl and Kennard, 1970). In practice, a value of k is chosen by computing $\hat{\theta}_1, \ldots, \hat{\theta}_p$ for a range of k values between 0 and 1 and plotting the results against k. The resulting graph is known as the *ridge trace* and is used to select an appropriate value for k. Guidelines for choosing k are given in the following example.

---

[9] Because the ridge method tends to shrink the estimates of the regression coefficients toward zero (see, e.g., Vinod and Ullah, 1981), ridge estimators are sometimes generically referred to as *shrinkage estimators*.

11.12 ESTIMATION BY THE RIDGE METHOD

A method for detecting collinearity that comes out of ridge analysis deals with the instability in the estimated coefficients resulting from slight changes in the estimation data. The instability may be observed in the *ridge trace*. The ridge trace is a simultaneous graph of the regression coefficients, $\hat{\theta}_1, \ldots, \hat{\theta}_p$, plotted against k for various values of k such as 0.001, 0.002, and so on. Figure 11.4 is the ridge trace for the IMPORT data. The graph is constructed from Table 11.13, which has the ridge estimated coefficients for 29 values of k ranging from 0 to 1. Typically, the values of k are chosen to be concentrated near the low end of the range. If the estimated coefficients show large fluctuations for small values of k, instability has been demonstrated and collinearity is probably at work.

What is evident from the trace or equivalently from Table 11.13 is that the estimated values of the coefficients θ_1 and θ_3 are quite unstable for small values of k. The estimate of θ_1 changes rapidly from an implausible negative value of -0.339 to a stable value of about 0.43. The estimate of θ_3 goes from 1.303 to stabilize at about 0.50. The coefficient of $\tilde{X}_2$ (STOCK), θ_2 is unaffected by the collinearity and remains stable throughout at about 0.21.

The next step in the ridge analysis is to select a value of k and to obtain the corresponding estimates of the regression coefficients. If collinearity is a serious problem, the ridge estimators will vary dramatically as k is slowly increased from zero. As k increases, the coefficients will eventually stabilize. Since k is a bias parameter, it is desirable to select the smallest value of k for which stability occurs since the size of k is directly related to the amount

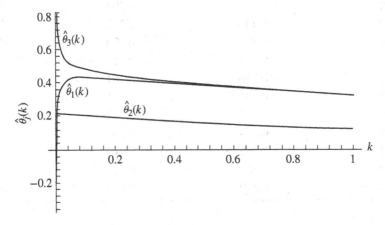

Figure 11.4 Ridge trace: IMPORT data (1949–1959).

Table 11.13 Ridge Estimates $\hat{\theta}_j(k)$, as Functions of the Ridge Parameter k, for the IMPORT Data (1949–1959)

k	$\hat{\theta}_1(k)$	$\hat{\theta}_2(k)$	$\hat{\theta}_3(k)$
0.000	−0.339	0.213	1.303
0.001	−0.117	0.215	1.080
0.003	0.092	0.217	0.870
0.005	0.192	0.217	0.768
0.007	0.251	0.217	0.709
0.009	0.290	0.217	0.669
0.010	0.304	0.217	0.654
0.012	0.328	0.217	0.630
0.014	0.345	0.217	0.611
0.016	0.359	0.217	0.597
0.018	0.370	0.216	0.585
0.020	0.379	0.216	0.575
0.022	0.386	0.216	0.567
0.024	0.392	0.215	0.560
0.026	0.398	0.215	0.553
0.028	0.402	0.215	0.548
0.030	0.406	0.214	0.543
0.040	0.420	0.213	0.525
0.050	0.427	0.211	0.513
0.060	0.432	0.209	0.504
0.070	0.434	0.207	0.497
0.080	0.436	0.206	0.491
0.090	0.436	0.204	0.486
0.100	0.436	0.202	0.481
0.200	0.426	0.186	0.450
0.300	0.411	0.173	0.427
0.400	0.396	0.161	0.408
0.500	0.381	0.151	0.391
0.600	0.367	0.142	0.376
0.700	0.354	0.135	0.361
0.800	0.342	0.128	0.348
0.900	0.330	0.121	0.336
1.000	0.319	0.115	0.325

of bias introduced. Several methods have been suggested for the choice of k. These methods include:

1. *Fixed Point.* Hoerl et al. (1975) suggest estimating k by

$$k = \frac{p\hat{\sigma}^2(0)}{\sum_{j=1}^{p}[\hat{\theta}_j(0)]^2}, \qquad (11.40)$$

where $\hat{\theta}_1(0), \ldots, \hat{\theta}_p(0)$ are the least squares estimates of $\theta_1, \ldots, \theta_p$ when the model in (11.36) is fitted to the data (i.e., when $k = 0$), and $\hat{\sigma}^2(0)$ is the corresponding residual mean square.

2. *Iterative Method.* Hoerl and Kennard (1976) propose the following iterative procedure for selecting k: Start with the initial estimate of k in (11.40). Denote this value by k_0. Then, calculate

$$k_1 = \frac{p\hat{\sigma}^2(0)}{\sum_{j=1}^{p}[\hat{\theta}_j(k_0)]^2}. \qquad (11.41)$$

Then use k_1 to calculate k_2 as

$$k_2 = \frac{p\hat{\sigma}^2(0)}{\sum_{j=1}^{p}[\hat{\theta}_j(k_1)]^2}. \qquad (11.42)$$

Repeat this process until the difference between two successive estimates of k is negligible.

3. *Ridge Trace.* The behavior of $\hat{\theta}_j(k)$ as a function of k is easily observed from the ridge trace. The value of k selected is the smallest value for which all the coefficients $\hat{\theta}_j(k)$ are stable. In addition, at the selected value of k, the residual sum of squares should remain close to its minimum value. The variance inflation factors,[10] $\text{VIF}_j(k)$, should also get down to less than 10. (Recall that a value of 1 is a characteristic of an orthogonal system and a value less than 10 would indicate a noncollinear or stable system.)

4. *Other Methods.* Many other methods for estimating k have been suggested in the literature. See, for example, Marquardt (1970),

---

[10] The formula for $\text{VIF}_j(k)$ is given in the Appendix to this chapter.

Mallows (1973), Goldstein and Smith (1974), McDonald and Galarneau (1975), Lawless and Wang (1976), Dempster et al. (1977), Wahba et al. (1979), Hoerl and Kennard (1981), Masuo (1988), Khalaf and Shukur (2005), and Dorugade and Kashid (2010). The appeal of the ridge trace, however, lies in its graphical representation of the effects that collinearity has on the estimated coefficients.

For the IMPORT data, the fixed point formula in (11.40) gives

$$k = \frac{3 \times 0.0012}{(-0.339)^2 + (0.213)^2 + (1.303)^2} = 0.0019. \tag{11.43}$$

The iterative method gives the following sequence: $k_0 = 0.0019, k_1 = 0.0035, k_2 = 0.0046, k_3 = 0.0050$, and $k_4 = 0.0051$. So, it converges after three iterations to $k = 0.005$. The ridge trace in Figure 11.4 (see also Table 11.13) appears to stabilize for k around 0.04. We therefore have three estimates of k (0.0019, 0.005, and 0.04).

From Table 11.13, we see that at any of these values the improper negative sign on the estimate of θ_1 has disappeared and the coefficient has stabilized (at 0.359 for $k = 0.016$ and at 0.42 for $k = 0.04$). From Table 11.14, we see that the sum of squared residuals, SSE(k), has only increased from 0.081 at $k = 0$ to 0.108 at $k = 0.016$, and to 0.117 at $k = 0.04$. Also, the variance inflation factors, VIF$_1(k)$ and VIF$_3(k)$, decreased from about 185 to values between 1 and 4. It is clear that values of k in the interval (0.016, 0.04) appear to be satisfactory.

The estimated coefficients from the model stated in standardized and original variables units are summarized in Table 11.15. The original coefficient $\hat{\beta}_j$ is obtained from the standardized coefficient $\hat{\theta}_j$ using (11.24). For example, $\hat{\beta}_1$ is calculated by

$$\hat{\beta}_{1j} = (s_y/s_1)\hat{\theta}_1 = (4.5437/29.9995)(0.4196) = 0.0635.$$

Thus, the resulting model in terms for the original variables fitted by the Ridge method using $k = 0.04$ is

$$\text{IMPORT} = -8.5537 + 0.0635 \text{DOPROD}$$
$$+ 0.5859 \text{STOCK} + 0.1156 \text{CONSUM}.$$

The equation gives a plausible representation of the relationship. Note that the final equation for these data is not particularly different from the result obtained by using the first two principal components (see Table 11.9), although the two computational methods appear to be very different.

Table 11.14 Residual Sum of Squares, SSE(k), and Variance Inflation Factors, VIF$_j$(k), as Functions of the Ridge Parameter k, for the IMPORT Data (1949–1959)

k	SSE(k)	VIF$_1$(k)	VIF$_2$(k)	VIF$_3$(k)
0.000	0.0810	186.11	1.02	186.00
0.001	0.0837	99.04	1.01	98.98
0.003	0.0911	41.80	1.00	41.78
0.005	0.0964	23.00	0.99	22.99
0.007	0.1001	14.58	0.99	14.57
0.009	0.1027	10.09	0.98	10.09
0.010	0.1038	8.60	0.98	8.60
0.012	0.1056	6.48	0.98	6.48
0.014	0.1070	5.08	0.97	5.08
0.016	0.1082	4.10	0.97	4.10
0.018	0.1093	3.39	0.97	3.39
0.020	0.1102	2.86	0.96	2.86
0.022	0.1111	2.45	0.96	2.45
0.024	0.1118	2.13	0.95	2.13
0.026	0.1126	1.88	0.95	1.88
0.028	0.1132	1.67	0.95	1.67
0.030	0.1139	1.50	0.94	1.50
0.040	0.1170	0.98	0.93	0.98
0.050	0.1201	0.72	0.91	0.72
0.060	0.1234	0.58	0.89	0.58
0.070	0.1271	0.49	0.87	0.49
0.080	0.1310	0.43	0.86	0.43
0.090	0.1353	0.39	0.84	0.39
0.100	0.1400	0.35	0.83	0.35
0.200	0.2052	0.24	0.69	0.24
0.300	0.2981	0.20	0.59	0.20
0.400	0.4112	0.18	0.51	0.18
0.500	0.5385	0.17	0.44	0.17
0.600	0.6756	0.15	0.39	0.15
0.700	0.8191	0.14	0.35	0.14
0.800	0.9667	0.13	0.31	0.13
0.900	1.1163	0.12	0.28	0.12
1.000	1.2666	0.11	0.25	0.11

Table 11.15 OLS and Ridge Estimates of the Regression Coefficients for IMPORT Data (1949–1959)

	OLS ($k = 0$)		Ridge ($k = 0.04$)	
Variable	Standardized Coefficients	Original Coefficients	Standardized Coefficients	Original Coefficients
Constant	0	−10.1280	0	−8.5537
DOPROD	−0.3393	−0.0514	0.4196	0.0635
STOCK	0.2130	0.5869	0.2127	0.5859
CONSUM	1.3027	0.2868	0.5249	0.1156
	$R^2 = 0.992$		$R^2 = 0.988$	

11.13 RIDGE REGRESSION: SOME REMARKS

Ridge regression provides a tool for judging the stability of a given body of data for analysis by least squares. In highly collinear situations, as has been pointed out, small changes (perturbations) in the data cause very large changes in the estimated regression coefficients. Ridge regression will reveal this condition. Least squares regression should be used with caution in these situations. Ridge regression provides estimates that are more robust than least squares estimates for small perturbations in the data. The method will indicate the sensitivity (or the stability) of the least squares coefficients to small changes in the data.

The ridge estimators are stable in the sense that they are not affected by slight variations in the estimation data. Because of the smaller mean square error property, values of the ridge estimated coefficients are expected to be closer than the OLS estimates to the true values of the regression coefficients. Also, forecasts of the response variable corresponding to values of the predictor variables not included in the estimation set tend to be more accurate.

The estimation of the bias parameter k is rather subjective. There are many methods for estimating k but there is no consensus as to which method is preferable. Regardless of the method of choice for estimating the ridge parameter k, the estimated parameter can be affected by the presence of outliers in the data. Therefore a careful checking for outliers should accompany any method for estimating k to ensure that the obtained estimate is not unduly influenced by outliers in the data.

As with the principal components method, the criteria for deciding when the ridge estimators are superior to the OLS estimators depend on the values of the true regression coefficients in the model. Although these values cannot be known, we still suggest that ridge analysis is useful in cases where extreme collinearity is suspected. The ridge coefficients can suggest an alternative interpretation of the data that may lead to a better understanding of the process under study.

Another practical problem with ridge regression is that it has not been implemented in some statistical packages. If a statistical package does not have a routine for ridge regression, ridge regression estimates can be obtained from the standard least squares package by using a slightly altered data set. Specifically, the ridge estimates of the regression coefficients can be obtained from the regression of Y^* on $X_1^*, \ldots, X_p^*$. The new response variable Y^* is obtained by augmenting $\tilde{Y}$ by p new fictitious observations, each of which is equal to zero. Similarly, the new predictor variable X_j^* is obtained by augmenting $\tilde{X}_j$ by p new fictitious observations, each of which is equal to zero except the one in the jth position which is equal to $\sqrt{k}$, where k is the chosen value of the ridge parameter. It can be shown that the ridge estimates $\hat{\theta}_1(k), \ldots, \hat{\theta}_p(k)$ are obtained by the least squares regression of Y^* on $X_1^*, \ldots, X_p^*$ without having a constant term in the model.

11.14 SUMMARY

Both alternative estimation methods, ridge regression and principal components regression, provide additional information about the data being analyzed. We have seen that the eigenvalues of the correlation matrix of predictor variables play an important role in detecting collinearity and in analyzing its effects. The regression estimates produced by these methods are biased but may be more accurate than OLS estimates in terms of mean square error. It is impossible to evaluate the gain in accuracy for a specific problem since a comparison of the two methods to OLS requires knowledge of the true values of the coefficients. Nevertheless, when severe collinearity is suspected, we recommend that at least one set of estimates in addition to the OLS estimates be calculated. The estimates may suggest an interpretation of the data that were not previously considered.

There is no strong theoretical justification for using principal components or ridge regression methods. We recommend that the methods be used in the presence of severe collinearity as a visual diagnostic tool for judging the suitability of the data for least squares analysis. When principal components or ridge regression analysis reveal the instability of a particular data set, the analyst should first consider using least squares regression on a reduced set of variables (as indicated in Chapter 10). If least squares regression is still unsatisfactory (high VIFs, coefficients with wrong signs, large condition number), only then should principal components or ridge regression be used.

11.15 BIBLIOGRAPHIC NOTES

The principal components techniques used in this chapter are derived in most books on multivariate statistical analysis. It should be noted that principal components analysis involves only the predictor variables. The analysis is aimed at characterizing and identifying dependencies (if they exist) among the predictor variables. For a comprehensive discussion of principal components, the reader is referred to Johnson and Wichern (1992) or Seber (1984). Several statistical software packages are now commercially available to carry out the analysis described in this chapter.

More recently, Jensen and Ramirez (2008) cast some doubt about the ability of ridge estimators to actually improve the condition of an ill-conditioned linear system and provide stable estimated regression coefficients and smaller variance inflation factors. They introduce *Surrogate Ridge Regression* as an improvement of Ridge Regression estimators. A brief description of the Surrogate Ridge Estimators is given in Hadi (2011) and in the Appendix to this chapter.

EXERCISES

11.1 Use the Advertising data in the file Advertising.csv at the Book's Website:

(a) Verify that the estimated coefficients and their standard errors, which are obtained using (11.14), are the same as those given in Table 11.3.

(b) Compute the five PCs in (11.12).

(c) Verify the regression results in Table 11.4.

11.2 Suppose we fit the model

$$Y = \beta_0 + \beta_1 X_1 + \beta_2 X_2 + \beta_3 X_3 + \varepsilon, \qquad (11.44)$$

to a set of data, where each of the three variables has a mean of 0 and a variance of 1. The three eigenvalues of the correlation matrix of the three predictor variables are 1.93, 1.06, and 0.01. The corresponding

Table 11.16 Three Eigenvectors of the Correlation Matrix of the Three Predictors in Model (11.44)

	V_1	V_2	V_3
X_1	0.500	−0.697	0.514
X_2	0.484	0.717	0.501
X_3	0.718	0.002	−0.696

Table 11.17 Regression Output from the Regression of Y on the Principal Components C_1, C_2, and C_3

ANOVA Table

Source	Sum of Squares	df	Mean Square	F-Test
Regression	86.6542	3	28.8847	225
Residual	12.3458	96	0.128602	

Coefficients Table

Variable	Coefficient	s.e.	t-Test	p-Value
C_1	0.67	0.03	25.9	0.0001
C_2	−0.02	0.03	−0.56	0.5782
C_3	−0.56	0.37	−1.53	0.1291

eigenvectors are given in Table 11.16. Table 11.17 shows a computer output obtained when regressing Y on the principal components C_1, C_2, and C_3.

(a) Compute the least squares estimate of β_0 when fitting the model in (11.44) to the data.

(b) Is there evidence of collinearity in the predictor variables? Explain.

(c) What is R^2 when Y is regressed on X_1, X_2, and X_3?

(d) What is the formula used to obtain C_1?

(e) Derive the principal components predicted equation $\hat{Y}_{PC}$ of the model in (11.44).

11.3 Refer to the eigenvectors given in Table 10.11.

(a) How many principal components can one construct for this data set?

(b) Write the formulas for two of these principal components.

11.4 Longley's (1967) data set is a classic example of collinear data. The data consist of a response variable Y and six predictor variables $X_1, \ldots, X_6$. The data can be found in the file longley.csv at the Book's Website. The initial model

$$Y = \beta_0 + \beta_1 X_1 + \ldots + \beta_6 X_6 + \varepsilon, \tag{11.45}$$

in terms of the original variables, can be written in terms of the standardized variables as

$$\tilde{Y} = \theta_1 \tilde{X}_1 + \ldots + \theta_6 \tilde{X}_6 + \varepsilon'. \tag{11.46}$$

(a) Fit the model (11.46) to the data using least squares. What conclusion can you draw from the data?

(b) From the results you obtained from the model in (11.46), obtain the least squares estimated regression coefficients in model (11.45).

(c) Now fit the model in (11.45) to the data using least squares and verify that the obtained results are consistent with those obtained above.

(d) Compute the correlation matrix of the six predictor variables and the corresponding scatter plot matrix. Do you see any evidence of collinearity?

(e) Compute the corresponding PCs, their sample variances, and the condition number. How many different sets of collinearity exist in the data? What are the variables involved in each set?

(f) Based on the number of PCs you choose to retain, obtain the PC estimates of the coefficients in (11.45) and (11.46).

(g) Using the ridge method, construct the ridge trace. What value of k do you recommend to be used in the estimation of the parameters in (11.45) and (11.46)? Use the chosen value of k and compute the ridge estimates of the regression coefficients in (11.45) and (11.46).

(h) Compare the estimates you obtained by the three methods. Which one would you recommend? Explain.

11.5 Repeat Exercise 11.4 using Hald's data discussed in Section 11.10 but using the original response variable Y and the four predictors $X_1, \ldots, X_4$. The data can be found in the file Hald.csv at the Book's Website.

11.6 From your analysis of the Longley and Hald data sets, do you observe the sort of problems pointed out in Section 11.10? Explain.

11.7 Download the data set in the file Collinear.data.csv at the Book's Website. The data which consist of a dependent variable Y and six predictor variables, $X_1, X_2, \ldots, X_6$. Analyze the data set for collinearity. In particular.

(a) Compute the condition number for the X-variables. Is there any evidence of collinearity?

(b) Compute all principal components (PCs) and regress Y on all PCs. Which PCs are significant?

(c) Construct the scatter plot of the first two PCs. What would the slope of the least squares regression line describing the relationship between these two PCs be? Why?

(d) How many sets of collinearity exist in the data?

(e) Which of the variables are involved in each set?

(f) What is the relationship among the variables in each set of collinear variables?

(g) How many principal components would you use to deal with collinearity in this case?

(h) Which model would you recommend to describe the relationship between Y and the other predictors variables?

Appendix: 11.A Principal Components

In this appendix we present the principal components approach to the detection of collinearity using matrix notation.

11.A.1 The Model

The regression model can be expressed as

$$\mathbf{Y} = \mathbf{Z}\boldsymbol{\theta} + \boldsymbol{\varepsilon}, \tag{11.A.1}$$

where $\mathbf{Y}$ is an $n \times 1$ vector of observations on the response variable,

$$\mathbf{Z} = (\mathbf{Z}_1, \ldots, \mathbf{Z}_p)$$

is an $n \times p$ matrix of n observations on p predictor variables, $\boldsymbol{\theta}$ is a $p \times 1$ vector of regression coefficients, and $\boldsymbol{\varepsilon}$ is an $n \times 1$ vector of random errors.

It is assumed that $E(\varepsilon) = \mathbf{0}$, $E(\varepsilon\varepsilon^T) = \sigma^2\mathbf{I}$, where $\mathbf{I}$ is the identity matrix of order n. It is also assumed, without loss of generality, that $\mathbf{Y}$ and $\mathbf{Z}$ have been centered and scaled so that $\mathbf{Z}^T\mathbf{Z}$ and $\mathbf{Z}^T\mathbf{Y}$ are matrices of correlation coefficients.[11]

There exist square matrices, Λ and $\mathbf{V}$ satisfying[12]

$$\mathbf{V}^T(\mathbf{Z}^T\mathbf{Z})\mathbf{V} = \Lambda \quad \text{and} \quad \mathbf{V}^T\mathbf{V} = \mathbf{V}\mathbf{V}^T = \mathbf{I}. \tag{11.A.2}$$

The matrix Λ is diagonal with the ordered eigenvalues of $\mathbf{Z}^T\mathbf{Z}$ on the diagonal. These eigenvalues are denoted by $\lambda_1 \geq \lambda_2 \geq \ldots \geq \lambda_p$. The columns of $\mathbf{V}$ are the normalized eigenvectors corresponding to $\lambda_1, \ldots, \lambda_p$. Since $\mathbf{V}\mathbf{V}^T = \mathbf{I}$, the regression model in (11.A.1) can be restated in terms of the PCs as

$$\mathbf{Y} = \mathbf{Z}\mathbf{V}\mathbf{V}^T\theta + \varepsilon = \mathbf{C}\alpha + \varepsilon, \tag{11.A.3}$$

where

$$\mathbf{C} = \mathbf{Z}\mathbf{V} \quad \text{and} \quad \alpha = \mathbf{V}^T\theta. \tag{11.A.4}$$

The matrix $\mathbf{C}$ contains p columns $\mathbf{C}_1, \ldots, \mathbf{C}_p$, each of which is a linear function of the predictor variables $\mathbf{Z}_1, \ldots, \mathbf{Z}_p$. The columns of $\mathbf{C}$ are orthogonal and are referred to as principal components (PCs) of the predictor variables $\mathbf{Z}_1, \ldots, \mathbf{Z}_p$. The columns of $\mathbf{C}$ satisfy $\mathbf{C}_j^T\mathbf{C}_j = \lambda_j$ and $\mathbf{C}_i^T\mathbf{C}_j = 0$ for $i \neq j$.

The PCs and the eigenvalues may be used to detect and analyze collinearity in the predictor variables. The restatement of the regression model given in (11.A.3) is a reparameterization of (11.A.1) in terms of orthogonal predictor variables. The λ's may be viewed as sample variances of the PCs. If $\lambda_i = 0$, all observations on the ith PC are also zero. Since the jth PC is a linear function of $\mathbf{Z}_1, \ldots, \mathbf{Z}_p$, when $\lambda_j = 0$ an exact linear dependence exists among the predictor variables. It follows that when λ_j is small (approximately equal to zero) there is an approximate linear relationship among the predictor variables. That is, a small eigenvalue is an indicator of collinearity. In addition, from (11.A.4) we have

$$\mathbf{C}_j = \sum_{i=1}^{p} v_{ij}\mathbf{Z}_i,$$

---

[11] Note that Z_j is obtained by transforming the original predictor variable X_j by

$$z_{ij} = \frac{x_{ij} - \bar{x}_j}{\sqrt{\sum(x_{ij} - \bar{x}_j)^2}}.$$

Thus, Z_j is centered and scaled to have unit length, that is, $\sum z_{ij}^2 = 1$.

[12] See, for example, Strang (1988) or Hadi (1996).

which identifies the exact form of the linear relationship that causes the collinearity.

11.A.2 Precision of Linear Functions of $\hat{\theta}$

Denoting $\hat{\alpha}$ and $\hat{\theta}$ as the least squares estimators for α and θ, respectively, it can be shown that $\hat{\alpha} = \mathbf{V}^T\hat{\theta}$, and conversely, $\hat{\theta} = \mathbf{V}\hat{\alpha}$. With $\hat{\alpha} = (\mathbf{C}^T\mathbf{C})^{-1}\mathbf{C}^T\mathbf{Y}$, it follows that the variance–covariance matrix of $\hat{\alpha}$ is $\mathbf{V}(\hat{\alpha}) = \sigma^2\Lambda^{-1}$, and the corresponding matrix for $\hat{\theta}$ is $\mathbf{V}(\hat{\theta}) = \sigma^2\mathbf{V}\Lambda^{-1}\mathbf{V}^T$. Let $\mathbf{L}$ be an arbitrary $p \times 1$ vector of constants. The linear function $\delta = \mathbf{L}^T\theta$ has least squares estimator $\hat{\delta} = \mathbf{L}^T\hat{\theta}$ and variance

$$\text{Var}(\hat{\delta}) = \sigma^2 \mathbf{L}^T \mathbf{V} \Lambda^{-1} \mathbf{V}^T \mathbf{L}. \tag{11.A.5}$$

Let $\mathbf{V}_j$ be the jth column of $\mathbf{V}$. Then $\mathbf{L}$ can be represented as $\mathbf{L} = \sum_{j=1}^{p} r_j \mathbf{V}_j$, for appropriately chosen constants $r_1, \ldots, r_p$. Then (11.A.5) becomes $\text{Var}(\hat{\delta}) = \sigma^2 \mathbf{R}^T \Lambda^{-1} \mathbf{R}$ or, equivalently,

$$\text{Var}(\hat{\delta}) = \sigma^2 \sum_{j=1}^{p} \frac{r_j^2}{\lambda_j}, \tag{11.A.6}$$

where Λ^{-1} is the inverse of Λ.

To summarize, the variance of $\hat{\delta}$ is a linear combination of the reciprocals of the eigenvalues. It follows that $\hat{\delta}$ will have good precision either if none of the eigenvalues are near zero or if r_j^2 is at most the same magnitude as λ_j when λ_j is small. Furthermore, it is always possible to select a vector, $\mathbf{L}$, and thereby a linear function of $\hat{\theta}$, so that the effect of one or a few small eigenvalues is eliminated and $\mathbf{L}^T\hat{\theta}$ has a small variance. Refer to Silvey (1969) for a more complete development of these concepts.

Appendix: 11.B Ridge Regression

In this appendix we present the Ridge regression method in matrix notation.

11.B.1 The Model

The regression model can be expressed as in (11.A.1). The least squares estimator for θ is $\hat{\theta} = (\mathbf{Z}^T\mathbf{Z})^{-1}\mathbf{Z}^T\mathbf{Y}$. It can be shown that

$$E[(\hat{\theta} - \theta)^T(\hat{\theta} - \theta)] = \sigma^2 \sum_{j=1}^{p} \lambda_j^{-1}, \tag{11.B.1}$$

where $\lambda_1 \geq \lambda_2 \geq \ldots \geq \lambda_p$ are the eigenvalues of $\mathbf{Z}^T\mathbf{Z}$. The left-hand side of (11.B.1) is called the *total mean square error*. It serves as a composite measure of the squared distance of the estimated regression coefficients from their true values.

11.B.2 Effect of Collinearity

It was argued in Chapter 10 and in Appendix 10.A that collinearity is synonymous with small eigenvalues. It follows from (11.B.1) that when one or more of the λ's are small, the total mean square error of $\hat{\theta}$ is large, suggesting imprecision in the least squares estimation method. The ridge regression approach is an attempt to construct an alternative estimator that has a smaller total mean square error value.

11.B.3 Ridge Regression Estimators

Hoerl and Kennard (1970) suggest a class of estimators indexed by a parameter $k > 0$. The estimator is (for a given value of k)

$$\hat{\theta}(k) = (\mathbf{Z}^T\mathbf{Z} + k\mathbf{I})^{-1}\mathbf{Z}^T\mathbf{Y} = (\mathbf{Z}^T\mathbf{Z} + k\mathbf{I})^{-1}\mathbf{Z}^T\mathbf{Z}\hat{\theta}. \tag{11.B.2}$$

The expected value of $\hat{\theta}(k)$ is

$$E[\hat{\theta}(k)] = (\mathbf{Z}^T\mathbf{Z} + k\mathbf{I})^{-1}\mathbf{Z}^T\mathbf{Z}\theta \tag{11.B.3}$$

and the variance–covariance matrix is

$$\mathrm{Var}[\hat{\theta}(k)] = (\mathbf{Z}^T\mathbf{Z} + k\mathbf{I})^{-1}\mathbf{Z}^T\mathbf{Z}(\mathbf{Z}^T\mathbf{Z} + k\mathbf{I})^{-1}\sigma^2. \tag{11.B.4}$$

The variance inflation factor, $\mathrm{VIF}_j(k)$, as a function of k is the jth diagonal element of the matrix $(\mathbf{Z}^T\mathbf{Z} + k\mathbf{I})^{-1}\mathbf{Z}^T\mathbf{Z}(\mathbf{Z}^T\mathbf{Z} + k\mathbf{I})^{-1}$.

The residual sum of squares can be written as

$$\begin{aligned}\mathrm{SSE}(k) &= (\mathbf{Y} - \mathbf{Z}\hat{\theta}(k))^T(\mathbf{Y} - \mathbf{Z}\hat{\theta}(k)) \\ &= (\mathbf{Y} - \mathbf{Z}\hat{\theta})^T(\mathbf{Y} - \mathbf{Z}\hat{\theta}) + (\hat{\theta}(k) - \hat{\theta})^T\mathbf{Z}^T\mathbf{Z}(\hat{\theta}(k) - \hat{\theta}).\end{aligned} \tag{11.B.5}$$

The total mean square error is

$$\begin{aligned}\mathrm{TMSE}(k) &= E[(\hat{\theta}(k) - \theta)^T(\hat{\theta}(k) - \theta)] \\ &= \sigma^2\,\mathrm{trace}[(\mathbf{Z}^T\mathbf{Z} + k\mathbf{I})^{-1}\mathbf{Z}^T\mathbf{Z}(\mathbf{Z}^T\mathbf{Z} + k\mathbf{I})^{-1}]\end{aligned}$$

$$+ k^2 \theta^T (Z^T Z + kI)^{-2} \theta$$

$$= \sigma^2 \sum_{j=1}^{p} \lambda_j (\lambda_j + k)^{-2} + k^2 \theta^T (Z^T Z + kI)^{-2} \theta. \quad (11.B.6)$$

Note that the first term on the right-hand side of (11.B.6) is the sum of the variances of the components of $\hat{\theta}(k)$ (total variance) and the second term is the square of the bias. Hoerl and Kennard (1970) prove that there exists a value of $k > 0$ such that

$$E[(\hat{\theta}(k) - \theta)^T (\hat{\theta}(k) - \theta)] < E[(\hat{\theta} - \theta)^T (\hat{\theta} - \theta)],$$

that is, the mean square error of the ridge estimator, $\hat{\theta}(k)$, is less than the mean square error of the OLS estimator, $\hat{\theta}$. Hoerl and Kennard (1970) suggest that an appropriate value of k may be selected by observing the ridge trace and some complementary summary statistics for $\hat{\theta}(k)$ such as $SSE(k)$ and $VIF_j(k)$. The value of k selected is the smallest value for which $\hat{\theta}(k)$ is stable. In addition, at the selected value of k, the residual sum of squares should remain close to its minimum value, and the variance inflation factors are less than 10, as discussed in Chapter 10.

Ridge estimators have been generalized in several ways. They are sometimes generically referred to as *shrinkage estimators*, because these procedures tend to shrink the estimates of the regression coefficients toward zero. To see one possible generalization, consider the regression model restated in terms of the principal components, $C = (C_1, \ldots, C_p)$, discussed in the Appendix to this chapter. The general model takes the form

$$Y = C\alpha + \varepsilon, \quad (11.B.7)$$

where

$$C = ZV, \quad \alpha = V^T \theta, \quad (11.B.8)$$

$$V^T Z^T Z V = \Lambda, \quad V^T V = VV^T = I,$$

and

$$\Lambda = \begin{pmatrix} \lambda_1 & 0 & 0 & \cdots & 0 & 0 \\ 0 & \lambda_2 & 0 & \cdots & 0 & 0 \\ \vdots & \vdots & \vdots & \ddots & \vdots & \vdots \\ 0 & 0 & 0 & \cdots & \lambda_{p-1} & 0 \\ 0 & 0 & 0 & \cdots & 0 & \lambda_p \end{pmatrix}, \quad \lambda_1 \geq \lambda_2 \geq \cdots \geq \lambda_p,$$

WORKING WITH COLLINEAR DATA

is a diagonal matrix consisting of the ordered eigenvalues of $Z^T Z$. The total mean square error in (11.B.6) becomes

$$\text{TMSE}(k) = E[(\hat{\theta}(k) - \theta)^T(\hat{\theta}(k) - \theta)]$$

$$= \sigma^2 \sum_{j=1}^{p} \frac{\lambda_j}{(\lambda_j + k)^2} + \sum_{j=1}^{p} \frac{k^2 \alpha_j^2}{(\lambda_j + k)^2}, \quad (11.\text{B}.9)$$

where $\alpha^T = (\alpha_1, \alpha_2, \ldots, \alpha_p)$. Instead of taking a single value for k, we can consider several different values k, say $k_1, k_2, \ldots, k_p$. We consider separate ridge parameters (i.e., shrinkage factors) for each of the regression coefficients. The quantity k, instead of being a scalar, is now a vector and denoted by $\mathbf{k}$. The total mean square error given in (11.B.9) now becomes

$$\text{TMSE}(\mathbf{k}) = E[(\hat{\theta}(\mathbf{k}) - \theta)^T(\hat{\theta}(\mathbf{k}) - \theta)]$$

$$= \sigma^2 \sum_{j=1}^{p} \frac{\lambda_j}{(\lambda_j + k_j)^2} + \sum_{j=1}^{p} \frac{k_j^2 \alpha_j^2}{(\lambda_j + k_j)^2}. \quad (11.\text{B}.10)$$

The total mean square error given in (11.B.10) is minimized by taking $k_j = \sigma^2/\alpha_j^2$. An iterative estimation procedure is suggested. At Step 1, k_j is computed by using ordinary least squares estimates for σ^2 and α_j. Then a new value of $\hat{\alpha}(\mathbf{k})$ is computed,

$$\hat{\alpha}(\mathbf{k}) = (C^T C + K)^{-1} C^T Y,$$

where K is a diagonal matrix with diagonal elements $k_1, \ldots, k_p$ from Step 1. The process is repeated until successive changes in the components of $\hat{\alpha}(\mathbf{k})$ are negligible. Then, using (11.B.8), the estimate of θ is

$$\hat{\theta}(\mathbf{k}) = V \hat{\alpha}(\mathbf{k}). \quad (11.\text{B}.11)$$

The two ridge-type estimators (one value of k, several values of k) defined previously, as well as other related alternatives to ordinary least squares estimation, are discussed by Dempster et al. (1977). The different estimators are compared and evaluated by Monte Carlo techniques. In general, the choice of the best estimation method for a particular problem depends on the specific model and data. Dempster et al. (1977) hint at an analysis that could be used to identify the best estimation method for a given set of data. At the present time, our preference is for the simplest version of the ridge method, a single ridge parameter k, chosen after an examination of the ridge trace.

Appendix: 11.C Surrogate Ridge Regression

We give here a very brief description of Surrogate Ridge regression introduced by Jensen and Ramirez (2008) who cast some doubt about the ability of ridge estimators to actually improve the condition of an ill-conditioned linear system and provide stable estimated regression coefficients and smaller variance inflation factors. Note that the ridge estimator in (11.B.2) is the solution of the linear system

$$(\mathbf{Z}^T\mathbf{Z} + k\mathbf{I}_p)\beta = \mathbf{Z}^T\mathbf{Y}, \tag{11.C.1}$$

The condition number of the matrix $(\mathbf{Z}^T\mathbf{Z} + k\mathbf{I}_p)$ on the left-hand side of (11.C.1) is $\sqrt{(\lambda_1 + k)/(\lambda_p + k)}$, which is smaller than the condition number of $(\mathbf{Z}^T\mathbf{Z}$, which is $\sqrt{\lambda_1/\lambda_p})$. Thus adding k to each of the diagonal elements of $\mathbf{Z}^T\mathbf{Z}$ improves its condition. But the matrix $\mathbf{Z}$ on the right-hand side of (11.C.1) remains ill-conditioned. To also improve the condition of the right-hand side of (11.C.1), Jensen and Ramirez (2008) propose replacing the ill-conditioned regression model in (11.A.1) by the *surrogate* but less ill-conditioned model

$$\mathbf{Y} = \mathbf{Z}_k\beta + \varepsilon, \tag{11.C.2}$$

where $\mathbf{Z}_k = \mathbf{U}(\mathbf{\Lambda} + k\mathbf{I}_p)^{1/2}\mathbf{V}^T$, the matrices $\mathbf{U}$ and $\mathbf{V}$ are obtained from the *singular-value decomposition* of $\mathbf{Z} = \mathbf{U}\mathbf{D}\mathbf{V}^T$ [see, e.g., Golub and van Loan (1989)] with $\mathbf{U}^T\mathbf{U} = \mathbf{V}^T\mathbf{V} = \mathbf{I}_p$, and $\mathbf{D}$ is a diagonal matrix containing the corresponding ordered singular values of $\mathbf{Z}$. Note that the square of the singular values of $\mathbf{X}$ are the eigenvalues of $\mathbf{Z}^T\mathbf{Z}$, that is, $\mathbf{D}^2 = \mathbf{\Lambda}$. Because $\mathbf{Z}_k^T\mathbf{Z}_k = \mathbf{Z}^T\mathbf{Z} + k\mathbf{I}_p$, the least squares estimator of the regression coefficients in (11.C.2) is the solution of the linear system

$$(\mathbf{Z}^T\mathbf{Z} + k\mathbf{I}_p)\beta = \mathbf{Z}_k^T\mathbf{Y}, \tag{11.C.3}$$

which is given by

$$\hat{\beta}_s(k) = (\mathbf{Z}^T\mathbf{Z} + k\mathbf{I}_p)^{-1}\mathbf{Z}_k^T\mathbf{Y}. \tag{11.C.4}$$

Jensen and Ramirez (2008) study the properties of the surrogate ridge regression estimator, $\hat{\beta}_s(k)$, in (11.C.4), and using a case study they demonstrate that the surrogate estimator is more conditioned than the classical ridge estimator, $\hat{\beta}(k)$, in (11.B.2). For example, they observe that (a) the condition of the variance of $||\hat{\beta}_s(k)||$ is monotonically increasing in k and (b) the maximum variance inflation factor is monotonically decreasing in k. These properties do not hold for the classical ridge estimator, $\hat{\beta}(k)$.

CHAPTER 12

VARIABLE SELECTION PROCEDURES

12.1 INTRODUCTION

In our discussion of regression problems so far we have assumed that the variables that go into the equation were chosen in advance. Our analysis involved examining the equation to see whether the functional specification was correct, and whether the assumptions about the error term were valid. The analysis presupposed that the set of variables to be included in the equation had already been decided. In many applications of regression analysis, however, the set of variables to be included in the regression model is not predetermined, and it is often the first part of the analysis to select these variables. There are some occasions when theoretical or other considerations determine the variables to be included in the equation. In those situations the problem of variable selection does not arise. But in situations where there is no clear-cut theory, the problem of selecting variables for a regression equation becomes an important one.

The problems of variable selection and the functional specification of the equation are linked to each other. The questions to be answered while formulating a regression model are: Which variables should be included, and in what form should they be included; that is, should they enter the equation as an original variable X or as some transformed variable such as X^2, $\log X$,

---

Regression Analysis By Example Using R, Sixth Edition. Ali S. Hadi and Samprit Chatterjee
© 2024 John Wiley & Sons, Inc. Published 2024 by John Wiley & Sons, Inc.
Companion website: www.wiley.com/go/hadi/regression_analysis_6e

or a combination of both? Although ideally the two problems should be solved simultaneously, we shall for simplicity propose that they be treated sequentially. We first determine the variables that will be included in the equation and after that investigate the exact form in which the variables enter it. This approach is a simplification, but it makes the problem of variable selection more tractable. Once the variables that are to be included in the equation have been selected, we can apply the methods described in the earlier chapters to arrive at the actual form of the equation.

12.2 FORMULATION OF THE PROBLEM

We have a response variable Y and q predictor variables $X_1, X_2, \ldots, X_q$. A linear model that represents Y in terms of q variables is

$$y_i = \beta_0 + \sum_{j=1}^{q} \beta_j x_{ij} + \varepsilon_i, \tag{12.1}$$

where β_j are parameters and ε_i represents random disturbances. Instead of dealing with the full set of variables (particularly when q is large) we might delete a number of variables and construct an equation with a subset of variables. This chapter is concerned with determining which variables are to be retained in the equation. Let us denote the set of variables retained by $X_1, X_2, \ldots, X_p$ and those deleted by $X_{p+1}, X_{p+2}, \ldots, X_q$. Let us examine the effect of variable deletion under two general conditions:

1. The model that connects Y to the X's has all β's $(\beta_0, \beta_1, \ldots, \beta_q)$ nonzero.
2. The model has $\beta_0, \beta_1, \ldots, \beta_p$ nonzero, but $\beta_{p+1}, \beta_{p+2}, \ldots, \beta_q$ zero.

Suppose that instead of fitting (12.1) we fit the subset model

$$y_i = \beta_0 + \sum_{j=1}^{p} \beta_j x_{ij} + \varepsilon_i. \tag{12.2}$$

We shall describe the effect of fitting the model to the full and partial set of X's under the two alternative situations described previously. In short, what are the effects of including variables in an equation when they should be properly left out (because the population regression coefficients are zero) and the effect of leaving out variables when they should be included (because the population regression coefficients are not zero)? We will examine the effect of deletion of variables on the estimates of parameters and the predicted values of Y. The solution to the problem of variable selection becomes a little clearer once the effects of retaining unessential variables or the deletion of essential variables in an equation are known.

12.3 CONSEQUENCES OF VARIABLES DELETION

Denote the estimates of the regression parameters by $\hat{\beta}_0^*, \hat{\beta}_1^*, \ldots, \hat{\beta}_q^*$ when the model (12.1) is fitted to the full set of variables $X_1, X_2, \ldots, X_q$. Denote the estimates of the regression parameters by $\hat{\beta}_0, \hat{\beta}_1, \ldots, \hat{\beta}_p$ when the model (12.2) is fitted. Let $\hat{y}_i^*$ and $\hat{y}_i$ be the predicted values from the full and partial set of variables corresponding to an observation $(x_{i1}, x_{i2}, \ldots, x_{iq})$. The results can now be summarized as follows (a summary using matrix notation is given in the Appendix to this chapter): $\hat{\beta}_0, \hat{\beta}_1, \ldots, \hat{\beta}_p$ are biased estimates of $\beta_0, \beta_1, \ldots, \beta_p$ unless the remaining β's in the model $(\beta_{p+1}, \beta_{p+2}, \ldots, \beta_q)$ are zero or the variables $X_1, X_2, \ldots, X_p$ are orthogonal to the variable set $(X_{p+1}, X_{p+2}, \ldots, X_q)$. The estimates $\hat{\beta}_0^*, \hat{\beta}_1^*, \ldots, \hat{\beta}_p^*$ have less precision than $\hat{\beta}_0, \hat{\beta}_1, \ldots, \hat{\beta}_p$; that is,

$$\text{Var}(\hat{\beta}_j^*) \geq \text{Var}(\hat{\beta}_j), \quad j = 0, 1, \ldots, p.$$

The variance of the estimates of regression coefficients for variables in the reduced equation is not greater than the variances of the corresponding estimates for the full model. Deletion of variables decreases or, more correctly, never increases the variances of estimates of the retained regression coefficients. Since $\hat{\beta}_j$ are biased and $\hat{\beta}_j^*$ are not, a better comparison of the precision of estimates would be obtained by comparing the mean square errors of $\hat{\beta}_j$ with the variances of $\hat{\beta}_j^*$. The mean squared errors (MSE) of $\hat{\beta}_j$ will be smaller than the variances of $\hat{\beta}_j^*$, only if the deleted variables have regression coefficients smaller in magnitude than the standard deviations of the estimates of the corresponding coefficients. The estimate of σ^2, based on the subset model, is generally biased upward, that is, $E(\hat{\sigma}^2) > \sigma^2$.

Let us now look at the effect of deletion of variables on prediction. The prediction $\hat{y}_i$ is biased unless the deleted variables have zero regression coefficients, or the set of retained variables are orthogonal to the set of deleted variables. The variance of a predicted value from the subset model is smaller than or equal to the variance of the predicted value from the full model; that is,

$$\text{Var}(\hat{y}_i) \leq \text{Var}(\hat{y}_i^*).$$

The conditions for $\text{MSE}(\hat{y}_i)$ to be smaller than $\text{Var}(\hat{y}_i^*)$ are identical to the conditions for $\text{MSE}(\hat{\beta}_j)$ to be smaller than $\text{Var}(\hat{\beta}_j^*)$, which we have already stated. For further details, refer to Chatterjee and Hadi (1988).

The rationale for variable selection can be outlined as follows: Even though the variables deleted have nonzero regression coefficients, the regression coefficients of the retained variables may be estimated with smaller variance from the subset model than from the full model. The same result also holds for the variance of a predicted response. The price paid for deleting variables is in the introduction of bias in the estimates. However, there are conditions (as we have described above), when the MSE of the biased estimates will be smaller than the variance of their unbiased estimates; that is, the gain in precision is not offset by the square of the bias. On the other hand, if some of the retained variables are extraneous or unessential, that is, have zero coefficients or coefficients whose magnitudes are smaller than the standard deviation of the estimates, the inclusion of these variables in the equation leads to a loss of precision in estimation and prediction.

The reader is referred to Sections 4.5, 5.13, and 5.14 for further elaboration on the interpretation of regression coefficients and the role of variables in regression modeling.

12.4 USES OF REGRESSION EQUATIONS

A regression equation has many uses. These are broadly summarized below.

12.4.1 Description and Model Building

A regression equation may be used to describe a given process or as a model for a complex interacting system. The purpose of the equation may be purely descriptive, to clarify the nature of this complex interaction. For this use there are two conflicting requirements: (1) to account for as much of the variation as possible, which points in the direction for inclusion of a large number of variables; and (2) to adhere to the principle of parsimony, which suggests that we try, for ease of understanding and interpretation, to describe the process with as few variables as possible. In situations where description is the prime goal, we try to choose the smallest number of predictor variables that accounts for the most substantial part of the variation in the response variable.

12.4.2 Estimation and Prediction

A regression equation is sometimes constructed for prediction. From the regression equation we want to predict the value of a future observation or estimate the mean response corresponding to a given observation. When a regression equation is used for this purpose, the variables are selected with an eye toward minimizing the MSE of prediction.

12.4.3 Control

A regression equation may be used as a tool for control. The purpose for constructing the equation may be to determine the magnitude by which the value of a predictor variable must be altered to obtain a specified value of the response (target) variable. Here the regression equation is viewed as a response function, with Y as the response variable. For control purposes it is desired that the coefficients of the variables in the equation be measured accurately; that is, the standard errors of the regression coefficients are small.

These are the broad uses of a regression equation. Occasionally, these functions overlap and an equation is constructed for some or all of these purposes. The main point to be noted is that the purpose for which the regression equation is constructed determines the criterion that is to be optimized in its formulation. It follows that a subset of variables that may be best for one purpose may not be best for another. The concept of the "best" subset of variables to be included in an equation always requires additional qualification.

Before discussing actual selection procedures we make two preliminary remarks. First, it is not usually meaningful to speak of the "best set" of variables to be included in a multiple regression equation. There is no unique "best set" of variables. A regression equation can be used for several purposes. The set of variables that may be best for one purpose may not be best for another. The purpose for which a regression equation is constructed should be kept in mind in the variable selection process. We shall show later that the purpose for which an equation is constructed determines the criteria for selecting and evaluating the contributions of different variables.

Second, since there is no best set of variables, there may be several subsets that are adequate and could be used in forming an equation. A good variable selection procedure should point out these several sets rather than generate a so-called single "best" set. The various sets of adequate variables throw light on the structure of data and help us in understanding the underlying process. In fact, the process of variable selection should be viewed as an intensive analysis of the correlational structure of the predictor variables and how they individually and jointly affect the response variable under study. These two points influence the methodology that we present in connection with variable selection.

12.5 CRITERIA FOR EVALUATING EQUATIONS

To judge the adequacy of various fitted equations we need a criterion. Several have been proposed in the statistical literature. We describe the two that we consider most useful. An exhaustive list of criteria is found in Hocking (1976).

12.5.1 Residual Mean Square

One measure that is used to judge the adequacy of a fitted equation is the residual mean square (RMS). With a p-term equation (includes a constant and $p-1$ variables), the RMS is defined as

$$\text{RMS}_p = \frac{\text{SSE}_p}{n-p} = \hat{\sigma}_p^2, \qquad (12.3)$$

where SSE_p is the residual sum of squares for a p-term equation. Between two equations, the one with the smaller RMS is usually preferred, especially if the objective is forecasting.

It is clear that RMS_p is related to the square of the multiple correlation coefficient R_p^2 and the square of the adjusted multiple correlation coefficient R_{ap}^2 which have already been described (Chapter 4) as measures for judging the adequacy of fit of an equation. Here we have added a subscript to R^2 and R_a^2 to denote their dependence on the number of terms in an equation. The relationship between these quantities are given by

$$R_p^2 = 1 - (n-p)\frac{\text{RMS}_p}{(\text{SST})} \qquad (12.4)$$

and

$$R_{ap}^2 = 1 - (n-1)\frac{\text{RMS}_p}{(\text{SST})}, \qquad (12.5)$$

where

$$\text{SST} = \sum (y_i - \bar{y})^2.$$

Note that R_{ap}^2 is more appropriate than R_p^2 when comparing models with different number of predictors because R_{ap}^2 adjusts (penalizes) for the number of predictor variables in the model.

12.5.2 Mallows C_p

We pointed out earlier that predicted values obtained from a regression equation based on a subset of variables are generally biased. To judge the performance of an equation we should consider the mean square error of the predicted value rather than the variance. The standardized total mean squared error of prediction for the observed data is measured by

$$J_p = \frac{1}{\sigma^2} \sum_{i=1}^{n} \text{MSE}(\hat{y}_i), \tag{12.6}$$

where $\text{MSE}(\hat{y}_i)$ is the mean squared error of the ith predicted value from a p-term equation, and σ^2 is the variance of the random errors. The $\text{MSE}(\hat{y}_i)$ has two components, the variance of prediction arising from estimation and a bias component arising from the deletion of variables.

To estimate J_p, Mallows (1973) uses the statistic

$$C_p = \frac{\text{SSE}_p}{\hat{\sigma}^2} + (2p - n), \tag{12.7}$$

where $\hat{\sigma}^2$ is an estimate of σ^2 and is usually obtained from the linear model with the full set of q variables. It can be shown that the expected value of C_p is p when there is no bias in the fitted equation containing p terms. Consequently, the deviation of C_p from p can be used as a measure of bias. The C_p statistic therefore measures the performance of the variables in terms of the standardized total mean square error of prediction for the observed data points irrespective of the unknown true model. It takes into account both the bias and the variance. Subsets of variables that produce values of C_p that are close to p are the desirable subsets. The selection of "good" subsets is done graphically. For the various subsets a graph of C_p is plotted against p. The line $C_p = p$ is also drawn on the graph. Sets of variables corresponding to points close to the line $C_p = p$ are the good or desirable subsets of variables to form an equation. The use of C_p plots is illustrated and discussed in more detail in the example that is given in Section 12.10. A very thorough treatment of the C_p statistic is given in Daniel and Wood (1980).

12.5.3 Information Criteria

Variable selection in the regression context can be viewed as a model selection problem. The Information criteria that we now describe arose first in the general problem of model selection. The Akaike (1973) Information Criterion (AIC) in selecting a model tries to balance the conflicting demands of accuracy (fit) and simplicity (small number of variables). This is the principle of parsimony already discussed in Section 4.10.2. AIC for a p-term equation (a constant, and $p-1$ variables) is given by

$$\text{AIC}_p = n\ln(\text{SSE}_p/n) + 2p. \qquad (12.8)$$

The models with smaller AIC are preferred.

We can see from (12.8) that for two models with similar SSE, AIC penalizes the model that has a larger number of variables. The numerical value of AIC for a single model is not very meaningful or descriptive. AIC can be used, however, to rank the models on the basis of their twin criteria of fit and simplicity. Models with AIC not differing by 2 should be treated as equally adequate. Larger differences in AIC indicate significant difference between the quality of the models. The one with the lower AIC should be adopted.

A great advantage of AIC is that it allows us to compare non-nested models. A group of models are nested if they can be obtained from a larger model as special cases (see Section 4.10). We cannot perform an F-Test, for example, to compare the adequacy of a model based on (X_1, X_2, X_3) with one based on (X_4, X_5). The choice of these two sets of variables may be dictated by the nature of the problem at hand. The AIC will allow us to make such comparisons but not the F-Test described earlier.

To compare models by AIC we must have complete data (no missing values). The AIC must be calculated on the same set of observations. If there are many missing values for some variables, application of AIC may be inefficient because observations in which some variables were missing will be dropped.

Several modifications of AIC have been suggested. One popular variation called Bayes Information Criterion (BIC), originally proposed by Schwarz (1978), is defined as

$$\text{BIC}_p = n\ln(\text{SSE}_p/n) + p(\ln n). \qquad (12.9)$$

The difference between AIC and BIC is in the severity of penalty for p. The penalty is far more severe in BIC when $n > 8$. This tends to control the overfitting (resulting in a choice of larger p) tendency of AIC.

Another modification of AIC to avoid overfitting is the bias corrected version, AIC^c, proposed by Hurvich and Tsai (1989), which is given by

$$\text{AIC}^c_p = \text{AIC}_p + \frac{2(p+2)(p+3)}{n-p-3}. \qquad (12.10)$$

The correction to AIC in (12.10) is small for large n and moderate p. The correction is large when n is small and p large. One should never fit a large and complex model with a small number of observations. In general the correction to AIC will be minor, and we will not discuss AIC^c further. To guard against overfitting in our analysis we will examine BIC.

12.6 COLLINEARITY AND VARIABLE SELECTION

In discussing variable selection procedures, we distinguish between two broad situations:

1. The predictor variables are not collinear; that is, there is no strong evidence of collinearity.
2. The predictor variables are collinear; that is, the data are highly multicollinear.

Depending on the correlation structure of the predictor variables, we propose different approaches to the variable selection procedure. If the data analyzed are not collinear, we proceed in one manner, and if collinear, we proceed in another.

As a first step in variable selection procedure we recommend calculating the variance inflation factors (VIFs) or the eigenvalues of the correlation matrix of the predictor variables. If none of the VIFs are greater than 10, collinearity is not a problem. Further, as we explained in Chapter 10, the presence of small eigenvalues indicates collinearity. If the condition number[1] is larger than 15, the variables are collinear. We may also look at the sum of the reciprocals of the eigenvalues. If any of the individual eigenvalues are less than 0.01, or the sum of the reciprocals of the eigenvalues is greater than, say, five times the number of predictor variables in the problem, we say that the variables are collinear. If the conditions above do not hold, the variables are regarded as noncollinear.

12.7 EVALUATING ALL POSSIBLE EQUATIONS

The first procedure described is very direct and applies equally well to both collinear and noncollinear data. The procedure involves fitting all possible

---

[1] Recall from Chapter 10 that the condition number is defined by $\kappa = \sqrt{\lambda_{max}/\lambda_{min}}$, where λ_{max} and λ_{min} are the maximum and minimum eigenvalues of the matrix of correlation coefficients.

subset equations to a given body of data. With q variables the total number of equations fitted is 2^q (including an equation that contains all the variables and another that contains no variables). The latter is simply $\hat{y}_i = \bar{y}$, which is obtained from fitting the model $Y = \beta_0 + \varepsilon$. This method clearly gives an analyst the maximum amount of information available concerning the nature of relationships between Y and the set of X's. However, the number of equations and supplementary information that must be looked at may be prohibitively large. Even with only six predictor variables, there are 64 (2^6) equations to consider; with seven variables the number grows to 128 (2^7), neither feasible nor practical. An efficient way of using the results from fitting all possible equations is to pick out the three "best" (on the basis of R^2, C_p, RMS, or the information criteria outlined earlier) equations containing a specified number of variables. This smaller subset of equations is then analyzed to arrive at the final model. These regressions are then carefully analyzed by examining the residuals for outliers, autocorrelation, or the need for transformations before deciding on the final model. The various subsets that are investigated may suggest interpretations of the data that might have been overlooked in a more restricted variable selection approach.

When the number of variables is large, the evaluation of all possible equations may not be practically feasible. Certain shortcuts have been suggested (Furnival and Wilson, 1974; La Motte and Hocking, 1970) which do not involve computing the entire set of equations while searching for the desirable subsets. But with a large number of variables these methods still involve a considerable amount of computation. There are variable selection procedures that do not require the evaluation of all possible equations. Employing these procedures will not provide the analyst with as much information as the fitting of all possible equations, but it will entail considerably less computation and may be the only available practical solution. These are discussed in Section 12.8. These procedures are quite efficient with noncollinear data. We do not, however, recommend them for collinear data.

12.8 VARIABLE SELECTION PROCEDURES

For cases when there are a large number of potential predictor variables, a set of procedures that does not involve computing of all possible equations has been proposed. These procedures have the feature that the variables are introduced or deleted from the equation one at a time, and involve examining only a subset of all possible equations. With q variables these procedures will involve evaluation of at most $q + 1$ equations, as contrasted with the evaluation of 2^q equations necessary for examining all possible equations. The procedures can be classified into two broad categories: (a) the *forward selection* (FS) procedure, and (b) the *backward elimination* (BE) procedure. There is also a very popular modification of the FS procedure called the *stepwise* method. The three procedures are described and compared below.

12.8.1 Forward Selection Procedure

The forward selection procedure starts with an equation containing no predictor variables, only a constant term. The first variable included in the equation is the one which has the highest absolute value of the simple correlation with the response variable Y. If the regression coefficient of this variable is significantly different from zero it is retained in the equation, and a search for a second variable is made. The variable that enters the equation as the second variable is one which has the highest absolute value of the correlation with Y, both Y and the candidate predictor variable have been adjusted for the effect of the first variable, that is, the variable with the highest absolute value of the simple correlation coefficient with the residuals from Step 1. The significance of the regression coefficient of the second variable is then tested. If the regression coefficient is significant, a search for a third variable is made in the same way. The procedure is terminated when the last variable entering the equation has an insignificant regression coefficient or all the variables are included in the equation. The significance of the regression coefficient of the last variable introduced in the equation is judged by the standard t-Test computed from the latest equation. Most forward selection algorithms use a low t cutoff value for testing the coefficient of the newly entered variable; consequently, the forward selection procedure goes through the full set of variables and provides us with $q + 1$ possible equations.

12.8.2 Backward Elimination Procedure

The backward elimination procedure starts with the full equation and successively drops one variable at a time. The variables are dropped on the basis of their contribution to the reduction of error sum of squares. The first variable deleted is the one with the smallest contribution to the reduction of error sum of squares. This is equivalent to deleting the variable which has the smallest absolute value of the t-Test in the equation. If all the t-Tests are significant, the full set of variables is retained in the equation. Assuming that there are one or more variables that have insignificant t-Tests, the procedure operates by dropping the variable with the smallest insignificant t-Test. The equation with the remaining $q-1$ variables is then fitted and the t-Tests for the new regression coefficients are examined. The procedure is terminated when all the t-Tests are significant or all variables have been deleted. In most backward elimination algorithms the cutoff value for the t-Test is set high so that the procedure runs through the whole set of variables, that is, starting with the q-variable equation and ending up with an equation containing only the constant term. The backward elimination procedure involves fitting at most $q+1$ regression equations.

12.8.3 Stepwise Method

The stepwise method is essentially a forward selection procedure but with the added proviso that at each stage the possibility of deleting a variable, as in backward elimination, is considered. In this procedure a variable that entered in the earlier stages of selection may be eliminated at later stages. The calculations made for inclusion and deletion of variables are the same as FS and BE procedures. Often, different levels of significance are assumed for inclusion and exclusion of variables from the equation.

AIC and BIC both can be used for setting up stepwise procedures (forward selection and backward elimination). For forward selection one starts with a constant as the fitting term and adds variables to the model. The procedure is terminated, when addition of a variable causes no reduction of AIC (BIC). In the backward procedure, we start with the full model (containing all the variables) and drop variables successively. The procedure is terminated when dropping a variable does not lead to any further reduction in the criteria.

The stepwise procedure based on information criteria differs in a major way from the procedures based on the t-statistic that gauges the significance of a variable. The information-based procedures are driven by all the variables in the model. The termination of the procedure is based solely on the decrease of the criterion, and not on the statistical significance of the entering or departing variable.

Most of the currently available software do not automatically produce AIC or BIC. They all, however, provide SSE, from which it is easy to compute (12.8) and (12.9) the information criteria.

12.9 GENERAL REMARKS ON VARIABLE SELECTION METHODS

The variable selection procedures discussed above should be used with caution. These procedures should not be used mechanically to determine the "best" variables. The order in which the variables enter or leave the equation in variable selection procedures should not be interpreted as reflecting the relative importance of the variables. If these caveats are kept in mind, the variable selection procedures are useful tools for variable selection in noncollinear situations. All three procedures will give nearly the same selection of variables with noncollinear data. They entail much less computing than that in the analysis of all possible equations.

Several stopping rules have been proposed for the variable selection procedures. A stopping rule that has been reported to be quite effective is as follows:

- In FS: Stop if minimum t-Test is less than 1.
- In BE: Stop if minimum absolute value of the t-Test is greater than 1.

In the following example we illustrate the effect of different stopping rules in variable selection.

We recommend the BE procedure over FS procedure for variable selection. One obvious reason is that in the BE procedure the equation with the full variable set is calculated and available for inspection even though it may not be used as the final equation. Although we do not recommend the use of variable selection procedures in a collinear situation, the BE procedure is better able to handle collinearity than the FS procedure (Mantel, 1970).

In an application of variable selection procedures several equations are generated, each equation containing a different number of variables. The various equations generated can then be evaluated using a statistic such as C_p, RMS, AIC, or BIC. The residuals for the various equations should also be examined. Equations with unsatisfactory residual plots are rejected. Only a total and comprehensive analysis will provide an adequate selection of variables and a useful regression equation. This approach to variable selection is illustrated by the following example.

12.10 A STUDY OF SUPERVISOR PERFORMANCE

To illustrate variable selection procedures in a noncollinear situation, consider the Supervisor Performance data discussed in Section 4.3. A regression equation was needed to study the qualities that led to the characterization of good supervisors by the people being supervised. The equation is to be constructed in an attempt to understand the supervising process and the relative importance of the different variables. In terms of the use for the

Table 12.1 Correlation Matrix for the Supervisor Performance Data

	X_1	X_2	X_3	X_4	X_5	X_6
X_1	1.000					
X_2	0.558	1.000				
X_3	0.597	0.493	1.000			
X_4	0.669	0.445	0.640	1.000		
X_5	0.188	0.147	0.116	0.377	1.000	
X_6	0.225	0.343	0.532	0.574	0.283	1.000

regression equation, this would imply that we want accurate estimates of the regression coefficients, in contrast to an equation that is to be used only for prediction. The variables in the problem are given in Table 4.2. The data can be found in the file Supervisor.csv at the Book's Website.[2]

The VIFs resulting from regressing Y on $X_1, X_2, \ldots, X_6$ are

$$\text{VIF}_1 = 2.7, \quad \text{VIF}_2 = 1.6, \quad \text{VIF}_3 = 2.3,$$

$$\text{VIF}_4 = 3.1, \quad \text{VIF}_5 = 1.2, \quad \text{VIF}_6 = 2.0.$$

The range of the VIFs (1.2 to 3.1) shows that collinearity is not a problem for these data. The same picture emerges if we examine the eigenvalues of the correlation matrix of the data (Table 12.1). The eigenvalues of the correlation matrix are

$$\lambda_1 = 3.169, \quad \lambda_2 = 1.006, \quad \lambda_3 = 0.763,$$

$$\lambda_4 = 0.553, \quad \lambda_5 = 0.317, \quad \lambda_6 = 0.192.$$

The sum of the reciprocals of the eigenvalues is 12.8. Since none of the eigenvalues are small (the condition number is 4.1) and the sum of the reciprocals of the eigenvalues is only about twice the number of variables, we conclude that the data in the present example are not seriously collinear and we can apply the variable selection procedures just described.

The result of forward selection procedure is given in Table 12.2. For successive equations we show the variables present, the RMS, and the value of the C_p statistic. The column labeled Rank shows the rank of the subset obtained by FS relative to best subset (on the basis of RMS) of the same size. The value of p is the number of predictor variables in the equation, including a constant term. Two stopping rules are used:

---

[2] http://www.aucegypt.edu/faculty/hadi/RABE6

A STUDY OF SUPERVISOR PERFORMANCE

Table 12.2 Variables Selected by the Forward Selection Method

Variables in Equation	min($|t|$)	$\hat{\sigma}_p$	C_p	p	Rank	AIC	BIC
X_1	7.74	6.993	1.41	2	1	118.63	121.43
$X_1 X_3$	1.57	6.817	1.11	3	1	118.00	122.21
$X_1 X_3 X_6$	1.29	6.734	1.60	4	1	118.14	123.74
$X_1 X_3 X_6 X_2$	0.59	6.820	3.28	5	1	119.73	126.73
$X_1 X_3 X_6 X_2 X_4$	0.47	6.928	5.07	6	1	121.45	129.86
$X_1 X_3 X_6 X_2 X_4 X_5$	0.26	7.068	7.00	7	—	123.36	133.17

1. Stop if the minimum absolute value of the t-Test is less than $t_{0.05}(n-p)$.
2. Stop if the minimum absolute value of the t-Test is less than 1.

The first rule is more stringent and terminates with variables X_1 and X_3. The second rule is less stringent and terminates with variables X_1, X_3, and X_6.

The results of applying the BE procedure are presented in Table 12.3. They are identical in structure to Table 12.2. For the BE we will use the stopping rules:

1. Stop if the minimum absolute value of the t-Test is greater than $t_{0.05}(n-p)$.
2. Stop if the minimum absolute value of the t-Test is greater than 1.

With the first stopping rule the variables selected are X_1 and X_3. With the second stopping rule the variables selected are X_1, X_3, and X_6. The FS and BE give identical equations for this problem, but this is not always the case (an example is given in Section 12.12). To describe the supervisor performance, the equation

$$Y = 13.58 + 0.62 X_1 + 0.31 X_3 - 0.19 X_6 \qquad (12.11)$$

Table 12.3 Variables Selected by Backward Elimination Method

Variables in Equation	min($|t|$)	$\hat{\sigma}_p$	C_p	p	Rank	AIC	BIC
$X_1 X_2 X_3 X_4 X_5 X_6$	0.26	7.068	7.00	7	—	123.36	133.17
$X_1 X_2 X_3 X_4 X_6$	0.47	6.928	5.07	6	1	121.45	129.86
$X_1 X_2 X_3 X_6$	0.59	6.820	3.28	5	1	119.73	126.73
$X_1 X_3 X_6$	1.29	6.734	1.60	4	1	118.14	123.74
$X_1 X_3$	1.57	6.817	1.11	3	1	118.00	122.21
X_1	7.74	6.993	1.41	2	1	118.63	121.43

VARIABLE SELECTION PROCEDURES

Table 12.4 Values of C_p Statistic (All Possible Equations)

Variables	C_p	Variables	C_p	Variables	C_p	Variables	C_p
1	1.41	1 5	3.41	1 6	3.33	1 5 6	5.32
2	44.40	2 5	45.62	2 6	46.39	2 5 6	47.91
1 2	3.26	1 2 5	5.26	1 2 6	5.22	1 2 5 6	7.22
3	26.56	3 5	27.94	3 6	24.82	3 5 6	25.02
1 3	1.11	1 3 5	3.11	1 3 6	1.60	1 3 5 6	3.46
2 3	26.96	2 3 5	28.53	2 3 6	24.62	2 3 5 6	25.11
1 2 3	2.51	1 2 3 5	4.51	1 2 3 6	3.28	1 2 3 5 6	5.14
4	30.06	4 5	31.62	4 6	27.73	4 5	29.50
1 4	3.19	1 4 5	5.16	1 4 6	4.70	1 4 5 6	6.69
2 4	29.20	2 4 5	30.82	2 4 6	25.91	2 4 5 6	27.74
1 2 4	4.99	1 2 4 5	6.97	1 2 4 6	6.63	1 2 4 5 6	8.61
3 4	23.25	3 4 5	25.23	3 4 6	16.50	3 4 5 6	18.42
1 3 4	3.09	1 3 4 5	5.09	1 3 4 6	3.35	1 3 4 5 6	5.29
2 3 4	24.56	2 3 4 5	26.53	2 3 4 6	17.57	2 3 4 5 6	19.51
1 2 3 4	4.49	1 2 3 4 5	6.48	1 2 3 4 6	5.07	1 2 3 4 5 6	7
5	57.91	6	57.95	5 6	58.76		

is chosen. The residual plots (not shown) for this equation are satisfactory. Since the present problem has only six variables, the total number of equations that can be fitted which contain at least one variable is 63. The C_p values for all 63 equations are shown in Table 12.4. The C_p values are plotted against p in Figure 12.1.[3] The best subsets of variables based on C_p values are given in Table 12.5.

It is seen that the subsets selected by C_p are different from those arrived at by the variable selection procedures as well as those selected on the basis of residual mean square. This anomaly suggests an important point concerning the C_p statistic that the reader should bear in mind. For applications of the C_p statistic, an estimate of σ^2 is required. Usually, the estimate of σ^2 is obtained from the residual sum of squares from the full model. If the full model has a large number of variables with no explanatory power (i.e., population regression coefficients are zero), the estimate of σ^2 from the residual sum of squares for the full model would be large. The loss in degrees of freedom for the divisor would not be balanced by a reduction in the error sum of squares.

---

[3] The R code for producing this and all other graphs in this book can be found at the Book's Website at http://www.aucegypt.edu/faculty/hadi/RABE6.

A STUDY OF SUPERVISOR PERFORMANCE

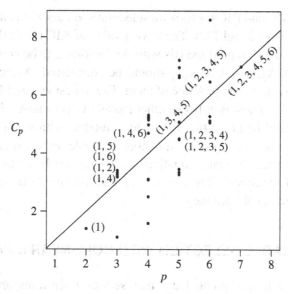

Figure 12.1 Supervisor's Performance data: scatter plot of C_p versus p for subsets with $C_p < 10$.

Table 12.5 Variables Selected on the Basis of C_p Statistic

Variables in Equation	min($|t|$)	$\hat{\sigma}_p$	C_p	p	Rank	AIC	BIC
X_1	7.74	6.993	1.41	2	1	118.63	121.43
$X_1 X_4$	0.47	7.093	3.19	3	2	120.38	124.59
$X_1 X_4 X_6$	0.69	7.163	4.70	4	5	121.84	127.45
$X_1 X_3 X_4 X_5$	0.07	7.080	5.09	5	6	121.97	127.97
$X_1 X_2 X_3 X_4 X_5$	0.11	7.139	6.48	6	4	123.24	131.65
$X_1 X_2 X_3 X_4 X_5 X_6$	0.26	7.068	7.00	7	—	123.36	133.17

If $\hat{\sigma}^2$ is large, then the value of C_p is small. For C_p to work properly, a good estimate of σ^2 must be available. When a good estimate of σ^2 is not available, C_p is of only limited usefulness. In our present example, the RMS for the full model with six variables is larger than the RMS for the model with three variables X_1, X_3, X_6. Consequently, the C_p values are distorted and not very useful in variable selection in the present case. The type of situation we have described can be spotted by looking at the RMS for different values of p. RMS will at first tend to decrease with p, but increase at later stages. This behavior indicates that the latter variables are not contributing significantly to the reduction of error sum of squares. Useful application of C_p requires a parallel monitoring of RMS to avoid distortions.

Values of AIC and BIC for forward selection and backward elimination are given in Tables 12.2 and 12.3. The lowest value of AIC (118.00) is obtained for X_1 and X_3. If we regard models with AIC within 2 to be equivalent, then X_1, X_1X_3, $X_1X_3X_6$, and $X_1X_3X_6X_2$ should be considered. Among these four candidate models we can pick one of them. The lowest value of BIC (121.43) is attained by X_1. There is only one other model (X_1X_3) whose BIC lies within 2 units. It should be noted that BIC selects models with smaller number of variables because of its penalty function. Variable selection should not be done mechanically. In many situations there may not be a "best model" or a "best set of variables". The aim of the analysis should be to identify all models of high equal adequacy.

12.11 VARIABLE SELECTION WITH COLLINEAR DATA

In Chapter 10 it was pointed out that serious distortions are introduced in standard analysis with collinear data. Consequently, we recommend a different set of procedures for selecting variables in these situations. Collinearity is indicated when the correlation matrix has one or more small eigenvalues. With a small number of collinear variables we can evaluate all possible equations and select an equation by methods that have already been described. But with a larger number of variables this method is not feasible.

Two different approaches to the problem have been proposed. The first approach tries to break down the collinearity of the data by deleting variables. The collinear structure present in the variables is revealed by the eigenvectors corresponding to the very small eigenvalues (see Chapters 10 and 11). Once the collinearities are identified, a set of variables can then be deleted to produce a reduced noncollinear data set. We can then apply the methods described earlier. The second approach uses ridge regression as the main tool. We assume that the reader is familiar with the basic terms and concepts of ridge regression (Chapter 11). The first approach (by judicious dropping of correlated variables) is the one that is almost always used in practice.

12.12 THE HOMICIDE DATA

In a study investigating the role of firearms in accounting for the rising homicide rate in Detroit, data were collected for the years 1961–1973. The data are reported in Gunst and Mason (1980, p. 360). The response variable (the homicide rate) and the predictor variables believed to influence or be

THE HOMICIDE DATA

Table 12.6 Homicide Data: Description of Variables

Variable	Symbol	Description
1	FTP	Number of full-time police per 100,000 population
2	UEMP	Percent of the population unemployed
3	M	Number of manufacturing workers (in thousands)
4	LIC	Number of handgun licenses issued per 100,000 population
5	GR	Number of handgun registration issued per 100,000 population
6	CLEAR	Percent of homicides cleared by arrest
7	W	Number of white males in the population
8	NMAN	Number of nonmanufacturing workers (in thousands)
9	G	Number of government workers (in thousands)
10	HE	Average hourly earnings
11	WE	Average weekly earnings
12	H	Number of homicides per 100,000 population

related to the rise in the homicide rate are defined in Table 12.6. The data can be found in the file Homicide.csv at the Book's Website.

We use these data to illustrate the danger of mechanical variable selection procedures, such as the FS and BE, in collinear situations. We are interested in fitting the model

$$H = \beta_0 + \beta_1 G + \beta_2 M + \beta_3 W + \varepsilon. \tag{12.12}$$

In terms of the centered and scaled version of the variables, the model becomes

$$\tilde{H} = \theta_1 \tilde{G} + \theta_2 \tilde{M} + \theta_3 \tilde{W} + \varepsilon'. \tag{12.13}$$

The OLS results are shown in Table 12.7. Can the number of predictor variables in this model be reduced? If the standard assumptions hold, the

Table 12.7 Homicide Data: The OLS Results from Fitting Model (12.13)

Variable	Coefficient	s.e.	t-Test	VIF
G	0.235	0.345	0.68	42
M	−0.405	0.090	−4.47	3
W	−1.025	0.378	−2.71	51
$n = 13$	$R^2 = 0.975$	$R_a^2 = 0.966$	$\hat{\sigma} = 0.0531$	df = 9

Table 12.8 Homicide Data: The Estimated Coefficients, Their t-Tests, and the Adjusted Squared Multiple Correlation Coefficient, R_a^2

	Model						
Variable	(a)	(b)	(c)	(d)	(e)	(f)	(g)
G: Coeff.	0.96			1.15	0.87	0.24	
t-Test	11.10			11.90	1.62	0.68	
M: Coeff.		0.55		−0.27		−0.40	−0.43
t-Test		2.16		−2.79		−4.47	−5.35
W: Coeff.			−0.95		−0.09	−1.02	−1.28
t-Test			−9.77		−0.17	−2.71	−15.90
R_a^2	0.91	0.24	0.89	0.95	0.90	0.97	0.97

small absolute value of the t-Test for the variable G (0.68) would indicate that the corresponding regression coefficient is insignificant and G can be omitted from the model. Let us now apply the forward selection and the backward elimination procedures to see which variables are selected. The regression output that we need to implement the two methods on the standardized versions of the variables are summarized in Table 12.8. In this table we give the estimated coefficients, their t-Tests, and the adjusted squared multiple correlation coefficient R_a^2 for each model for comparison purposes.

The first variable to be selected by the FS is G because it has the largest t-Test among the three models that contain a single variable [Models (a)–(c) in Table 12.8]. Between the two candidates for the two-variable models [Models (d) and (e)], Model (d) is better than Model (e). Therefore, the second variable to enter the equation is M. The third variable to enter the equation is W [Model (f)] because it has a significant t-Test. Note, however, the dramatic change of the significance of G in Models (a), (d), and (f). It was highly significant coefficient in Models (a) and (d), but became insignificant in Model (f). Collinearity is a suspect!

The BE method starts with the three-variable Model (f). The first variable to leave is G (because it has the lowest absolute value of the t-Test), which leads to Model (g). Both M and W in Model (g) have significant t-Tests and the BE procedure terminates.

Observe that the first variable eliminated by the BE (G) is the same as the first variable selected by the FS. That is, the variable G, which was selected by the FS as the most important of the three variables, was regarded by the BE as the least important! Among other things, the reason for this anomalous result is collinearity. The eigenvalues of the correlation matrix, $\lambda_1 = 2.65$,

$\lambda_2 = 0.343$, and $\lambda_3 = 0.011$, give a large condition number ($\kappa = 15.6$). Two of the three variables (G and W) have large VIF (42 and 51). The sum of the reciprocals of the eigenvalues is also very large (96). In addition to collinearity, since the observations were taken over time (for the years 1961–1973), we are dealing with time series data here. Consequently, the error terms can be autocorrelated (see Chapter 9). Examining the pairwise scatter plots of the data will reveal other problems with the data.

This example shows clearly that automatic applications of variable selection procedure in multicollinear data can lead to the selection of a wrong model. In Sections 12.13 and 12.14 we make use of ridge regression for the process of variable selection in multicollinear situations.

12.13 VARIABLE SELECTION USING RIDGE REGRESSION

One of the goals of ridge regression is to produce a regression equation with stable coefficients. The coefficients are stable in the sense that they are not affected by slight variations in the estimation data. The objectives of a good variable selection procedure are (a) to select a set of variables that provides a clear understanding of the process under study, and (b) to formulate an equation that provides accurate forecasts of the response variable corresponding to values of the predictor variables not included in the study. It is seen that the objectives of a good variable selection procedure and ridge regression are very similar and, consequently, one (ridge regression) can be employed to accomplish the other (variable selection).

The variable selection is done by examining the ridge trace, a plot of the ridge regression coefficients against the ridge parameter k. For a collinear system, the characteristic pattern of ridge trace has been described in Chapter 11. The ridge trace is used to eliminate variables from the equation. The guidelines for elimination are

1. Eliminate variables whose coefficients are stable but small. Since ridge regression is applied to standardized data, the magnitude of the various coefficients is directly comparable.

2. Eliminate variables with unstable coefficients that do not hold their predicting power, that is, unstable coefficients that tend to zero.

3. Eliminate one or more variables with unstable coefficients. The variables remaining from the original set, say p in number, are used to form the regression equation.

At the end of each of the above steps, we refit the model that includes the remaining variables before we proceed to the next step.

The subset of variables remaining after elimination should be examined to see if collinearity is no longer present in the subset. We illustrate this procedure by an example.

12.14 SELECTION OF VARIABLES IN AN AIR POLLUTION STUDY

McDonald and Schwing (1973) present a study that relates total mortality to climate, socioeconomic, and pollution variables. Fifteen predictor variables selected for the study are listed in Table 12.9. The response variable is the total age-adjusted mortality from all causes. We will not comment on the epidemiological aspects of the study, but merely use the data as an illustrative example for variable selection. A very detailed discussion of the problem is presented by McDonald and Schwing in their article and we refer the interested reader to it for more information.

Table 12.9 Description of Variables, Means, and Standard Deviations, SD ($n = 60$)

Variable	Description	Mean	SD
X_1	Mean annual precipitation (inches)	37.37	9.98
X_2	Mean January temperature (degrees Fahrenheit)	33.98	10.17
X_3	Mean July temperature (degrees Fahrenheit)	74.58	4.76
X_4	Percent of population over 65 years of age	8.80	1.46
X_5	Population per household	3.26	0.14
X_6	Median school years completed	10.97	0.85
X_7	Percent of housing units that are sound	80.91	5.14
X_8	Population per square mile	3876.05	1454.10
X_9	Percent of nonwhite population	11.87	8.92
X_{10}	Percent employment in white-collar jobs	46.08	4.61
X_{11}	Percent of families with income under $3000	14.37	4.16
X_{12}	Relative pollution potential of hydrocarbons	37.85	91.98
X_{13}	Relative pollution potential of oxides of nitrogen	22.65	46.33
X_{14}	Relative pollution potential of sulfur dioxide	53.77	63.39
X_{15}	Percent relative humidity	57.67	5.37
Y	Total age-adjusted mortality from all causes.	940.36	62.21

The original data were not available to us before the Fourth Edition of this book was published. But we obtained the data before the Fifth Edition. The original data can be found in the file Air.Pollution.csv at the Book's Website. In the Fourth Edition we did the analysis starting with the correlation matrix of the response and the 15 predictor variables and mentioned that it is not a good practice to perform the analysis based only on the correlation matrix because without the original data we will not be able to perform diagnostics checking which is necessary in any thorough analysis of the data. We can now perform some diagnostic checks regarding the validity of the standard assumptions of the linear regression model. We have examined some of the diagnostic plots discussed in Chapter 5 and found one outlier and a couple of high-leverage points, but luckily, they are not consequential; that is, the results with them and without them are not substantially different. As can be expected from the nature of the variables, some of them are highly correlated with each other. The evidence of collinearity is clearly seen if we examine the eigenvalues of the correlation matrix (not shown). The eigenvalues of the correlation matrix of the 15 predictor variables are

$$\lambda_1 = 4.5284, \quad \lambda_6 = 0.9604, \quad \lambda_{11} = 0.1664,$$

$$\lambda_2 = 2.7548, \quad \lambda_7 = 0.6127, \quad \lambda_{12} = 0.1270,$$

$$\lambda_3 = 2.0545, \quad \lambda_8 = 0.4720, \quad \lambda_{13} = 0.1140,$$

$$\lambda_4 = 1.3484, \quad \lambda_9 = 0.3709, \quad \lambda_{14} = 0.0460,$$

$$\lambda_5 = 1.2232, \quad \lambda_{10} = 0.2164, \quad \lambda_{15} = 0.0049.$$

There are two very small eigenvalues; the largest eigenvalue is more than 930 times larger than the smallest eigenvalue. The sum of the reciprocals of the eigenvalues is 265, which is more than 17 times the number of variables. The data show strong evidence of collinearity.

The initial OLS results from fitting a linear model to the centered and scaled data are given in Table 12.10. Although the model has a high R^2, some of the estimated coefficients have small absolute value of the t-Tests. In the presence of collinearity, a small absolute value of the t-Test does not necessarily mean that the corresponding variable is not important. The small absolute value of the t-Test might be due of variance inflation because of the presence of collinearity. As can be seen in Table 12.10, VIF_{12} and VIF_{13} are very large.

Table 12.10 OLS Regression Output for the Air Pollution Data (15 Predictor Variables)

Variable	Coefficient	s.e.	t-Test	VIF
X_1	0.306	0.148	2.06	4.11
X_2	−0.317	0.181	−1.75	6.14
X_3	−0.237	0.146	−1.63	3.97
X_4	−0.213	0.200	−1.07	7.47
X_5	−0.232	0.152	−1.53	4.31
X_6	−0.233	0.161	−1.45	4.86
X_7	−0.054	0.146	−0.37	3.99
X_8	0.084	0.094	0.89	1.66
X_9	0.640	0.190	3.36	6.78
X_{10}	−0.014	0.123	−0.11	2.84
X_{11}	−0.011	0.216	−0.05	8.72
X_{12}	−0.994	0.726	−1.37	98.64
X_{13}	0.998	0.749	1.33	104.98
X_{14}	0.088	0.150	0.59	4.23
X_{15}	0.009	0.101	0.09	1.91
$n = 60$	$R^2 = 0.765$	$R_a^2 = 0.685$	$\hat{\sigma} = 0.56$	df = 44

The ridge trace for the 15 regression coefficients is shown in Figures 12.2–12.4. Each figure shows five curves. If we put all 15 curves, the graph would be quite cluttered and the curves would be difficult to trace. To make the three graphs comparable, the scale is kept the same for all graphs. From the ridge trace, we see that some of the coefficients are quite unstable and some are small regardless of the value of the ridge parameter k.

We now follow the guidelines suggested for the selection of variables in multicollinear data. Following the first criterion we eliminate variables 7, 8, 10, 11, and 15. These variables all have fairly stable coefficients, as shown by the flatness of their ridge traces, but are very small. Although variable 14 has a small coefficient at $k = 0$ (see Table 12.10), its value increases sharply as k increases from zero. So, it should not be eliminated at this point.

We now repeat the analysis using the 10 remaining variables: 1, 2, 3, 4, 5, 6, 9, 12, 13, and 14. The corresponding OLS results are given in Table 12.11. There is still evidence of collinearity. The largest eigenvalue, $\lambda_1 = 3.378$, is about 619 times the smallest value $\lambda_{10} = 0.005$. The two VIFs for variable 12 and 13 are still high. The corresponding ridge traces are shown in Figures 12.5 and 12.6. Variable 14 continues to have a small coefficient at $k = 0$ but it

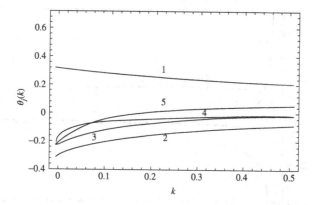

Figure 12.2 Air Pollution data: ridge traces for $\hat{\theta}_1, \ldots, \hat{\theta}_5$ (the 15-variable model).

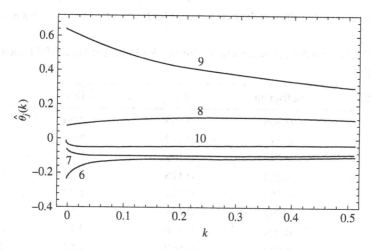

Figure 12.3 Air Pollution data: ridge traces for $\hat{\theta}_6, \ldots, \hat{\theta}_{10}$ (the 15-variable model).

increases as k increases from zero. So, it should be kept in the model at this stage. None of the other 9 variables satisfy the first criterion.

The second criterion suggests eliminating variables with unstable coefficients that tend to zero. Examination of the ridge traces in Figures 12.5 and 12.6 shows that variables 12 and 13 fall in this category.

The OLS results for the remaining eight variables are shown in Table 12.12. Collinearity has disappeared. Now, the largest and smallest eigenvalues are 2.886 and 0.094, which give a small condition number ($\kappa = 5.5$). The sum of the reciprocals of the eigenvalues is 23.5, about three times the number of variables. All values of VIF are less than 10. Since the retained variables are not collinear, we can now apply the variables selection methods for

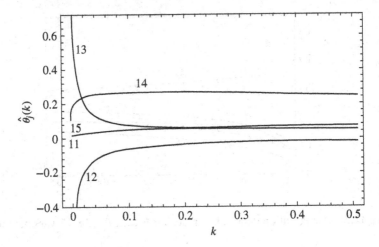

Figure 12.4 Air Pollution data: ridge traces for $\hat{\theta}_{11}, \ldots, \hat{\theta}_{15}$ (the 15-variable model).

Table 12.11 OLS Regression Output for the Air Pollution Data (10 Predictor Variables)

Variable	Coefficient	s.e.	t-Test	VIF
X_1	0.306	0.135	2.26	3.75
X_2	−0.345	0.119	−2.91	2.88
X_3	−0.245	0.108	−2.26	2.39
X_4	−0.224	0.175	−1.28	6.22
X_5	−0.268	0.137	−1.97	3.81
X_6	−0.292	0.103	−2.85	2.15
X_9	0.664	0.140	4.75	3.99
X_{12}	−1.017	0.659	−1.54	88.86
X_{13}	1.018	0.674	1.51	92.99
X_{14}	0.096	0.127	0.76	3.30
$n = 60$	$R^2 = 0.760$	$R_a^2 = 0.711$	$\hat{\sigma} = 0.537$	df = 49

noncollinear data discussed in Sections 12.7 and 12.8. This is left as an exercise for the reader.

An alternative way of analyzing these Air Pollution data is as follows: The collinearity in the original 15 variables is actually a simple case of collinearity; it involves only two variables (12 and 13). So, the analysis can proceed by eliminating any one of the two variables. The reader can verify that the remaining 14 variables are not collinear. The standard variables

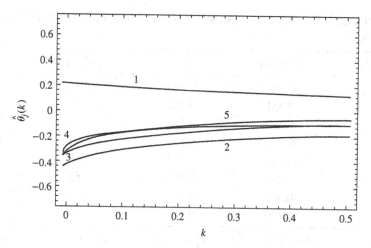

Figure 12.5 Air Pollution data: ridge traces for $\hat{\theta}_1, \ldots, \hat{\theta}_5$ (the 10-variable model).

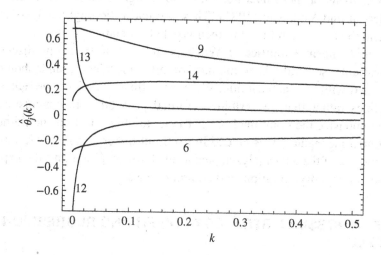

Figure 12.6 Air Pollution data: ridge traces for $\hat{\theta}_6$, $\hat{\theta}_9$, $\hat{\theta}_{12}$, $\hat{\theta}_{13}$ and $\hat{\theta}_{14}$ (the 10-variable model).

selection procedures for noncollinear data can now be utilized. We leave this as an exercise for the reader.

In our analysis of the Air Pollution data, we did not use the third criterion, but there are situations where this criterion is needed. We should note that ridge regression was used successfully in this example as a tool for variable selection. Because the variables selected at an intermediate stage were found to be noncollinear, the standard OLS was utilized.

Table 12.12 OLS Regression Output for the Air Pollution Data (Eight Predictor Variables)

Variable	Coefficient	s.e.	t-Test	VIF
X_1	0.331	0.120	2.76	2.91
X_2	−0.351	0.106	−3.31	2.28
X_3	−0.217	0.104	−2.09	2.19
X_4	−0.155	0.163	−0.95	5.42
X_5	−0.221	0.134	−1.66	3.62
X_6	−0.270	0.102	−2.65	2.10
X_9	0.692	0.133	5.219	3.57
X_{14}	0.230	0.083	2.77	1.40
$n = 60$	$R^2 = 0.749$	$R_a^2 = 0.709$	$\hat{\sigma} = 0.539$	df $= 51$

An analysis of these data not using ridge regression has been given by Henderson and Velleman (1981). They present a thorough analysis of the data and the reader is referred to their article for details.

Some General Comments: We hope it is clear from our discussion that variable selection is a mixture of art and science, and should be performed with care and caution. We have outlined a set of approaches and guidelines rather than prescribing a formal procedure. In conclusion, we must emphasize the point made earlier that variable selection should not be performed mechanically as an end in itself but rather as an exploration into the structure of the data analyzed, and as in all true explorations, the explorer is guided by theory, intuition, and common sense.

12.15 A POSSIBLE STRATEGY FOR FITTING REGRESSION MODELS

In this concluding section of the chapter we outline a possible sequence of steps that may be used to fit a regression model satisfactorily. Let us emphasize at the beginning that there is no single correct approach. The reader may be more comfortable with a different sequence of steps and should feel free to follow such a sequence. In almost all cases the analysis described here will lead to meaningful interpretable models useful in real-life applications.

We assume that we have a response variable Y which we want to relate to some or all of a set of variables $X_1, X_2, \ldots, X_p$. The set, $X_1, X_2, \ldots, X_p$, is often generated from external subject matter considerations. The set of variables is often large and we want to come to an acceptable reduced set. Our objective is to construct a valid and viable regression model. A possible sequence of steps are

1. Examine the variables $(Y, X_1, X_2, \ldots, X_p)$ one at a time. This can be done by calculating the summary statistics, and also graphically by looking at histograms, dot plots, or box plots (see Chapter 5). The distributions of the values should not be too skewed, nor the range of the variables very large. Look for outliers (check for transcription errors). Make transformations to induce symmetry and reduce skewness. Logarithmic transformations are useful in this situation (see Chapter 7).

2. Construct pairwise scatter plots for each variable. When p, the number of predictor variables, is large, this may not be feasible. Pairwise scatter plots are quite informative on the relationship between two variables. A look at the correlation matrix will point out obvious collinearity problems. Delete redundant variables. Calculate the condition number of the correlation matrix to get an idea of the severity of the collinearity (Chapters 10 and 11).

3. Fit the full linear regression model. Delete variables with no significant explanatory power (insignificant t-Tests). For the reduced model, examine the residuals:

   (a) Check linearity. If none, make a transformation on the variable (Chapter 7).

   (b) Check for heteroscedasticity and autocorrelation (for time series data). If present, take appropriate action (Chapters 8 and 9).

   (c) Look for outliers, high-leverage points, and influential points. If present, take appropriate action (Chapter 5).

4. Examine if additional variables can be dropped without compromising the integrity of the model. Examine if new variables are to be brought into the model (added variable plots, residual plus component plots) (Chapters 5 and 12). Repeat Step 3. Monitor the fitting process by examining the information criteria (AIC or BIC). This is particularly relevant in examining non-nested models.

5. For the final fitted model, check variance inflation factors. Ensure satisfactory residual plots and no negative diagnostic messages (Chapters 4, 6, 7, and 10). If need be, repeat Step 4.

6. Attempt should then be made to validate the fitted model. When the amount of data is large, the model may be fitted by part of the

data and validated by the remainder of the data. Resampling methods such as bootstrap, jackknife, and cross-validation are also possibilities, particularly when the amount of data available is not large [see Efron (1982) and Diaconis and Efron (1983)].

The steps we have described are, in practice, often not done sequentially but implemented synchronously. The process described is an iterative process and it may be necessary to recycle through the outlined steps several times to arrive at a satisfactory model. They enumerate the factors that must be considered for constructing a satisfactory model.

One important component that we have not included in our outlined steps is the subject matter knowledge of the analyst in the area in which the model is constructed. This knowledge should always be incorporated in the model-building process. Incorporation of this knowledge will often accelerate the process of arriving at a satisfactory model because it will help considerably in the appropriate choice of variables and corresponding transformations. After all is said and done, statistical model building is an art. The techniques that we have described are the tools by which this task can be attempted methodically.

12.16 BIBLIOGRAPHIC NOTES

There is a vast amount of literature on variable selection scattered in statistical journals. A very comprehensive review with an extensive bibliography may be found in Hocking (1976). A detailed treatment on variable selection with special emphasis on C_p statistic is given in the book by Daniel and Wood (1980). Refinements on the application of C_p statistic are given by Mallows (1973). The variable selection procedures are discussed in the book by Draper and Smith (1998). Use of ridge regression in connection with variable selection is discussed by Hoerl and Kennard (1970) and McDonald and Schwing (1973).

EXERCISES

12.1 As we have seen in Section 12.14, the three noncollinear subsets of predictor variables below have emerged. Apply one or more variable selection methods to each subset and compare the resulting final models:

(a) The subset of eight variables: 1, 2, 3, 4, 5, 6, 9, and 14.

(b) The subset of 14 variables obtained after omitting variable 12.

(c) The subset of 14 variables obtained after omitting variable 13.

12.2 The estimated regression coefficients in Table 12.10 correspond to the standardized versions of the variables because they are computed using the correlation matrix of the response and predictor variables. Using the means and standard deviations of the variables in Table 12.9, write the estimated regression equation in terms of the original variables (before centering and scaling).

12.3 In the Homicide data discussed in Section 12.12, we observed that when fitting the model in (12.13), the FS and BE methods give contradictory results. In fact, there are several other subsets in the data (not necessarily with three predictor variables) for which the FS and BE methods give contradictory results. Find one or more of these subsets.

12.4 Use the variable selection methods, as appropriate, to find one or more subsets of the predictor variables, in the file Homicide.csv at the Book's Website, that best account for the variability in the response variable H.

12.5 Property Valuation: Scientific mass appraisal is a technique in which linear regression methods applied to the problem of property valuation. The objective in scientific mass appraisal is to predict the sale price of a home from selected physical characteristics of the building and taxes (local, school, county) paid on the building. Twenty-four observations were obtained from *Multiple Listing* (Vol. 87) for Erie, PA, which is designated as Area 12 in the directory. These data were originally presented by Narula and Wellington (1977). The list of variables are given in Table 12.13. The data can be found in the file Property.Valuation.csv at the Book's Website.

Answer the following questions, in each case justifying your answer by appropriate analyses.

(a) In a fitted regression model that relates the sale price to taxes and building characteristics, would you include all the variables?

(b) A veteran real estate agent has suggested that local taxes, number of rooms, and age of the house would adequately describe the sale price. Do you agree?

(c) A real estate expert who was brought into the project reasoned as follows: The selling price of a home is determined by its desirability

Table 12.13 List of Variables for Data in the file Property.Valuation.csv at the Book's Website

Variable	Definition
Y	Sale price of the house in thousands of dollars
X_1	Taxes (local, county, school) in thousands of dollars
X_2	Number of bathrooms
X_3	Lot size (in thousands of square feet)
X_4	Living space (in thousands of square feet)
X_5	Number of garage stalls
X_6	Number of rooms
X_7	Number of bedrooms
X_8	Age of of the home (years)
X_9	Number of fireplaces

and this is certainly a function of the physical characteristic of the building. This overall assessment is reflected in the local taxes paid by the homeowner; consequently, the best predictor of sale price is the local taxes. The building characteristics are therefore redundant in a regression equation which includes local taxes. An equation that relates sale price solely to local taxes would be adequate. Examine this assertion by examining several models. Do you agree? Present what you consider to be the most adequate model or models for predicting sale price of homes in Erie, PA.

12.6 Refer to the data in the file Gasoline.Consumption.csv at the Book's Website.

(a) Would you include all the variables to predict the gasoline consumption of the cars? Explain, giving reasons.

(b) Six alternative models have been suggested:

  (i) Regression of Y on X_1
  (ii) Regression of Y on X_{10}
  (iii) Regression of Y on X_1 and X_{10}
  (iv) Regression of Y on X_2 and X_{10}
  (v) Regression of Y on X_8 and X_{10}
  (vi) Regression of Y on X_8 and X_5, and X_{10}.

Among these regression models, which would you choose to predict the gasoline consumption of automobiles? Can you suggest a better model?

(c) Plot Y against X_1, X_2, X_8, and X_{10} (one at a time). Do the plots suggest that the relationship between Y and the 11 predictor variables may not be linear?

(d) The gasoline consumption was determined by driving each car with the same load over the same track (a road length of about 123 miles). Instead of using Y (miles per gallon), it was suggested that we consider a new variable, $W = 100/Y$ (gallons per hundred miles). Plot W against X_1, X_2, X_8, and X_{10} and examine if the relationship between W and the 11 predictor variables is more linear than that between Y and the 11 predictor variables.

(e) Repeat Part (b) using W in place of Y. What are your conclusions?

(f) Regress Y on X_{13}, where $X_{13} = X_8/X_{10}$.

(g) Write a brief report describing your findings. Make a recommendation on the model to be used for predicting gasoline consumption of cars.

12.7 Refer to the data in the file Presidential.Election.csv at the Book's Website and, as in Exercise 10.9, consider fitting a model relating V to all the variables (including a time trend representing year of election) plus as many interaction terms involving two or three variables as you possibly can.

(a) Starting with the model in Exercise 10.9(a). Apply two or more variable selection methods to choose the best model(s) that might be expected to perform best in predicting future presidential elections.

(b) Repeat the above exercise starting with the model in Exercise 10.9(d).

(c) Which one of the models obtained above would you prefer?

(d) Use your chosen model to predict the proportion of votes expected to be obtained by a presidential candidate in the United States presidential elections in the years 2000, 2004, and 2008.

(e) Which one of the above three predictions would you expect to be more accurate than the other two? Explain.

(f) The result of the 2000 presidential election was not known at the time this edition went to press. If you happen to be reading this book after the election of the year 2000 and beyond, were your predictions in Exercise correct?

12.8 Cigarette Consumption Data: Consider the Cigarette Consumption data described in Exercise 4.17 and given in the file Cigarette. Consumption.csv at the Book's Website. The organization wanted to construct a regression equation that relates statewide cigarette consumption (per capita basis) to various socioeconomic and demographic variables, and to determine whether these variables were useful in predicting the consumption of cigarettes.

(a) Construct a linear regression model that explains the per capita sale of cigarettes in a given state. In your analysis, pay particular attention to outliers. See if the deletion of an outlier affects your findings. Look at residual plots before deciding on a final model. You need not include all the variables in the model if your analysis indicates otherwise. Your objective should be to find the smallest number of variables that describes the state sale of cigarettes meaningfully and adequately.

(b) Write a report describing your findings.

Appendix: 12.A Effects of Incorrect Model Specifications

In this appendix we discuss the effects of an incorrect model specification on the estimates of the regression coefficients and predicted values using matrix notation. Define the following matrix and vectors:

$$\mathbf{X} = \begin{bmatrix} x_{10} & x_{11} & \cdots & x_{1p} & x_{1(p+1)} & \cdots & x_{1q} \\ x_{20} & x_{21} & \cdots & x_{2p} & x_{2(p+1)} & \cdots & x_{2q} \\ \vdots & \vdots & \ddots & \vdots & \vdots & \ddots & \vdots \\ x_{n0} & x_{n1} & \cdots & x_{np} & x_{n(p+1)} & \cdots & x_{nq} \end{bmatrix}, \quad \mathbf{Y} = \begin{bmatrix} y_1 \\ \vdots \\ y_n \end{bmatrix},$$

$$\boldsymbol{\beta} = \begin{bmatrix} \beta_0 \\ \beta_1 \\ \vdots \\ \beta_p \\ \hline \beta_{p+1} \\ \vdots \\ \beta_q \end{bmatrix}, \quad \boldsymbol{\varepsilon} = \begin{bmatrix} \varepsilon_1 \\ \varepsilon_2 \\ \vdots \\ \varepsilon_n \end{bmatrix},$$

where $x_{i0} = 1$ for $i = 1, \ldots, n$. The matrix $\mathbf{X}$, which has n rows and $q + 1$ columns, is partitioned into two submatrices $\mathbf{X}_p$ and $\mathbf{X}_r$, of dimensions $n \times (p + 1)$ and $n \times r$, where $r = q - p$. The vector β is similarly partitioned into β_p and β_r, which have $p + 1$ and r components, respectively.

The full linear model containing all q variables is given by

$$\mathbf{Y} = \mathbf{X}\beta + \varepsilon = \mathbf{X}_p \beta_p + \mathbf{X}_r \beta_r + \varepsilon, \qquad (12.A.1)$$

where ε_i's are independently normally distributed errors with zero means and unit variance.

The linear model containing only p variables (i.e., an equation with $p + 1$ terms) is

$$\mathbf{Y} = \mathbf{X}_p \beta_p + \varepsilon. \qquad (12.A.2)$$

Let us denote the least squares estimate of β obtained from the full model (12.A.1) by $\hat{\beta}^*$, where

$$\hat{\beta}^* = \begin{pmatrix} \hat{\beta}_p^* \\ \hat{\beta}_r^* \end{pmatrix} = (\mathbf{X}^T \mathbf{X})^{-1} \mathbf{X}^T \mathbf{Y}.$$

The estimate $\hat{\beta}_p$ of β_p obtained from the subset model (12.A.2) is given by

$$\hat{\beta}_p = (\mathbf{X}_p^T \mathbf{X}_p)^{-1} \mathbf{X}_p^T \mathbf{Y}.$$

Let $\hat{\sigma}_q^2$ and $\hat{\sigma}_p^2$ denote the estimates of σ^2 obtained from (12.A.1) and (12.A.2), respectively. Then it follows that

$$\hat{\sigma}_q^2 = \frac{\mathbf{Y}^T \mathbf{Y} - \hat{\beta}^{*T} \mathbf{X}^T \mathbf{Y}}{n - q - 1} \quad \text{and} \quad \hat{\sigma}_p^2 = \frac{\mathbf{Y}^T \mathbf{Y} - \hat{\beta}_p^T \mathbf{X}_p^T \mathbf{Y}}{n - p - 1}.$$

It is known from standard theory that $\hat{\beta}^*$ and $\hat{\sigma}_q^2$ are unbiased estimates of β and σ^2. It can be shown that $E(\hat{\beta}_p) = \beta_p + \mathbf{A}\beta_r$, where

$$\mathbf{A} = (\mathbf{X}_p^T \mathbf{X}_p)^{-1} \mathbf{X}_p^T \mathbf{X}_r.$$

Further,

$$\mathrm{Var}(\hat{\beta}_p) = (\mathbf{X}_p^T \mathbf{X}_p)^{-1} \sigma^2,$$
$$\mathrm{Var}(\hat{\beta}^*) = (\mathbf{X}^T \mathbf{X})^{-1} \sigma^2,$$

and

$$\mathrm{MSE}(\hat{\beta}_p) = (\mathbf{X}_p^T \mathbf{X}_p)^{-1} \sigma^2 + \mathbf{A}\beta_r \beta_r^T \mathbf{A}^T.$$

We can summarize the properties of $\hat{\boldsymbol{\beta}}_p$ and $\hat{\boldsymbol{\beta}}_p^*$ as follows:

1. The $\hat{\boldsymbol{\beta}}_p$ is a biased estimate of $\boldsymbol{\beta}_p$ unless (1) $\boldsymbol{\beta}_r = 0$ or (2) $\mathbf{X}_p^T \mathbf{X}_r = 0$.

2. The matrix $\text{Var}(\hat{\boldsymbol{\beta}}_p^*) - \text{Var}(\hat{\boldsymbol{\beta}}_p)$ is positive semidefinite; that is, variances of the least squares estimates of regression coefficients obtained from the full model are larger than the corresponding variances of the estimates obtained from the subset model. In other words, the deletion of variables always results in smaller variances for the estimates of the regression coefficients of the remaining variables.

3. If the matrix $\text{Var}(\hat{\boldsymbol{\beta}}_r^*) - \boldsymbol{\beta}_r \boldsymbol{\beta}_r^T$ is positive semidefinite, then the matrix $\text{Var}(\hat{\boldsymbol{\beta}}_p^*) - \text{MSE}(\hat{\boldsymbol{\beta}}_p)$ is positive semidefinite. This means that the least squares estimates of regression coefficients obtained from the subset model have smaller mean square error than estimates obtained from the full model when the variables deleted have regression coefficients that are smaller than the standard deviation of the estimates of the coefficients.

4. $\hat{\sigma}_p^2$ is generally biased upward as an estimate of σ^2.

To see the effect of model misspecification on prediction, let us examine the prediction corresponding to an observation, say $\mathbf{x}^T = (\mathbf{x}_p^T : \mathbf{x}_r^T)$. Let $\hat{y}^*$ denote the predicted value corresponding to $\mathbf{x}^T$ when the full set of variables are used. Then $\hat{y}^* = \mathbf{x}^T \hat{\boldsymbol{\beta}}^*$ with mean $\mathbf{x}^T \boldsymbol{\beta}$ and prediction variance $\text{Var}(\hat{y}^*)$:

$$\text{Var}(\hat{y}^*) = \sigma^2 (1 + \mathbf{x}^T (\mathbf{X}^T \mathbf{X})^{-1} \mathbf{x}).$$

On the other hand, if the subset model (12.A.2) is used, the estimated predicted value $\hat{y} = \mathbf{x}_p^T \hat{\boldsymbol{\beta}}_p$ with mean

$$E(\hat{y}) = \mathbf{x}_p^T \boldsymbol{\beta}_p + \mathbf{x}_p^T \mathbf{A} \boldsymbol{\beta}_r$$

and prediction variance

$$\text{Var}(\hat{y}) = \sigma^2 (1 + \mathbf{x}_p^T (\mathbf{X}_p^T \mathbf{X}_p)^{-1} \mathbf{x}_p).$$

The prediction mean square error is given by

$$\text{MSE}(\hat{y}) = \sigma^2 (1 + \mathbf{x}_p^T (\mathbf{X}_p^T \mathbf{X}_p)^{-1} \mathbf{x}_p) + (\mathbf{x}_p^T \mathbf{A} \boldsymbol{\beta}_r - \mathbf{x}_r^T \boldsymbol{\beta}_r)^2.$$

The properties of $\hat{y}^*$ and $\hat{y}$ can be summarized as follows:

1. $\hat{y}$ is biased unless $\beta_r = 0$ or $\mathbf{X}_p^T\mathbf{X}_r = \mathbf{0}$ (i.e., $\mathbf{X}_p$ and $\mathbf{X}_r$ are orthogonal).
2. $\text{Var}(\hat{y}^*) \geq \text{Var}(\hat{y})$.
3. If the matrix $\text{Var}(\hat{\beta}_r^*) - \beta_r\beta_r^T$ is positive semidefinite, then $\text{Var}(\hat{y}^*) \geq \text{MSE}(\hat{y})$.

The significance and interpretation of these results in the context of variable selection are given in the main body of the chapter.

CHAPTER 13

LOGISTIC REGRESSION

13.1 INTRODUCTION

In our discussion of regression analysis so far the response variable Y has been regarded as a continuous quantitative variable. The predictor variables, however, have been both quantitative as well as qualitative. Indicator variables, which we have described earlier, fall into the second category. There are situations, however, where the response variable is qualitative. In this chapter we present methods for dealing with this situation. The methods presented in this chapter are very different from the method of least squares considered in the previous chapters.

Consider a procedure in which individuals are selected on the basis of their scores in a battery of tests. After five years the candidates are classified as "good" or "poor." We are interested in examining the ability of the tests to predict the job performance of the candidates. Here the response variable, performance, is dichotomous. We can code "good" as 1 and "poor" as 0, for example. The predictor variables are the scores in the tests.

In a study to determine the risk factors for cancer, health records of several people were studied. Data were collected on several variables, such

---

Regression Analysis By Example Using R, Sixth Edition. Ali S. Hadi and Samprit Chatterjee
© 2024 John Wiley & Sons, Inc. Published 2024 by John Wiley & Sons, Inc.
Companion website: www.wiley.com/go/hadi/regression_analysis_6e

as age, gender, smoking, diet, and the family's medical history. The response variable was the person had cancer ($Y = 1$) or did not have cancer ($Y = 0$).

In the financial community, the "health" of a business is of primary concern. The response variable is solvency of the firm (bankrupt = 0, solvent =1), and the predictor variables are the various financial characteristics associated with the firm. Situations where the response variable is a dichotomous variable are quite common and occur extensively in statistical applications.

13.2 MODELING QUALITATIVE DATA

The qualitative data with which we are dealing, the binary response variable, can always be coded as having two values, 0 or 1. Rather than predicting these two values we try to model the probabilities that the response takes one of these two values. The limitation of the previously considered standard linear regression model is obvious.

We illustrate this point by considering a simple regression problem in which we have only one predictor. The same considerations hold for the multiple regression case. Let π denote the probability that $Y = 1$ when $X = x$. If we use the standard linear model to describe π, then our model for the probability would be

$$\pi = \Pr(Y = 1 | X = x) = \beta_0 + \beta_1 x. \tag{13.1}$$

Since π is a probability it must lie between 0 and 1. The linear function given in (13.1) is unbounded and hence cannot be used to model probability. There is another reason why ordinary least squares method is unsuitable. The response variable Y is a binomial random variable; consequently, its variance will be a function of π and depends on X. The assumption of equal variance (homoscedasticity) does not hold. We could use the weighted least squares, but there are problems with that approach. The values of π are not known. In order to use weighted least squares approach, we will have to start with an initial guess for the value of π and then iterate. Instead of this complex method we will describe an alternative method for modeling probabilities.

13.3 THE LOGIT MODEL

The relationship between the probability π and X can often be represented by a *logistic response function*. It resembles an *S*-shaped curve, a sketch of which is given in Figure 13.1. The probability π initially increases slowly with

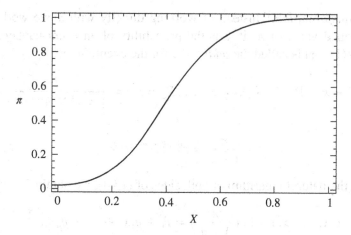

Figure 13.1 Logistic response function.

increase in X, then the increase accelerates, finally stabilizes, but does not increase beyond 1. Intuitively this makes sense. Consider the probability of a questionnaire being returned as a function of cash reward, or the probability of passing a test as a function of the time put in studying for it.

The shape of the S-curve given in Figure 13.1 can be reproduced if we model the probabilities as follows:

$$\pi = \Pr(Y = 1 | X = x) = \frac{e^{\beta_0 + \beta_1 x}}{1 + e^{\beta_0 + \beta_1 x}}, \qquad (13.2)$$

where e is the base of the natural logarithm. The probabilities here are modeled by the distribution function (cumulative probability function) of the logistic distribution. There are other ways of modeling the probabilities that would also produce the S-curve. The cumulative distribution of the normal curve has also been used. This gives rise to the *probit* model. We will not discuss the probit model here, as we consider the logistic model simpler and superior to the probit model.

The logistic model can be generalized directly to the situation where we have several predictor variables. The probability π is modeled as

$$\pi = \Pr(Y = 1 | X_1 = x_1, \ldots, X_p = x_p)$$
$$= \frac{e^{\beta_0 + \beta_1 x_1 + \beta_2 x_2 + \cdots + \beta_p x_p}}{1 + e^{\beta_0 + \beta_1 x_1 + \cdots + \beta_p x_p}}. \qquad (13.3)$$

Equation (13.3) is called the *logistic regression function*. It is nonlinear in the parameters $\beta_0, \beta_1, \ldots, \beta_p$. However, it can be linearized by the *logit transformation*.[1] Instead of working directly with π we work with a transformed value of π. If π is the probability of an event happening, the ratio $\pi/(1-\pi)$ is called the *odds ratio* for the event. Since

$$1 - \pi = \Pr(Y = 0 | X_1 = x_1, \ldots, X_p = x_p) = \frac{1}{1 + e^{\beta_0 + \beta_1 x_1 + \cdots + \beta_p x_p}},$$

then

$$\frac{\pi}{1-\pi} = e^{\beta_0 + \beta_1 x_1 + \cdots + \beta_p x_p}. \tag{13.4}$$

Taking the natural logarithm of both sides of (13.4), we obtain

$$g(x_1, \ldots, x_p) = \ln\left(\frac{\pi}{1-\pi}\right) = \beta_0 + \beta_1 x_1 + \cdots + \beta_p x_p. \tag{13.5}$$

The logarithm of the odds ratio is called the *logit*. It can be seen from (13.5) that the logit transformation produces a linear function of the parameters $\beta_0, \beta_1, \ldots, \beta_p$. Note also that while the range of values of π in (13.3) is between 0 and 1, the range of values of $\ln[\pi/(1-\pi)]$ is between $-\infty$ and $+\infty$, which makes the logits (the logarithm of the odds ratio) more appropriate for linear regression fitting.

Modeling the response probabilities by the logistic distribution and estimating the parameters of the model given in (13.3) constitutes fitting a logistic regression. In logistic regression the fitting is carried out by working with the logits. The logit transformation produces a model that is linear in the parameters. The method of estimation used is the *maximum likelihood* method. The maximum likelihood estimates are obtained numerically, using an iterative procedure. Unlike least squares fitting, no closed-form expression exists for the estimates of the parameters. We will not go into the computational aspects of the problem but refer the reader to McCullagh and Nelder (1989), Seber (1984), Hosmer and Lemeshow (1989), and Dobson and Barnett (2008).

To fit a logistic regression in practice a computer program is essential. Most regression packages have a logistic regression option. After the fitting one looks at the same set of questions that are usually considered in linear regression. Questions about the suitability of the model, the variables to be retained, and goodness of fit are all considered. Tools used are not the usual R^2, t-, and F-Tests, the ones employed in least squares regression, but others which provide answers to these same questions. Hypothesis testing is done

---

[1] See Chapter 7 for transformation of variables.

EXAMPLE: ESTIMATING PROBABILITY OF BANKRUPTCIES   413

by different methods, since the method of estimation is maximum likelihood as opposed to least squares. Information criteria such as AIC and BIC can be used for model selection. Instead of SSE, the logarithm of the likelihood for the fitted model is used. An explicit formula is given in Section 13.6.

13.4 EXAMPLE: ESTIMATING PROBABILITY OF BANKRUPTCIES

Detecting ailing financial and business establishments is an important function of audit and control. Systematic failure to do audit and control can lead to grave consequences, such as the savings-and-loan fiasco of the 1980s in the United States. The data we analyze here consist of some of the operating financial ratios of 33 firms that went bankrupt after 2 years and 33 that remained solvent during the same period. The data can be found in the file Financal.Ratios.csv at the Book's Website.[2]

A multiple logistic regression model is fitted using variables X_1, X_2, and X_3. The output from fitting the model is given in Table 13.1. Three financial ratios were available for each firm:

$$X_1 = \frac{\text{Retained Earnings}}{\text{Total Assets}},$$

$$X_2 = \frac{\text{Earnings Before Interest and Taxes}}{\text{Total Assets}},$$

$$X_3 = \frac{\text{Sales}}{\text{Total Assets}}.$$

The response variable is defined as

$$Y = \begin{cases} 0, & \text{if bankrupt after 2 years,} \\ 1, & \text{if solvent after 2 years.} \end{cases}$$

Table 13.1 has a certain resemblance to the standard regression output. Some of the output serve similar functions. We now describe and interpret the output obtained from fitting a logistic regression. If π denotes the probability of a firm remaining solvent after 2 years, the fitted logit is given by

$$\hat{g}(x_1, \ldots, x_p) = -10.15 + 0.33\, x_1 + 0.18\, x_2 + 5.09\, x_3. \tag{13.6}$$

---

[2] http://www.aucegypt.edu/faculty/hadi/RABE6

Table 13.1 Output from the Logistic Regression Using X_1, X_2, and X_3

Variable	Coeff.	s.e.	Z-Test	p-Value	Odds Ratio	95% C.I. Lower	Upper
Constant	−10.15	10.84	−0.94	0.35			
X_1	0.33	0.30	1.10	0.27	1.39	0.77	2.51
X_2	0.18	0.11	1.69	0.09	1.20	0.97	1.48
X_3	5.09	5.08	1.00	0.32	161.98	0.01	3.43×10^6
Log-Likelihood = −2.906			G = 85.683		df = 3	p-Value 0.000	

This corresponds to the fitted regression equation in standard analysis. Here instead of predicting Y we obtain a model to predict the logits, $\log[\pi/(1-\pi)]$. From the logits, after transformation, we can get the predicted probabilities. The constant and the coefficients are read directly from the second column in the table. The standard errors (s.e.) of the coefficients are given in the third column. The fourth column headed by Z is the ratio of the coefficient and the standard deviation. The Z is sometimes referred to as the Wald Statistic (Test). The Z corresponding to the coefficient of X_2 is obtained from dividing 0.181 by 0.107. In the standard regression this would be the *t*-Test. This ratio for the logistic regression has a normal distribution as opposed to a *t*-distribution that we get in linear regression. The fifth column gives the *p*-value corresponding to the observed Z value and should be interpreted like any *p*-value (see Chapters 3 and 4). These *p*-values are used to judge the significance of the coefficient. Values smaller than 0.05 would lead us to conclude that the coefficient is significantly different from 0 at the 5% significance level. From the *p*-values in Table 13.1, we see that none of the variables individually are significant for predicting the logits of the observations.

In the standard regression output the regression coefficients have a simple interpretation. The regression coefficient of the *j*th predictor variable X_j is the expected change in Y for unit change in X_j when other variables are held fixed. The coefficient of X_2 in (13.6) is the expected change in the logit for unit change in X_2 when the other variables are held fixed. The coefficients of a logistic regression fit have another interpretation that is of major practical importance. Keeping X_1 and X_3 fixed, for unit increase in X_2 the relative odds of

$$\frac{\Pr(\text{Firm solvent after 2 years})}{\Pr(\text{Firm bankrupt})}$$

is multiplied by $e^{\hat{\beta}_2} = e^{0.181} = 1.198$, that is, there is an increase of 20%. These values for each of the variables is given in the sixth column headed by Odds Ratio. They represent the change in odds ratio for unit change of a particular variable, while the others are held constant. The change in odds ratio for unit change in variable X_j, while the other variables are held fixed, is $e^{\hat{\beta}_j}$. If X_j was a binary variable, taking values 1 or 0, then $e^{\hat{\beta}_j}$ would be the actual value of the odds ratio rather than the change in the value of the odds ratio.

The 95% confidence intervals of the odds ratios are given in the last two columns of the table. If the confidence interval does not contain the value 1, the variable has a significant effect on the odds ratio. If the interval is below 1, the variable lowers significantly the relative odds. On the other hand, if the interval lies above 1, the relative odds is significantly increased by the variable.

To see whether the variables collectively contribute in explaining the logits a test that examines whether the coefficients $\beta_1, \ldots, \beta_p$ are all zero is performed. This corresponds to the case in multiple regression analysis where we test whether all the regression coefficients can be taken to be zero. The statistic G given at the bottom of Table 13.1 performs that task. The statistic G has a chi-square distribution. The p-value is considerably smaller than 0.05 and indicates that the variables collectively influence the logits.

13.5 LOGISTIC REGRESSION DIAGNOSTICS

After fitting a logistic regression model certain diagnostic measures can be examined for the detection of outliers, high-leverage points, influential observations, and other model deficiencies. The diagnostic measures developed in Chapter 5 for the standard linear regression model can be adapted to the logistic regression model. Regression packages with a logistic regression option usually give various diagnostic measures. These include:

1. The estimated probabilities $\hat{\pi}_i$, $i = 1, \ldots, n$.

2. One or more types of residuals, for example, the *standardized deviance residuals*, DR_i, and the *standardized Personian residuals*, PR_i, $i = 1, \ldots, n$.

3. The *weighted leverages*, p_{ii}^*, which measure the potential effects of the observations in the predictor variables on the obtained logistic regression results.

4. The scaled difference in the regression coefficients when the ith observation is deleted: DBETA_i, $i = 1, \ldots, n$.

5. The change in the chi-squared statistics G when the ith observation is deleted: DFG_i, $i = 1, \ldots, n$.

The formulas and derivations of these measures are beyond the scope of this book. The interested reader is referred to Pregibon (1981), Landwehr et al. (1984), Hosmer and Lemeshow (1989), and the references therein. The above measures, however, can be used in the same way as the corresponding measures obtained from a linear fit (Chapter 5). For example, the following graphical displays can be examined:

1. The scatter plot of DR_i versus $\hat{\pi}_i$
2. The scatter plot of PR_i versus $\hat{\pi}_i$
3. The index plots of DR_i, DBETA_i, DG_i, and p^*_{ii}.

As an illustrative example using the Bankruptcy data, the index plots of DR_i, DBETA_i, and DG_i obtained from the fitted logistic regression model in (13.6) are shown in Figures 13.2, 13.3, and 13.4,[3] respectively. It can easily be seen from these graphs that observations 9, 14, 52, and 53 are unusual and that they may have undue influence on the logistic regression results. We leave it as an exercise for the reader to determine if their deletion would make a significant difference in the results and the conclusion drawn from the analysis.

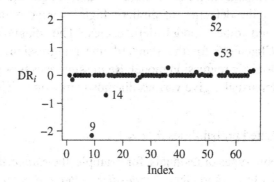

Figure 13.2 Bankruptcy data: Index plot of DR_i, the standardized deviance residuals.

---

[3] The R code for producing these and all other graphs in this book can be found at the Book's Website at http://www.aucegypt.edu/faculty/hadi/RABE6.

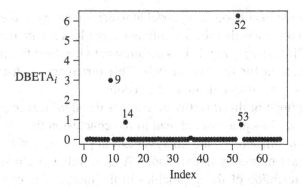

Figure 13.3 Bankruptcy data: Index plot of $DBETA_i$, the scaled difference in the regression coefficients when the ith observation is deleted.

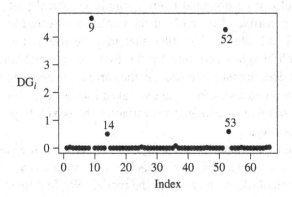

Figure 13.4 Bankruptcy data: Index plot of DG_i, the change in the chi-squared statistics G when the ith observation is deleted.

13.6 DETERMINATION OF VARIABLES TO RETAIN

In the analysis of the Bankruptcy data we have determined so far that the variables X_1, X_2, and X_3 collectively have explanatory power. Do we need all three variables? This is analogous to the problem of variable selection in multiple regression that was discussed in Chapter 12. Instead of looking at the reduction in the error sum of squares we look at the change in the likelihood (more precisely, the logarithm of the likelihood) for the two fitted models. The reason for this is that in logistic regression the fitting criterion is the maximum likelihood, whereas in least squares it is the least sum of squares. Let $L(p)$ denote the logarithm of the likelihood when we have a model with p variables and a constant. Similarly, let $L(p + q)$ be the logarithm of the likelihood for a model in which we have $p + q$ variables and a constant. To see whether the q additional variables contribute significantly we look at $2[L(p + q) - L(p)]$. This quantity is twice the difference between the log-likelihood for the two models. This difference is distributed as a chi-square variable with q degrees of freedom.

The magnitude of this quantity determines the significance of the test. A small value of chi-square would lead to the conclusion that the q variables do not add significantly to the improvement in prediction of the logits, and is therefore not necessary in the model. A large value of chi-square would call for the retention of the q variables in the model. The critical value is determined by the significance level of the test. This test procedure is valid when n, the number of observations available for fitting the model, is large.

An idea of the predictive power of a variable for possible inclusion in the logistic model can be obtained from a simple graphical plot. Side-by-side boxplots are constructed for each of the explanatory variable. Side-by-side boxplots will indicate the variables that may be useful for this purpose. Variables with boxplots different for the two groups are likely candidates. Note that this does not take into account the correlation between the variables. The formal procedure described above takes into account the correlations. With a large number of explanatory variables the boxplots provide a quick screening procedure.

In the Bankruptcy data we are analyzing, let us see if the variable X_3 can be deleted without degrading the model. We want to answer the question: Should the variable X_3 be retained in the model? We fit a logistic regression using X_1 and X_2. The results are given in Table 13.2. The log-likelihood for the model with X_1, X_2, and X_3 is -2.906, whereas with only X_1 and X_2 it is -4.736. Here $p = 2$ and $q = 1$, and $2[L(3) - L(2)] = 3.66$. This is a chi-square variable with 1 degree of freedom. Using the R command

qchisq(0.05, 1, lower.tail = FALSE)

Table 13.2 Output From the Logistic Regression Using X_1 and X_2

Variable	Coefficient	s.e.	Z-Test	p-Value	Odds Ratio	95% C.I. Lower	Upper
Constant	−0.550	0.951	−0.58	0.563			
X_1	0.157	0.075	2.10	0.036	1.17	1.01	1.36
X_2	0.195	0.122	1.59	0.112	1.21	0.96	1.54
Log-Likelihood = −4.736		G = 82.024		df = 2	p-Value 0.000		

DETERMINATION OF VARIABLES TO RETAIN   419

Table 13.3 Output from the Logistic Regression Using X_1

Variable	Coefficient	s.e.	Z-Test	p-Value	Odds Ratio	95% C.I. Lower	Upper
Constant	−1.167	0.816	−1.43	0.153			
X_1	0.177	0.057	3.09	0.002	1.19	1.07	1.33
Log-Likelihood = −7.902		G = 75.692		df = 1		p-Value = 0.0000	

we find that the 5% critical value of the chi-square distribution with 1 degree of freedom to be 3.84. Since 3.66 < 3.84, we can conclude that the variable X_3 can be deleted without affecting the effectiveness of the model at the 5% level of significant. Equivalently, the p-value for this test is

pchisq(3.66, 1, lower.tail = FALSE) = 0.0557,

which is larger than the significance level 0.05, leading to the same conclusion.

Let us now see if we can delete X_2. The result of regressing Y on X_1 is given in Table 13.3. The resulting log-likelihood is −7.902. The test statistic, which we have described earlier, has a value of 6.332. This is distributed as a chi-square random variable with 1 degree of freedom. The 5% value, as we saw earlier, was 3.84. The analysis indicates that we should not delete X_2 from our model. The p-value for this test, as can be verified, is 0.019. To predict probabilities of bankruptcies of firms in our data we should include both X_1 and X_2 in our model.

The procedure that we have outlined above enables us to test any *nested model*. A set of models are said to be nested if they can be obtained from a larger model as special cases. The methodology is similar to that used in analyzing nested models in multiple regression. The only difference is that here our test statistic is based on the log of the likelihood instead of sum of squares.

The AIC and BIC criteria discussed in Section 12.5.3 can be used to judge the suitability of various logistic models, and thereby the desirability of retaining a variable in the model. In the context of p-term logistic regression, AIC and BIC are

$$\text{AIC} = -2(\text{Log-Likelihood of the Fitted Model}) + 2p, \qquad (13.7)$$

$$\text{BIC} = -2(\text{Log-Likelihood of the Fitted Model}) + p \log n, \qquad (13.8)$$

where p denotes the number of variables in the model. Table 13.4 shows AIC and BIC for all possible models. The best AIC model is the one that includes

Table 13.4 The AIC and BIC Criteria for Various Logistic Regression models

Variables	AIC	BIC
$X_1 X_2 X_3$	13.81	22.57
$X_1 X_2$	15.47	22.04
$X_1 X_3$	18.12	24.69
$X_2 X_3$	33.40	39.97
X_1	19.80	24.18
X_2	34.50	38.88
X_3	92.46	96.84
None	93.50	95.69

all three variables (lowest AIC). While BIC picks $X_1 X_2$ as the best model, the one containing all three variables is equally adequate. The BIC for the two top models differ by less than 2.

13.7 JUDGING THE FIT OF A LOGISTIC REGRESSION

The overall fit of a multiple regression model is judged, for example, by the value of R^2 from the fitted model. No such simple satisfactory measure exists for logistic regression. Some ad hoc measures have been proposed which are based on the ratio of likelihoods. Most of these are functions of the ratio of the likelihood for the model and the likelihood of the data under a binomial model. These measures are not particularly informative and we will consider a different approach.

The logistic regression equation attempts to model probabilities for the two values of Y (0 or 1). To judge how well the model is doing we will determine the number of observations in the sample that the model is classifying correctly. Our approach will be to fit the logistic model to the data and calculate the fitted logits. From the fitted logits we will calculate the fitted probabilities for each observation. If the fitted probability for an observation is greater than 0.5, we will assign it to Group 1 ($Y = 1$), and if less than 0.5 we will classify it in Group 0 ($Y = 0$). We will then determine what proportion of the data is classified correctly. A high proportion of correct classification will indicate to us that the logistic model is working well. A low proportion of correct classification will indicate poor performance.

Different cutoff values, other than 0.5, have been suggested in the literature. In most practical situations, without any auxiliary information, such as the relative cost of misclassification or the relative frequency of the two categories in the population, 0.5 is recommended as a cutoff value.

A slightly more problematical question is how high the correct classification probability has to be before logistic regression is thought to be effective. Suppose that in a sample of size n there are n_1 observations from Group 1, and n_2 from Group 2. If we classify all the observations into one group or the other, then we will get either n_1/n or n_2/n proportions of observations classified correctly. As a base level for correct classification we can take the $\max(n_1/n, n_2/n)$. The proportion of observation classified correctly by the logistic regression should be much higher than the base level for the logistic model to be deemed useful.

For the Bankruptcy data that we have been analyzing logistic regression performs very well. Using variables X_1 and X_2, we find that the model misclassifies one observation from the solvent group (observation number 36) and one observation from the bankruptcy group (observation number 9). The overall correct classification rate $(64/66) = 0.97$. This is considerably higher than the base level rate of 0.5.

The concept of overall correct classification for the observed sample to judge the adequacy of the logistic model that we have discussed has been generalized. This generalization is used to produce a statistic to judge the fit of the logistic model. It is sometimes called the *Concordance Index* and is denoted by C. This statistic is calculated by considering all possible pairs formed by taking one observation from each group. Each of the pairs is then classified by using the fitted model. The Concordance Index is the percent of all possible pairs that is classified correctly. Thus, C lies between 0.5 and 1. Values of C close to 0.5 show the logistic model performing poorly (no better than guessing). The value of C for the logistic model with $X_1X_2X_3$ is 0.99. Several currently available software compute the value of C.

The observed correct classification rate should be treated with caution. In practice, if this logistic regression was applied to a new set of observations from this population, it would be very unlikely to do as well. The classification probability has an upward bias. The bias arises due to the fact that the same data that were used to fit the model were used to judge the performance of the model. The model fitted to a given body of data is expected to perform well on the same body of data. The true measure of the performance of the logistic regression model for classification is the probability of classifying a future observation correctly and not a sample observation. This upward bias in the estimate of correct classification probability can be reduced by using resampling methods, such as jack-knife or bootstrap. These will not be discussed here. The reader is referred to Efron (1982) and Diaconis and Efron (1983).

13.8 THE MULTINOMIAL LOGIT MODEL

In our discussion of logistic regression we have so far assumed that the qualitative response variable assumes only two values, generically, 1 for success and 0 for failure. The logistic regression model can be extended to situations where the response variable assumes more than two values. In a study of the choice of mode of transportation to work, the response variable may be private automobile, car pool, public transport, bicycle, or walking. The response falls into five categories. There is no natural ordering of the categories. We might want to analyze how the choice is related to factors such as age, gender, income, distance traveled, and so forth. The resulting model can be analyzed by using slightly modified methods that were used in analyzing the dichotomous outcomes. This method is called the *multinomial (polytomous)* logistic regression.

The response categories are not ordered in the example described above. There are situations where the response categories are ordered. In an opinion survey, the response categories might be strongly agree, agree, no opinion, disagree, and strongly disagree. The response categories are naturally ordered. In a clinical trial the responses to a treatment could be classified as improved, no change, or worse. For these situations a different method called the *Proportional Odds Model* is used. We discuss it in Section 13.8.3.

13.8.1 Multinomial Logistic Regression

We have n independent observations with p explanatory variables. The qualitative response variable has k categories. To construct the logits in the multinomial case one of the categories is considered the base level and all the logits are constructed relative to it. Any category can be taken as the base level. We will take category k as the base level in our description of the method. Since there is no ordering, it is apparent that any category may be labeled k. Let π_j denote the multinomial probability of an observation falling in the jth category. We want to find the relationship between this probability and the p explanatory variables, $X_1, X_2, \ldots, X_p$. The multiple logistic regression model then is

$$\ln\left(\frac{\pi_j(x_i)}{\pi_k(x_i)}\right) = \beta_{0j} + \beta_{1j}x_{1i} + \beta_{2j}x_{2i} + \cdots + \beta_{pj}x_{pi}; \quad \begin{matrix} j = 1, 2, \ldots, (k-1), \\ i = 1, 2, \ldots, n. \end{matrix}$$

Since all the π's add to unity, this reduces to

$$\ln(\pi_j(x_i)) = \frac{\exp(\beta_{0j} + \beta_{1j}x_{1i} + \beta_{2j}x_{2i} + \cdots + \beta_{pj}x_{pi})}{1 + \sum_{j=1}^{k-1} \exp(\beta_{0j} + \beta_{1j}x_{1i} + \beta_{2j}x_{2i} + \cdots + \beta_{pj}x_{pi})},$$

for $j = 1, 2, \ldots, (k-1)$. The model parameters are estimated by the method of maximum likelihood. Statistical software is available to do this fitting. We illustrate the method by an example.

13.8.2 Example: Determining Chemical Diabetes

To determine the treatment and management of diabetes it is necessary to determine whether the patient has chemical diabetes or overt diabetes. The data we present here is from a study conducted to determine the nature of chemical diabetes. The measurements were taken on 145 nonobese volunteers who were subjected to the same regimen. Many variables were measured, but we consider only three of them. These are insulin response (IR), the steady-state plasma glucose (SSPG), which measures insulin resistance, and relative weight (RW). The diabetic status of each subject was recorded. The clinical classification (CC) categories were overt diabetes (a), chemical diabetes (b), and normal (c). The data set is found in Andrews and Herzberg (1985). More details of the study are found in Reaven and Miller (1979). The data can be found in the file Diabetes.csv at the Book's Website.

Side-by-side boxplots of the explanatory variables indicate that the distribution of IR and SSPG differs for the three categories. The distribution of RW on the other hand does not differ substantially for the three categories. The boxplots are shown in Figure 13.5. The results of fitting a multinomial logistic model using the variables IR, SSPG, and RW are given in Table 13.5. Each of the logistic models is given relative to normal patients.

We see that RW has insignificant values in each of the logit models. This is consistent with what we observed in the side-by-side boxplots. We now fit the multinomial logistic model with two variables, SSPG and IR. The results are given in Table 13.6.

Looking at Logit (1/3), we see that higher values of SSPG increase the odds of overt diabetes, while a decrease in IR reduces the same odds when compared to normal subjects. Looking at Logit (2/3), we see that the higher values of SSPG increase the odds of chemical diabetes when compared to the normal subjects. The IR value does not significantly affect the odds. This indicates the difference between chemical and overt diabetes and has implications for the treatment of the two conditions.

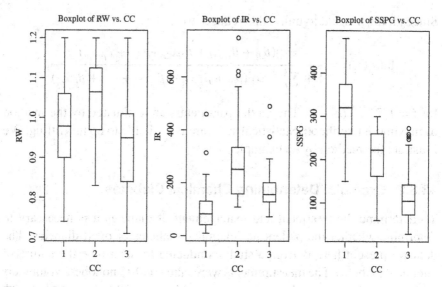

Figure 13.5 Side-by-side boxplots for the Diabetes data.

Table 13.5 Multinomial Logistic Regression Output with RW, SSPG, and IR (Base Level =3)

Variable	Coefficient	s.e.	Z-Test	p-Value	Odds Ratio	95% C.I. Lower	95% C.I. Upper
Logit 1: (2/3)							
Constant	−7.615	2.336	−3.26	0.001			
RW	3.473	2.446	1.42	0.156	32.23	0.27	3894.2
SSPG	0.016	0.005	3.29	0.001	1.02	1.01	1.03
IR	0.004	0.002	1.53	0.127	1.00	1.00	1.01
Logit 2: (1/3)							
Constant	−1.845	3.463	−0.53	0.594			
RW	−5.868	3.867	−1.52	0.129	0.00	0.00	5.53
SSPG	0.046	0.009	4.92	0.000	1.05	1.03	1.07
IR	−0.0134	0.005	−2.66	0.008	0.99	0.98	1.00
Log-Likelihood = −68.415	G = 159.369	df = 6	p-Value = 0.0000				

Although we have taken 3 as the base level, from our computation we can derive other comparisons. We can get Logit (1/2) from the relation

$$\text{Logit}(1/2) = \text{Logit}(1/3) - \text{Logit}(2/3). \qquad (13.9)$$

Table 13.6 Multinomial Logistic Regression Output with SSPG and IR (Base Level = 3)

Variable	Coefficient	s.e.	Z-Test	p-Value	Odds Ratio	95% C.I. Lower	95% C.I. Upper
Logit 1: (2/3)							
Constant	−4.549	0.771	−5.90	0.000			
SSPG	0.020	0.004	4.38	0.000	1.02	1.01	1.03
IR	0.003	0.002	1.42	0.155	1.00	1.00	1.01
Logit 2: (1/3)							
Constant	−7.111	1.688	−4.21	0.000			
SSPG	0.043	0.008	5.34	0.000	1.04	1.03	1.06
IR	−0.013	0.005	−2.89	0.004	0.99	0.98	1.00

Log-Likelihood = −72.029 $G = 152.141$ df = 4   p-Value = 0.0000

Table 13.7 Classification Table of Diabetes Data Using Multinomial Logistic Regression

CC	Predict 1	Predict 2	Predict 3	All
1	27	3	3	33
2	1	22	13	36
3	2	5	69	76
All	30	30	85	145

We can judge how well the multinomial logistic regression classifies the observations into different categories. The methodology is similar to binary logistic regression. An observation is classified to that category for which it has the highest estimated probability. The classification table for the multinomial logistic regression is given in Table 13.7.

One can see that 118 out of 145 subjects studied are classified correctly by this procedure. Thus, 81% of the observations are correctly classified. This is considerably higher than the maximum correct rate 59% (85/145), which would have been obtained if all the observations were put in one category. Multinomial logistic regression has performed well on this data. It is a powerful technique that should be used more extensively.

13.8.3 Ordinal Logistic Regression

The response variable in many studies, as has been pointed out earlier, can be qualitative and fall in more than two categories. The categories may sometimes be ordered. In a consumer satisfaction study, the responses might be highly satisfied, satisfied, dissatisfied, and highly dissatisfied. An analyst may want to study the socioeconomic and demographic factors that influence the response. The logistic model, slightly modified, can be used for this analysis. The logits here are based on the cumulative probabilities. Several logistic models can be based on the cumulative logits. We describe one of these, the *proportional odds model*.

Again, we have n independent observations with p predictors. The response variable falls into k categories $(1, 2, \ldots, k)$. The k categories are ordered. Let Y denote the response variable. The cumulative distribution for Y is

$$F_j(x_i) = \Pr(Y \leq j | X_i = x_{i1}, \ldots, X_p = x_{ip},); \quad j = 1, 2, \ldots, (k-1).$$

The proportional odds model is given by,

$$L_j(x_i) = \ln\left(\frac{F_j(x_i)}{1 - F_j(x_i)}\right) = \beta_{0j} + \beta_{1j}x_{1i} + \beta_{2j}x_{2i} + \cdots + \beta_{pj}x_{pi},$$

for $j = 1, 2, \ldots, (k-1)$. The cumulative logit has a simple interpretation. It can be interpreted as the logit for a binary response in which the categories from 1 to j is one category, and the remaining categories from $j+1$ to k is the second category. The model is fitted by the maximum likelihood method. Several statistical software packages will carry out this procedure. Increase in the value of a response variable with a positive β will increase the probability of being in a lower numbered category, all other variables remaining the same. The number of parameters estimated to describe the data is fewer in the ordinal than in the nominal model. For a more detailed discussion the reader is referred to Agresti (2002), Simonoff (2003), and Hoffmann (2016).

13.8.4 Example: Determining Chemical Diabetes Revisited

We will use the data on chemical diabetes considered in Section 13.8.2 to illustrate ordinal logistic regression. The clinical classifications in the previous categories are ordered, but we did not take it into consideration in our analysis. The progression of diabetes goes from normal (c), chemical (b), to overt diabetes (a). The classification states have a natural order and we

Table 13.8 Ordinal Logistic Regression Model (Proportional Odds) Using SSPG and IR

Variable	Coefficient	s.e.	Z-Test	p-Value	Odds Ratio	95% C.I. Lower	Upper
Constant 1	−6.794	0.872	−7.79	0.000			
Constant 2	−4.189	0.665	−6.30	0.000			
IR	−0.004	0.002	−2.30	0.021	1.00	0.99	1.00
SSPG	0.028	0.004	7.73	0.000	1.03	1.02	1.04

Log-Likelihood = −81.749   G = 132.700   df = 2   p-Value = 0.0000

will use them in our analysis. We will fit the proportional odds logit model. The results of the fit are given in Table 13.8.

The fit for the model is good. Both variables have a significant relationship to the group membership. The coefficient of SSPG is positive. This indicates that higher values of SSPG increase the probability of being in a lower numbered category, other factors being the same. The coefficient of IR is negative, indicating that higher values of this variable increase the probability of being in a higher numbered category, other factors remaining the same. The coefficient of concordance is high (0.90) showing the ability of the model to classify the group membership is high. In Table 13.9, we give the classification table for the ordinal logistic regression.

Of the 145 subjects ordinal logit regression classifies 114 subjects to their correct group. This gives the correct classification rate as 79%. This is comparable to the rate achieved by the multinomial logit model. It is generally expected that the ordinal model will do better than the multinomial model because of the additional information provided by the ordering of the categories. It should be also noted the ordinal logit model uses fewer

Table 13.9 Classification Table of Diabetes Data Using Multinomial Logistic Regression

CC	Predict 1	2	3	All
1	26	5	2	33
2	3	20	13	36
3	0	8	68	76
All	29	33	83	145

parameters than the multinomial model. In our example the ordinal model uses 4 parameters, while the nominal version uses 6. For a more detailed discussion the reader is referred to Agresti (2002) and Simonoff (2003).

13.9 CLASSIFICATION PROBLEM: ANOTHER APPROACH

The method of logistic regression has been used to model the probability that an observation belongs to one group given the measurements on several characteristics. We have described how the fitted logits could then be used for classifying an observation into one of two categories. A different statistical methodology is available if our primary interest is *classification*. When the sole interest is to predict the group membership of each observation, a statistical method called *discriminant analysis* is commonly used. Without discussing discriminant analysis here, we indicate a simple regression method that will accomplish the same task. The reader can find a discussion of discriminant analysis in McLachlan (1992), Rencher (1995), and Johnson (1998).

The essential idea in discriminant analysis is to find a linear combination of the predictor variables $X_1, \ldots, X_p$, such that the scores given by this linear combination separate the observations from the two groups as far as possible. One way that this separation can be accomplished is by fitting a multiple regression model to the data. The response variable is Y, taking values 0 and 1, and the predictors are $X_1, \ldots, X_p$. As has been pointed out earlier, some of the fitted values will be outside the range of 0 and 1. This does not matter here, as we are not trying to model probabilities, but only to predict group membership. We calculate the average of the predicted values of all the observations. If the predicted value for a given observation is greater than the average predicted value we assign that observation to the group which has $Y = 1$; if the predicted value is smaller than the average predicted value we assign it to the group with $Y = 0$. From this assignment we determine the

Table 13.10 Results from the OLS Regression of Y on X_1, X_2, X_3

Variable	Coefficient	s.e.	t-Test	p-Value
Constant	0.322	0.087	3.68	0.0005
X_1	0.003	0.001	3.76	0.0004
X_2	0.004	0.001	2.96	0.0044
X_3	0.149	0.045	3.28	0.0017
$n = 66$	$R^2 = 0.57$	$R_a^2 = 0.55$	$\hat{\sigma} = 0.3383$	df $= 62$

CLASSIFICATION PROBLEM: ANOTHER APPROACH

Table 13.11 Classification of Observations by Fitted Values

Row	Y	Fitted	Assigned	Row	Y	Fitted	Assigned
1	0	−0.00	0	34	1	0.72	1
2	0	0.48	0	35	1	0.82	1
3	0	−0.12	0	36	1	0.73	1
4	0	0.31	0	37	1	0.80	1
5	0	0.23	0	38	1	0.65	1
6	0	0.14	0	39	1	0.80	1
7	0	0.33	0	40	1	0.75	1
8	0	−0.32	0	41	1	0.76	1
9	0	0.52	1[a]	42	1	0.83	1
10	0	0.12	0	43	1	1.10	1
11	0	0.23	0	44	1	1.42	1
12	0	−0.07	0	45	1	0.86	1
13	0	−0.80	0	46	1	0.66	1
14	0	0.55	1[a]	47	1	0.81	1
15	0	0.03	0	48	1	0.58	1
16	0	−0.45	0	49	1	0.97	1
17	0	0.64	1[a]	50	1	1.03	1
18	0	0.45	0	51	1	0.77	1
19	0	0.44	0	52	1	0.48	0[a]
20	0	0.14	0	53	1	0.60	1
21	0	0.22	0	54	1	0.74	1
22	0	0.37	0	55	1	0.81	1
23	0	0.18	0	56	1	0.84	1
24	0	0.05	0	57	1	0.62	1
25	0	0.55	1[a]	58	1	0.81	1
26	0	0.56	1[a]	59	1	0.74	1
27	0	0.39	0	60	1	0.84	1
28	0	0.34	0	61	1	0.86	1
29	0	0.39	0	62	1	0.80	1
30	0	0.26	0	63	1	0.68	1
31	0	0.39	0	64	1	0.83	1
32	0	0.12	0	65	1	0.61	1
33	0	0.44	0	66	1	0.59	1

[a] Wrongly classified observations.

number of observations classified correctly in the sample. The variables used in this classification procedure are determined exactly by the same methods as those used for variable selection in multiple regression.

We illustrate this method by applying it to the Bankruptcy data that we have used earlier to illustrate least squares regression. Table 13.10 gives the OLS regression results using the three predictor variables X_1, X_2, and X_3. All three variables have significant regression coefficients and should be retained for classification equation.

Table 13.11 displays the observed Y, the predicted Y, and the assigned group for the Bankruptcy data. The average value of the predicted Y is 0.5. All observations with predicted value less than 0.5 are assigned to $Y = 0$, and those with predicted value greater than 0.5 are assigned to the group with $Y = 1$. The wrongly classified observations are marked with an asterisk. It is seen that five bankrupt firms are classified as solvent, and one solvent firm is classified as bankrupt. The logistic regression, it should be noted, classified only two observations wrongly. One solvent firm and one bankrupt firm were misclassified. For the Bankruptcy data presented in Table 13.1, the logistic regression performs better than the multiple regression in classifying the sample data. In general this is true. The logistic regression does not have to make the restrictive assumption of multivariate normality for the predictor variables. For classification problems we recommend the use of logistic regression. If a logistic regression package is not available, then the multiple regression approach may be tried.

EXERCISES

13.1 The diagnostic plots in Figures 13.2, 13.3, and 13.4 show three unusual observations in the Bankruptcy data. Fit a logistic regression model to the 63 observations without these three observations and compare your results with the results obtained in Section 13.5. Does the deletion of the three points cause a substantial change in the logistic regression results?

13.2 Examine the various logistic regression diagnostics obtained from fitting the logistic regression Y on X_1 and X_2 (Table 13.2) and determine if the data contain unusual observations.

13.3 The *O-rings* in the booster rockets used in space launching play an important part in preventing rockets from exploding. Probabilities of O-ring failures are thought to be related to temperature. A detailed discussion of the background of the problem is found in *The Flight of the Space Shuttle Challenger* in Chatterjee, Handcock, and Simonoff (1995, pp. 33–35). Each flight has six O-rings that could be potentially damaged in a particular flight. The data from 23 flights are given in the file Challenger.csv at the Book's Website. For each flight we have the number of O-rings damaged and the temperature of the launch.

(a) Fit a logistic regression connecting the probability of an O-ring failure with temperature. Interpret the coefficients.

(b) The data for Flight 18 that was launched when the launch temperature was 75 degrees Fahrenheit was thought to be problematic, and was deleted. Fit a logistic regression to the reduced data set. Interpret the coefficients.

(c) From the fitted model, find the probability of an O-ring failure when the temperature at launch was 31 degrees Fahrenheit. This was the temperature forecast for the day of the launching of the fatal *Challenger* flight on January 20, 1986.

(d) Would you have advised the launching on that particular day?

13.4 Field-goal-kicking data for the entire American Football League (AFL) and National Football League (NFL) for the 1969 season are given in Table 13.12 and can also be found at the Book's Website. Let $\pi(X)$ denote the probability of kicking a field goal from a distance of X yards.

Table 13.12 Field-Goal-Kicking Performances of the American Football League (AFL) and National Football League (NFL) for the 1969 Season. The Variable Z is an Indicator Variable Representing League

League	Distance	Success	Attempts	Z
NFL	14.5	68	77	0
NFL	24.5	74	95	0
NFL	34.5	61	113	0
NFL	44.5	38	138	0
NFL	52.0	2	38	0
AFL	14.5	62	67	1
AFL	24.5	49	70	1
AFL	34.5	43	79	1
AFL	44.5	25	82	1
AFL	52.0	7	24	1

Source: Morris and Rolph (1981, p. 200).

(a) For each of the leagues, fit the model

$$\pi(X) = \frac{e^{\beta_0 + \beta_1 X + \beta_2 X^2}}{1 + e^{\beta_0 + \beta_1 X + \beta_2 X^2}}.$$

(b) Let Z be an indicator variable representing the league, that is,

$$Z = \begin{cases} 1, & \text{for the AFL,} \\ 0, & \text{for the NFL.} \end{cases}$$

Fit a single model combining the data from both leagues by extending the model to include the indicator variable Z; that is, fit

$$\pi(X, Z) = \frac{e^{\beta_0 + \beta_1 X + \beta_2 X^2 + \beta_3 Z}}{1 + e^{\beta_0 + \beta_1 X + \beta_2 X^2 + \beta_3 Z}}.$$

(c) Does the quadratic term contribute significantly to the model?

(d) Are the probabilities of scoring field goals from a given distance the same for each league?

13.5 Using the data given in the file Health.Care.csv at the Book's Website (A description of the data is found in Section 1.3.8):

(a) Are rural facilities different from nonrural facilities? Use logistic regression to determine the best fitting model.

(b) How do the hospital characteristics affect patient care revenue? Use a sequential approach to arrive at the best model.

13.6 Using the data in the file Diabetes at the Book's Website:

(a) Show that inclusion of the variable RW does not result in a substantial improvement in the classification rate from the multinomial logistic model using IR and SSPG.

(b) Fit an ordinal logistic model using RW, IR, and SSPG to explain CC. Show that there is no substantial improvement in fit, and the correct classification rate from a model using only IR and SSPG.

CHAPTER 14

FURTHER TOPICS

14.1 INTRODUCTION

In this chapter we discuss two topics that have come up several times earlier, but we did not focus on them. We will be discussing generalized linear models (GLM) and robust regression. These are two vast topics that would require full-length books. We will give brief descriptions of the topics and provide examples that illustrate the concepts. GLM unifies the concept of linear model building, a primary activity of statistical analysts.

The importance of robust models in any statistical analysis cannot be overemphasized. The earlier chapters have provided us with methods for constructing robust models. In Section 14.5 we discuss methods that exclusively aim at robustness. The discussion on these two topics will not be exhaustive but reflect our personal experience and preferences.

14.2 GENERALIZED LINEAR MODEL

As in Chapter 4, given a response variable Y and p predictor variables X_1, X_2, ..., X_p, the linear regression model can be described as follows: An observation Y_i can be written as

$$Y_i = \beta_0 + \beta_1 X_{i1} + \beta_2 X_{i2} + \cdots + \beta_p X_{ip} + \varepsilon_i$$
$$= \mu_i + \varepsilon_i, \tag{14.1}$$

where μ_i is called the linear predictor and ε_i is a random error assumed to have a Gaussian (normal) distribution.

The GLM extends the linear regression model in two ways. The ε_i is assumed to have a distribution coming from the exponential family. The exponential family includes several standard distributions, in addition to the Gaussian. For example, it includes the binomial, Poisson, Gamma, and Inverse Gaussian distributions.

The second generalization is that the mean function μ_i is not necessarily the linear predictor, but some monotonic differentiable function of the linear predictor,

$$h(\mu_i) = \beta_0 + \beta_1 X_{i1} + \beta_2 X_{i2} + \cdots + \beta_p X_{ip}, \tag{14.2}$$

where $h(\mu)$ denotes the function that links μ to the linear predictor. The function relating the mean to the linear predictor is called the link function.

These two generalizations considerably increase the flexibility of linear models. The GLM can be used in situations where a linear regression model would not be appropriate. These models are fitted by the method of maximum likelihood. Most statistical packages have programs that can be used to fit and analyze generalized linear models. Applications of GLM using R are discussed in Dunn and Smyth (2018).

The GLM were first proposed by Nelder and Wedderburn (1972) and extensively developed by McCullagh and Nelder (1989). For computational details the reader should consult the references given above. A very accessible discussion is given in Simonoff (2003) and Fox (2016). A good theoretical foundation of generalized linear models is found in Agresti (2015). A recent book chapter about GLM is written by Dunn (2023).

The logistic regression, which we discussed in the previous chapter, is an example of GLM although we did not describe it in those terms. We now describe logistic regression as a GLM. The probability distribution of the random error was binomial, since there are only two outcomes. Instead of the mean, π_i, being the linear predictor, we took a function of π_i, namely, $\ln[\pi_i/(1-\pi_i)]$ as the linear predictor. The logistic regression model can now be described as a GLM from the binomial family with a logit link function. Another example of GLM is the Poisson regression model. This is discussed in the next section.

14.3 POISSON REGRESSION MODEL

Poisson regression models are appropriate when the response variable is count data. A researcher in the public health area may be interested in studying the number of hospitalizations of a group of people and the characteristics associated with these patients. Simonoff (2003) studies the number of tornado deaths in relation to the month, year, and the classification of the tornado's severity. In Section 7.4, we have analyzed injury accidents in airlines. These data can be analyzed by Poisson regression, because here we are dealing with count data. We analyzed these data earlier by using the square root transformation, which is an approximation to the exact method that we are now considering. Note that in these data sets the counts are small numbers, and small values are observed more frequently than large values.

The Poisson regression model can be described as follows: The random component has a Poisson distribution (see Section 7.4 for the Poisson distribution), and the mean is linked to the linear predictor by a logarithmic function

$$\ln(\mu_i) = \beta_0 + \beta_1 X_{i1} + \beta_2 X_{i2} + \cdots + \beta_p X_{ip}. \tag{14.3}$$

The test and the inferences on the Poisson model are carried out in the same way as the logit model (logistic regression). In some analysis instead of analyzing the number of cases (y) we may be interested in analyzing the rates of occurrence. Let y_i be the number observed out of a_i that are exposed to the risk. To construct a model for the rate we have only to modify the link function. The link function for the rate is

$$\ln(\mu_i) = \beta_0 + \beta_1 X_{i1} + \beta_2 X_{i2} + \cdots + \beta_p X_{ip} + \ln(a_i). \tag{14.4}$$

The quantity $\ln(a_i)$ is called the off-set, and the $\ln(\mu_i)$ now represents the logarithm of the mean rate of occurrence instead of the logarithm of the mean occurrence.

We now illustrate Poisson regression by an example.

14.4 INTRODUCTION OF NEW DRUGS

The number of new drugs (D) for 16 diseases brought to the U.S. market between 1992 and 2005 can be found in the file New.Drugs.csv. at the Book's Website. Also provided are the prevalence rates (P) of these diseases for 100,000 people. The money allocated for research by the National Institute of Health (M) during the year 1994 for a specific disease in millions of dollars is also given. This data set was kindly provided to us by Dr. Salomeh Keyhani of Mount Sinai Medical School. This is a part of a much larger database. We are using these three variables to illustrate the application of Poisson regression.

We are interested in studying the relationship between D (the response variable), with P and M. It should be noted that D is an integer variable with small values. A new drug coming to the market is a rare event. Poisson distribution is often used to model rare events. We now will fit a GLM with a Poisson random component and log link function. The result is given in Table 14.1

The large values of Wald's (Z-Test) show that the two variables are strongly related to the response variable D. The value of AIC[1] is 82.14. The fitted values ($\hat{D}$) are given in the last column of the data in the file New.Drugs.csv at the Book's Website. The agreement between observed and fitted values is satisfactory.

We will fit the data by least squares and compare this AIC value with that obtained from the least squares fit. The least squares fit is given in Table 14.2.

Only the M coefficient is significant in the least squares fit. As we have pointed out earlier, the linear regression model is not appropriate here as the response variable is count data. The value of AIC is 82.1 for the Poisson model compared to 88.2 for the linear regression model. The AIC indicates

Table 14.1 Output From the Poisson Regression Using P and M

Variable	Coefficient	s.e.	Z-Test	p-Value
Constant	0.8778	0.2074	4.233	0.0000
P	2.700×10^{-5}	9.508×10^{-6}	2.840	0.0045
M	1.998×10^{-3}	3.008×10^{-4}	6.642	0.0000
	Log-Likelihood = −9.721	df = 2	AIC = 82.14	

---

[1] See Section 12.5.3.

Table 14.2 Output From the Linear Regression Using P and M

Variable	Coefficient	s.e.	t-Test	p-Value
Constant	0.8362	1.379	0.607	0.5546
P	1.317×10^{-4}	8.973×10^{-5}	1.467	0.1661
M	0.01688	3.378×10^{-3}	4.996	0.0002
$R^2 = 0.671$	$R_a^2 = 0.620$	$\hat{\sigma} = 3.292$	df $= 13$	AIC $= 88.207$

that the Poisson model fits the data better than does the linear regression model. The pattern of residuals is also more satisfactory for the Poisson model. The Poisson regression model is more appropriate for these data.

14.5 ROBUST REGRESSION

The regression model fitted to data should be robust, in the sense that deletion of one or two observations should not cause a drastic change in the model. In Chapter 5 (particularly Sections 5.8 and 5.9), we have described how to detect these points. Our prescription was to delete these points to get a more stable and realistic model. We will now consider a method in which instead of deleting these points (which may be considered subjective), we reduce the impact of these points. There are several ways of getting a regression model which is robust. We describe a method which is simple and effective. The problem with the least squares is that the procedure gives too much weight to outliers and high-leverage points in the fitting. This has been illustrated extensively in Sections 5.8 and 5.9. The effect of these points can be reduced by down weighting these points in the fitting. We use weighted least squares (WLS), in which low weight is given to points with (a) high leverage and (b) large residuals. Since the weights are determined by the residuals, and as these change from iteration to iteration, the procedure is an iterative one. The explicit form of the weights and the procedure are given in Algorithm 14.1 below, where we use Q^j to denote the value of Q in the jth iteration step.

Algorithm 14.1

Input: An $n \times 1$ response vector $\mathbf{Y}$ and the corresponding $n \times p$ predictors matrix $\mathbf{X}$.

Output: A weighted least squares robust estimates of the regression coefficients and the corresponding residual vector.

Step 0: Compute the weighted least squares estimate of the regression coefficients when using $w_i^0 = 1/\max(p_{ii}, p/n)$ as a weight for the ith observation, where p_{ii} is the ith diagonal element of the projection matrix $\mathbf{P} = \mathbf{X}(\mathbf{X}^T\mathbf{X})^{-1}\mathbf{X}^T$. Let this estimate be denoted by $\hat{\boldsymbol{\beta}}^0$.

Step j: For $j = 1, 2, \ldots$, until convergence, compute

$$\mathbf{e}^{j-1} = \mathbf{Y} - \hat{\mathbf{Y}}^{j-1} = \mathbf{Y} - \mathbf{X}\hat{\boldsymbol{\beta}}^{j-1}, \tag{14.5}$$

which is the residuals of the fit at Step $j - 1$. Compute the new weights

$$w_i^j = \frac{(1 - p_{ii})^2}{\max(|e_i^{j-1}|, m_e^{j-1})}, \tag{14.6}$$

where m_e^{j-1} is the median of $(|e_1^{j-1}|, \ldots, |e_n^{j-1}|)$. Compute the weighted least squares estimate of the regression coefficients when using w_i^j as a weight for the ith observation. Let this estimate be denoted by $\hat{\boldsymbol{\beta}}^j$.

As can be seen from the weighting scheme, those points with high leverage (high p_{ii}), or with large residuals (e_i) get low weights. The details of this procedure can be found in Chatterjee and Mächler (1997).

We provide two examples to illustrate the procedure.

14.6 FITTING A QUADRATIC MODEL

We illustrate the problem with least squares fit by a simple artificial data set given in Table 14.3. There are 10 observations on two variables Y and X. The plot of the data in Figure 14.1[2] shows clearly a quadratic pattern. The least squares fit is given in Table 14.4.

Table 14.3 Data Illustrating Robust Regression

Y	X	Y	X
18.8	3.4	92.7	12.5
20.5	5.7	124.2	14.2
29.4	7.5	142.4	15.2
45.0	8.8	154.7	15.8
72.4	11.1	118.4	17.9

---

[2] The R code for producing this and all other graphs in this book can be found at the Book's Website at http://www.aucegypt.edu/faculty/hadi/RABE6.

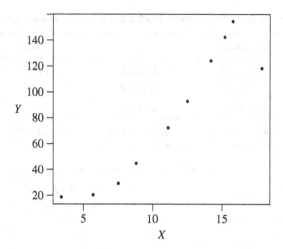

Figure 14.1 Scatter plot of Y versus X for the Data Set in Table 14.3.

The least squares fit shows that both the linear and quadratic terms are statistically insignificant. We will now fit the model by the robust regression method that we have outlined. The results are given in Table 14.5.

Figure 14.2 shows the least squares and the robust fits superposed on the scatter plot of Y and X. The robust fit tracks the data considerably better than the least squares fit. The least squares fit is pulled away from the main body of the data by the high-leverage outlier points in the top right and the bottom left. The robust fit does not suffer from this because such points are down weighted. Here this is visually obvious. In higher dimensions this would

Table 14.4 Least Squares Quadratic Fit for the Data Set in Table 14.3

ANOVA Table

Source	Sum of Squares	df	Mean Square	F-Test	p-Value
Regression	21206	2	10603	25.14	0.001
Residuals	2952	7	422		

Coefficients Table

Variable	Coefficient	s.e.	t-Test	p-Value
Constant	−28.77	37.69	−0.76	0.470
X	9.329	7.788	1.20	0.270
X^2	0.041	0.359	0.12	0.911
$n = 10$	$R^2 = 0.878$	$R_a^2 = 0.843$	$\hat{\sigma} = 20.5349$	df = 7

Table 14.5 Robust Regression Quadratic Fit for the Data Set in Table 14.3

Variable	Coefficient	s.e.	Z-Test	p-Value
Constant	15.2614	0.8003	19.07	0.000
X	−3.5447	0.1829	−19.38	0.000
X^2	0.7831	0.0099	79.10	0.000
n = 10	$R^2 = 0.837$	$R_a^2 = 0.780$	$\hat{\sigma} = 41.425$	df = 7

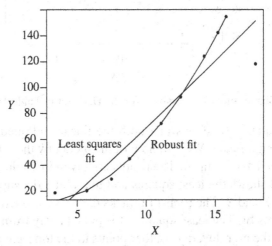

Figure 14.2 Least squares and robust fits superposed on the scatter plot of Y versus X for the data set in Table 14.3.

not be apparent, but the robust procedure would automatically take this into account. This is a commonly occurring situation.

We now illustrate robust regression by using real-life data.

14.7 DISTRIBUTION OF PCB IN U.S. BAYS

The Poly Chlorinated Biphenyl (PCB) is a health hazard found in water from industrial waste and city drainage. The concentration is measured in parts per billion. We do not include any bay or estuary in which no PCB was detected in the two years. We want to study the relationship of PCB level for the two years 1984 and 1985 in U.S. Bays and estuaries. The data is taken from *Environmental Quality* 1987–1988, published by the Council on Environmental Quality. The data can be found in the file PCB.csv at the Book's Website. An exhaustive description of the data along with a thorough analysis can be found in Chatterjee et al. (1995).

Table 14.6 Least Squares Regression of ln(PCB85) on ln(PCB84) for the Data in the file PCB.csv at the Book's Website

ANOVA Table

Source	Sum of Squares	df	Mean Square	F-Test	p-Value
Regression	59.605	1	59.605	87.96	0.000
Residual Error	18.296	27	0.678		

Coefficients Table

Variable	Coefficient	s.e.	t-Test	p-Value
Constant	1.001	0.315	3.17	0.004
ln(PCB84)	0.718	0.077	9.38	0.000

$n = 29$ $R^2 = 0.765$ $R_a^2 = 0.756$ $\hat{\sigma} = 0.823$ df = 27

To overcome the skewness of the data we will transform the data. We will work with the logarithm of the PCB concentrations. The result of the least squares fit is given in Table 14.6.

The fit is problematic. Two observations (Boston Harbor and Delaware Bay) require attention. Both are outliers with standardized residuals of −2.12 and 4.11, respectively. Boston Harbor is a high-leverage point with a large value of Cook's distance. Delaware Bay is not a high-leverage point, but has a high value of Cook's distance. These two points have a significant effect on the fit. We will examine the relationship between the PCB levels in two succeeding years when these two aberrant points are removed. The regression result with the two observations deleted is given in Table 14.7.

This fit has no problems, and is an acceptable description of the relationship of the PCB levels between the years 1984 and 1985. It should be pointed out that this relationship does not hold for Boston Harbor and Delaware Bay. These two bays present special conditions and should be investigated. The deletion of the two points from the fit gives us a better picture of the overall relationship of PCB levels for U.S. bays and estuaries.

Robust regression provides us an alternative approach. Using the robust regression algorithm outlined earlier, we get the following fitted model:

$$\ln(\text{PCB85}) = 0.175 + 0.927 \ln(\text{PCB84}). \qquad (14.7)$$

The standard errors of the two coefficients in (14.7) are 0.25 and 0.0056, respectively. The robust regression applied to the complete data gives results similar to those obtained by the diagnostic prescription of deleting the two

Table 14.7 Least Squares Regression of ln(PCB85) on ln(PCB84) for the Data Set in the file PCB.csv at the Book's Website, when Boston and Delaware Are Deleted

ANOVA Table

Source	Sum of Squares	df	Mean Square	F-Test	p-Value
Regression	64.712	1	64.712	908.24	0.000
Residual Error	1.781	25	0.071		

Coefficients Table

Variable	Coefficient	s.e.	t-Test	p-Value
Constant	0.093	0.122	0.76	0.456
ln(PCB84)	0.960	0.032	30.14	0.000
$n = 27$	$R^2 = 0.973$	$R_a^2 = 0.972$	$\hat{\sigma} = 0.267$	df = 25

observations. The weights given to the deleted points are very small compared to the other data points. This is done mechanically by the rules built into the algorithm. The robust procedure does not require a detailed analysis of regression diagnostics. The final weights used in the iteration point out the problematic observations for further investigation.

Figure 14.3 shows the least squares and robust fit superposed on the scatter plot. It is seen that the least squares line is influenced strongly by the outliers

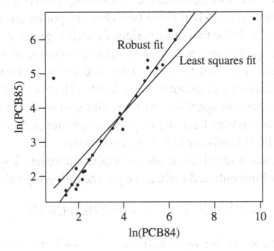

Figure 14.3 Least squares and robust fits superposed on scatter plot of ln(PCB85) versus ln(PCB84) for the data set in the file PCB.csv at the Book's Website.

and high-leverage points. The robust fit tracks the data more accurately without being unduly influenced by high-leverage points and outliers.

As can be seen from (14.6), in the presence of masking, observations in a high-leverage group tend to have large weights because they tend to have small values of p_{ii}. For this reason the Chatterjee–Mächler procedure is not very effective in the presence of masking. The algorithm we have given has been extended to cover problems of masking and swamping, but that is beyond the scope of this book. For details, the reader is referred to Billor et al. (2006).

Much work has been done on robust regression, but it is not widely used in practice. We hope this brief exposition will bring it to public attention.

EXERCISES

14.1 Use the data on injury incidents in airlines given in Table 7.6 to fit a Poisson regression model. Compare the three fits (least squares, transformed least squares, and Poisson), and decide which procedure provides the best description of the data.

14.2 Using the data on the distribution of PCB in U.S. bays and estuaries given in the file PCB.csv at the Book's Website, do a thorough analysis that relates the 1985 PCB levels to 1984 levels. Compare the results of your analysis to the robust fit given in the text.

14.3 Use the data set given in the file Supervisor.Performasnce.csv a the Book's Website and regress Y on X_1 and X_3 by least squares and the robust procedure. Verify that both procedures give similar results.

14.4 Use the data given in the file Magazine.Advertising.csv at the Book's Website and regress $\ln R$ on $\ln P$ using least squares. Observations 15, 22, 23, and 41 are problematic. Do these points have any special feature? Show that the robust fit for the full data set gives results comparable to the least squares results after deleting the four points.

14.5 Use the data on Field-Goal Kicking given in Table 13.12:

(a) Fit a Poison regression model to relate Success with the Distance from which the kick is taken. Use Attempts as offset.

(b) Fit a logistic model relating the probability of a successful kick to the distance from which the kick is taken.

(c) Show that the logistic model gives a better fit than the Poisson regression model.

REFERENCES

1. Agresti, A. (2002), *Categorical Data Analysis*, 2nd ed., New York: John Wiley & Sons.
2. Agresti, A. (2015), *Foundations of Linear and, Generalized Linear Models*, New York: John Wiley & Sons.
3. Akaike, H. (1973), "Information Theory and an Extension of Maximum Likelihood Principle," in *Second International Symposium on Information Theory* (B. N. Petrov and F. Caski, Eds.), Budapest: Akademia Kiado, 267–281.
4. Andrews, D. F. and Herzberg, A. M. (1985) *Data: A Collection of Problems from Many Fields for the Student and Research Worker*, New York: Springer-Verlag.
5. Anscombe, F. J. (1973), "Graphs in Statistical Analysis," *The American Statistician*, 27, 17–21.
6. Atkinson, A. C. (1985), *Plots, Transformations, and Regression: An Introduction to Graphical Methods of Diagnostic Regression Analysis*, Oxford: Clarendon Press.
7. Barnett, V. and Lewis, T. (1994), *Outliers in Statistical Data*, 3rd ed., New York: John Wiley & Sons.
8. Bartlett, G., Stewart, J., and Abrahamowicz, M. (1998), "Quantitative Sensory Testing of Peripheral Nerves," *Student: A Statistical Journal for Graduate Students*, 2, 289–301.

REFERENCES

9. Bates, D. M. and Watts, D. G. (1988), *Nonlinear Regression Analysis and Its Applications*, New York: John Wiley & Sons.
10. Becker, R. A., Cleveland, W. S., and Wilks, A. R. (1987), "Dynamic Graphics for Data Analysis," *Statistical Science*, 2, 4, 355–395.
11. Belsley, D. A. (1991), *Conditioning Diagnostics: Collinearity and Weak Data in Regression*, New York: John Wiley & Sons.
12. Belsley, D. A., Kuh, E., and Welsch, R. E. (1980), *Regression Diagnostics: Identifying Influential Data and Sources of Collinearity*, New York: John Wiley & Sons.
13. Billor, N., Chatterjee, S., and Hadi, A. S. (2006), "A Re-Weighted Least Squares Method for Robust Regression Estimation," *American Journal of Mathematical and Management Sciences*, 26, 229–252.
14. Birkes, D. and Dodge, Y. (1993), *Alternative Methods of Regression*, New York: John Wiley & Sons.
15. Boehmke, B. C. (2016), *Data Wrangling with R (Use R!)*, Switzerland: Springer International Publishing.
16. Box, G. E. P. and Pierce, D. A. (1970), "Distribution of Residual Autocorrelation in Autoregressive-Integrated Moving Average Time Series Models," *Journal of the American Statistical Association*, 64, 1509–1526.
17. Carroll, R. J. and Ruppert, D. (1988), *Transformation and Weighting in Regression*, London: Chapman and Hall.
18. Chambers, J. M., Cleveland, W. S., Kleiner, B., and Tukey, P. A. (1983), *Graphical Methods for Data Analysis*, Boston, MA: Duxbury Press.
19. Chatterjee, S. and Hadi, A. S. (1986), "Influential Observations, High Leverage Points, and Outliers in Linear Regression (with discussions)," *Statistical Science*, 1, 379–416.
20. Chatterjee, S. and Hadi, A. S. (1988), *Sensitivity Analysis in Linear Regression*, New York: John Wiley & Sons.
21. Chatterjee, S., Handcock, M. S., and Simonoff, J. S. (1995), *A Casebook for a First Course in Statistics and Data Analysis*, New York: John Wiley & Sons. pp. 33–35.
22. Chatterjee, S. and Mächler, M. (1997), "Robust Regression: A Weighted Least Squares Approach," *Communications in Statistics, Theory and Methods*, 26, 1381–1394.
23. Chi-Lu, C. and Van Ness, J. W. (1999), *Statistical Regression with Measurement Error*, London: Arnold.
24. Christensen, R. (1996), *Analysis of Variance, Design and Regression: Applied Statistical Methods*, New York: Chapman and Hall.
25. Coakley, C. W. and Hettmansperger, T. P. (1993), "A Bounded Influence, High Breakdown, Efficient Regression Estimator," *Journal of the American Statistical Association*, 88, 872–880.

26. Cochrane, D. and Orcutt, G. H. (1949), "Application of Least Squares Regression to Relationships Containing Autocorrelated Error Terms," *Journal of the American Statistical Association*, 44, 32–61.
27. Coleman, J. S., Cambell, E. Q., Hobson, C. J., McPartland, J., Mood, A. M., Weinfield, F. D., and York, R. L. (1966), *Equality of Educational Opportunity*, Washington, DC: U.S. Government Printing Office.
28. Conover, W. J. (1980), *Practical Nonparametric Statistics*, New York: John Wiley & Sons.
29. Cook, R. D. (1977), "Detection of Influential Observations in Linear Regression," *Technometrics*, 19, 15–18.
30. Cook, R. D. and Weisberg, S. (1982), *Residuals and Influence in Regression*, London: Chapman and Hall.
31. Cotton, R. (2013), *Learning R*, Sebastopol, CA: O'Reilly Media.
32. Cox, D. R. (1989), *The Analysis of Binary Data*, 2nd ed., London: Methuen.
33. Crawley, M. J. (2012), *The R Book*, 2nd ed., New York: John Wiley & Sons.
34. Dalgaard, P. (2008), *Introductory Statistics with R*, 2nd ed., New York: Springer.
35. Daniel, C. and Wood, F. S. (1980), *Fitting Equations to Data: Computer Analysis of Multifactor Data*, 2nd ed., New York: John Wiley & Sons.
36. Davies, T. M. (2016), *The Book of R: A First Course in Programming and Statistics*, San Francisco, CA: No Starch Press.
37. Dempster, A. P., Schatzoff, M., and Wermuth, N. (1977), "A Simulation Study of Alternatives to Ordinary Least Squares," *Journal of the American Statistical Association*, 72, 77–106.
38. Diaconis, P. and Efron, B. (1983), "Computer Intensive Methods in Statistics," *Scientific American*, 248, 116–130.
39. Dobson, A. J. and Barnett, A. G. (2008), *Introduction to Generalized Linear Models*, 3rd ed., Boca Raton, FL: Chapman and Hall.
40. Dodge, Y. and Hadi, A. S. (1999), "Simple Graphs and Bounds for the Elements of the Hat Matrix" *Journal of Applied Statistics*, 26, 817–823.
41. Dorugade, A. V. and Kashid, D. N. (2010) "Alternative Method for Choosing Ridge Parameter for Regression," *Applied Mathematical Sciences*, 4, 447–456.
42. Draper, N. R. and Smith, H. (1998), *Applied Regression Analysis*, 3rd ed., New York: John Wiley & Sons.
43. Dunn, P. K. (2023), "Generalized linear models", in *International Encyclopedia of Education*, 4th ed., (R. J. Tierney, F. Rizvi, and K. Ercikan Eds.), Elsevier, 583–589.
44. Dunn, P. K. and Smyth, G. K. (2018), *Generalized Linear Models With Examples in R*, New York: Springer.
45. Durbin, J. and Watson, G. S. (1950), "Testing for Serial Correlation in Least Squares Regression," *Biometrika*, 37, 409–428.

46. Durbin, J. and Watson, G. S. (1951), "Testing for Serial Correlation in Least Squares Regression, II," *Biometrika*, 38, 159–178.
47. Efron, B. (1982), *The Jacknife, the Bootstrap and Other Resampling Plans*, CBMS- National Science Monograph 38, Society of Industrial and Applied Mathematics.
48. Eisenhauer, J. G. (2003), "Regression Through the Origin," *Teaching Statistics*, 25, 76–80.
49. Ezekiel, M. (1924), "A Method for Handling Curvilinear Correlation for Any Number of Variables," *Journal of the American Statistical Association*, 19, 431–453.
50. Faraway, J. J. (2006), *Extending the Linear Model with R: Generalized Linear, Mixed Effects and Nonparametric Regression Models*, Boca Raton, FL, New York: Chapman & Hall/CRC.
51. Finney, D. J. (1964), *Probit Analysis*, London: Cambridge University Press.
52. Fox, J. (1984), *Linear Statistical Models and Related Methods*, New York: John Wiley & Sons.
53. Fox, J. (2016), *Applied Regression Analysis and Generalized Linear Models*, 3rd ed., Sage Publications.
54. Friedman, M. and Meiselman, D. (1963), "The Relative Stability of Monetary Velocity and the Investment Multiplier in the United States, 1897–1958," in *Commission on Money and Credit, Stabilization Policies*, Englewood Cliffs, NJ: Prentice-Hall.
55. Fuller, W. A. (1987), *Measurement Error Models*, New York: John Wiley & Sons.
56. Furnival, G. M. and Wilson, R. W., Jr. (1974), "Regression by Leaps and Bounds," *Technometrics*, 16, 499–512.
57. Gibbons, J. D. (1993), *Nonparametric Statistics: An Introduction*, Newbury Park, CA: Sage Publications.
58. Gillespie, C. and Lovelace, R. (2017) *Efficient R Programming: A Practical Guide to Smarter Programming*, Sebastopol, CA: O'Reilly Media.
59. Goldstein, M. and Smith, A. F. M. (1974), "Ridge-Type Estimates for Regression Analysis," *Journal of the Royal Statistical Society (B)*, 36, 284–291.
60. Golub, G. H. and van Loan, C. (1989), *Matrix Computations*, Baltimore, MD: Johns Hopkins.
61. Gray, J. B. (1986), "A Simple Graphic for Assessing Influence in Regression," *Journal of Statistical Computation and Simulation*, 24, 121–134.
62. Gray, J. B. and Ling, R. F. (1984), "K-Clustering as a Detection Tool for Influential Subsets in Regression (with Discussion)," *Technometrics*, 26, 305–330.
63. Graybill, F. A. (1976), *Theory and Application of the Linear Model*, Belmont, CA: Duxbury Press.
64. Graybill, F. A. and Iyer, H. K. (1994), *Regression Analysis: Concepts and Applications*, Belmont, CA: Duxbury Press.

65. Green, W. H. (1993), *Econometric Analysis*, 2nd ed., Saddle River, NJ: Prentice-Hall.
66. Grolemund, G. (2014), *Hands on Programming With R: Write Your Own Functions and Simulations*, Sebastopol, CA: O'Reilly Media.
67. Grosser, M., Bumann H., and Wickham, H. (2022) *Advanced R Solutions*, New York: Chapman & Hall/CRC.
68. Gunst, R. F. and Mason, R. L. (1980), *Regression Analysis and Its Application: A Data-Oriented Approach*, New York: Marcel Dekker.
69. Hadi, A. S. (1988), "Diagnosing Collinearity-Influential Observations," *Computational Statistics and Data Analysis*, 7, 143–159.
70. Hadi, A. S. (1992), "A New Measure of Overall Potential Influence in Linear Regression," *Computational Statistics and Data Analysis*, 14, 1–27.
71. Hadi, A. S. (1993), "Graphical Methods for Linear Models," Chapter c23 in *Handbook of Statistics: Computational Statistics*, (C. R. Rao, Ed.), Vol. 9, New York: North-Holland Publishing Company, 775–802.
72. Hadi, A. S. (1996), *Matrix Algebra as a Tool*, Belmont, CA: Duxbury Press.
73. Hadi, A. S. (2011), "Ridge and Surrogate Ridge Regressions," in *International Encyclopedia of Statistical Science*, (M. Lovric, Ed.), New York: Springer, Part 18, 1232–1234.
74. Hadi, A. S. and Ling, R. F. (1998), "Some Cautionary Notes on the Use of Principal Components Regression," *The American Statistician*, 52, 15–19.
75. Hadi, A. S. and Simonoff, J. S. (1993), "Procedures for the Identification of Multiple Outliers in Linear Models," *Journal of the American Statistical Association*, 88, 1264–1272.
76. Hadi, A. S. and Son, M. S. (1997), "Detection of Unusual Observations in Regression and Multivariate Data," Chapter 13 in *Handbook of Applied Economic Statistics* (A. Ullah and D. E. A. Giles, Eds.), New York: Marcel Dekker, 441–463.
77. Hadi, A. S. and Velleman, P. F. (1997), "Computationally Efficient Adaptive Methods for the Identification of Outliers and Homogeneous Groups in Large Data Sets," *Proceedings of the Statistical Computing Section, American Statistical Association*, 124–129.
78. Haith, D. A. (1976), "Land Use and Water Quality in New York Rivers," *Journal of the Environmental Engineering Division*, ASCE 102 (No. EEI. Proc. Paper 11902, Feb. 1976), 1–15.
79. Hamilton, D. J. (1987), "Sometimes $R^2 > r_{y \cdot x_1}^2 + r_{y \cdot x_2}^2$, Correlated Variables Are Not Always Redundant," *The American Statistician*, 41, 2, 129–132.
80. Hamilton, D. J. (1994), *Time Series Analysis*, Princeton, NJ: Princeton University Press.
81. Hampel, F. R., Ronchetti, E. M., Rousseeuw, P. J., and Stahel, W. A. (1986), *Robust Statistics: The Approach Based on Influence Functions*, New York: John Wiley & Sons.
82. Hand, D. J., Daly, F., Lunn, A. D., McConway, K. J., and Ostrowski, E. (1994), *A Handbook of Small Data Sets*, New York: Chapman and Hall.

83. Hawkins, D. M. (1980), *Identification of Outliers*, London: Chapman and Hall.
84. Henderson, H. V. and Velleman, P. F. (1981), "Building Multiple Regression Models Interactively," *Biometrics*, 37, 391–411.
85. Hildreth, C. and Lu, J. (1960), "Demand Relations with Autocorrelated Disturbances," *Technical Bulletin No. 276*, Michigan State University, Agricultural Experiment Station.
86. Hoaglin, D. C. and Welsch, R. E. (1978), "The Hat Matrix in Regression and ANOVA," *The American Statistician*, 32, 17–22.
87. Hocking, R. R., (1976), "The Analysis and Selection of Variables in Linear Regression," *Biometrics*, 32, 1–49.
88. Hoerl, A. E. (1959), "Optimal Solution of Many Variables," *Chemical Engineering Progress*, 55, 69–78.
89. Hoerl, A. E. and Kennard, R. W. (1970), "Ridge Regression: Biased Estimation for Nonorthogonal Problems," *Technometrics*, 12, 69–82.
90. Hoerl, A. E. and Kennard, R. W. (1976), "Ridge Regression: Iterative Estimation of the Biasing Parameter," *Communications in Statistics, Theory and Methods*, A5, 77–88.
91. Hoerl, A. E. and Kennard, R. W. (1981), "Ridge Regression – 1980: Advances, Algorithms, and Applications," *American Journal of Mathematical and Management Sciences*, 1, 5–83.
92. Hoerl, A. E., Kennard, R. W., and Baldwin, K. F. (1975), "Ridge Regression: Some Simulations," *Communications in Statistics, Theory and Methods*, 4, 105–123.
93. Hoffmann, J. P. (2016), *Regression Models for Categorical, Count, and Related Variables: An Applied Approach*, University of California Press.
94. Hollander, M. and Wollfe, D. A. (1999), *Nonparametric Statistical Methods*, New York: John Wiley & Sons.
95. Hosmer, D. W. and Lemeshow, S. (1989), *Applied Logistic Regression*, New York: John Wiley & Sons.
96. Huber, P. J. (1981), *Robust Statistics*, New York: John Wiley & Sons.
97. Huber, P. J. (1991), "Between Robustness and Diagnostics," in *Directions in Robust Statistics and Diagnostics* (W. Stahel and S. Weisberg, Eds.), New York: Springer-Verlag, 121–130.
98. Hurvich, C. M. and Tsai, C.-L. (1989), "Regression and Time Series Model Selection in Small Samples," *Biometrika*, 76, 297–307.
99. Iversen, G. R. (1976), *Analysis of Variance*, Beverly Hills, CA: Sage Publications.
100. Iversen, G. R. and Norpoth, H. (1987), *Analysis of Variance*, Beverly Hills, CA: Sage Publications.
101. Jensen, D. R. and Ramirez, D. E. (2008), "Anomalies in the Foundations of Ridge Regression," *International Statistical Review*, 76, 89–105.

REFERENCES

102. Jerison, H. J. (1973), *Evolution of the Brain and Intelligence*, New York: Academic Press.
103. Johnson, D. E. (1998), *Applied Multivariate Methods for Data Analysts*, Belmont, CA: Duxbury Press.
104. Johnson, R. A. and Wichern, D. W. (1992), *Applied Multivariate Statistical Analysis*, 3rd ed., Englewood Cliffs, NJ: Prentice-Hall.
105. Johnston, J. (1984), *Econometric Methods*, 2nd ed., New York: McGraw-Hill.
106. Kabacoff, R. L. (2015), *R in Action: Data Analysis and Graphics with R*, 2nd ed., Shelter Island, NY: Manning Publication Co.
107. Khalaf, G. and Shukur, G. (2005), "Choosing Ridge Parameter for Regression Problem," *Communications in Statistics, Theory and Methods*, 34, 1177–1182.
108. Kmenta, J. (1986), *Elements of Econometrics*, New York: Macmillan.
109. Krasker, W. S. and Welsch, R. E. (1982), "Efficient Bounded-Influence Regression Estimation," *Journal of the American Statistical Association*, 77, 595–604.
110. Krishnaiah, P. R. (Ed.) (1980), *Analysis of Variance*, New York: North-Holland Publishing Co.
111. La Motte, L. R. and Hocking, R. R. (1970), "Computational Efficiency in the Selection of Regression Variables," *Technometrics*, 12, 83–93.
112. Lander, J. P. (2014), *R for Everyone: Advanced Analytics and Graphics*, India: Pearson.
113. Landwehr, J., Pregibon, D., and Shoemaker, A. (1984), "Graphical Methods for Assessing Logistic Regression Models," *Journal of the American Statistical Association*, 79, 61–83.
114. Larsen, W. A. and McCleary, S. J. (1972), "The Use of Partial Residual Plots in Regression Analysis," *Technometrics*, 14, 781–790.
115. Lawless, J. F. and Wang, P. (1976), "A Simulation of Ridge and Other Regression Estimators," *Communications in Statistics, Theory and Methods*, A5, 307–323.
116. Lehmann, E. L. (1975), *Nonparametric Statistical Methods Based on Ranks*, New York: McGraw-Hill.
117. Lilja, D. J. and Linse, G. M. (2022), *Linear Regression Using R: An Introduction to Data Modeling*, 2nd ed., Minnesota, USA: University of Minnesota Libraries Publishing.
118. Lindman, H. R. (1992), *Analysis of Variance in Experimental Design*, New York: Springer-Verlag.
119. Longley, J. W. (1967), "An Appraisal of Least-Squares Programs from the Point of View of the User," *Journal of the American Statistical Association*, 62, 819–841.
120. Maindonald, J. and Braun, J. (2003), *Data Analysis and Graphics Using R*, Cambridge, UK: Cambridge University Press.

121. Malinvaud, E. (1968), *Statistical Methods of Econometrics*, Chicago: Rand McNally.
122. Mallows, C. L. (1973), "Some Comments on C_p," *Technometrics*, 15, 661–675.
123. Manly, B. F. J. (1986), *Multivariate Statistical Methods*, New York: Chapman and Hall.
124. Mantel, N. (1970), "Why Stepdown Procedures in Variable Selection," *Technometrics*, 12, 591–612.
125. Marquardt, D. W. (1970), "Generalized Inverses, Ridge Regression, Biased Linear Estimation and Nonlinear Estimation," *Technometrics*, 12, 591–612.
126. Masuo, N. (1988), "On the Almost Unbiased Ridge Regression Estimation," *Communications in Statistics, Simulation*, 17, 729–743.
127. Matloff, N. (2011), *The Art of R Programming: A Tour of Statistical Software Design*, San Francisco, CA: No Starch Press.
128. McCallum, B. T. (1970), "Artificial Orthogonalization in Regression Analysis," *Review of Economics and Statistics*, 52, 110–113.
129. McCullagh, P. and Nelder, J. A. (1989), *Generalized Linear Models*, 2nd ed., London: Chapman and Hall.
130. McCulloch, C. E. and Meeter, D. (1983), "Discussion of 'Outlier..........s'," by R. J. Beckman and R. D. Cook, *Technometrics*, 25, 119–163.
131. McDonald, G. C. and Galarneau, D. I. (1975), "A Monte Carlo Evaluation of Some Ridge Type Estimators," *Journal of the American Statistical Association*, 70, 407–416.
132. McDonald, G. C. and Schwing, R. C. (1973), "Instabilities of Regression Estimates Relating Air Pollution to Mortality," *Technometrics*, 15, 463–481.
133. McLachlan, G. J. (1992), *Discriminant Analysis and Statistical Pattern Recognition*, New York: John Wiley & Sons.
134. Montgomery, D. C., Peck, E. A., and Vining, G. G. (2012) *Introduction to Linear Regression Analysis*, 5th ed., New York: John Wiley & Sons.
135. Moore, D. S. and McCabe, G. P. (1993), *Introduction to the Practice of Statistics*, New York: W. H. Freeman and Company.
136. Morris, C. N. and Rolph, J. E. (1981), *Introduction to Data Analysis and Statistical Inference*, Englewood Cliffs, NJ: Prentice-Hall.
137. Mosteller, F. and Moynihan, D. F. (Eds.) (1972), *On Equality of Educational Opportunity*, New York: Random House.
138. Mosteller, F. and Tukey, J. W. (1977), *Data Analysis and Regression*, Reading, MA: Addison-Wesley.
139. Myers, R. H. (1990), *Classical and Modern Regression with Applications*, 2nd ed., Boston, MA: PWS-KENT Publishing Company.
140. Narula, S. C. and Wellington, J. F. (1977), "Prediction, Linear Regression, and the Minimum Sum of Relative Errors," *Technometrics*, 19, 2, 185–190.
141. Nelder, J. A. and Wedderburn, R. W. M. (1972), "Generalized Linear Models," *Journal of the Royal Statistical Society (A)*, 135, 370–384.

142. Pregibon, D. (1981), "Logistic Regression Diagnostics," *The Annals of Statistics*, 9, 705–724.
143. R Core Team (2022). R: A language and environment for statistical computing. R Foundation for Statistical Computing, Vienna, Austria. https://www.R-project.org.
144. Racine, J. and Hyndman, R. (2002), "Using R To Teach Econometrics," *Journal of Applied Econometrics*, 17, 175–189.
145. Rao, C. R. (1973), *Linear Statistical Inference and Its Applications*, New York: John Wiley & Sons.
146. Ratkowsky, D. A. (1983), *Nonlinear Regression Modeling: A Unified Practical Approach*, New York: Marcel Dekker.
147. Ratkowsky, D. A. (1990), *Handbook of Nonlinear Regression Models*, New York: Marcel Dekker.
148. Reaven, G. M. and Miller, R. G. (1979), "An Attempt to Define the Nature of Chemical Diabetes Using a Multidimensional Analysis," *Diabetologia*, 16, 17–24.
149. Rencher, A. C. (1995), *Methods of Multivariate Analysts*, New York: John Wiley & Sons.
150. Robert, C. P. and Casella, G., (2010), *Introducing Monte Carlo Methods with R*, New York, Springer.
151. Rousseeuw, P. J. and Leroy, A. M. (1987), *Robust Regression and Outlier Detection*, New York: John Wiley & Sons.
152. Scheffé, H. (1959), *The Analysis of Variance*, New York: John Wiley & Sons.
153. Schwarz, G. (1978), "Estimating the Dimensions of a Model," *Annals of Statistics*, 121, 461–464.
154. Searle, S. R. (1971), *Linear Models*, New York: John Wiley & Sons.
155. Seber, G. A. F. (1977), *Linear Regression Analysis*, New York: John Wiley & Sons.
156. Seber, G. A. F. (1984), *Multivariate Observations*, New York: John Wiley & Sons.
157. Seber, G. A. F. and Lee, A. J. (2003), *Linear Regression Analysis*, New York: John Wiley & Sons.
158. Seber, G. A. F. and Wild, C. J. (1989), *Nonlinear Regression*, New York: John Wiley & Sons.
159. Sen, A. and Srivastava, M. (1990), *Regression Analysis: Theory, Methods, and Applications*, New York: Springer-Verlag.
160. Simplilearn (2023), https://www.simplilearn.com/what-is-r-article. Accessed January 25, 2023.
161. Shumway, R. H. (1988), *Applied Statistical Time Series Analysis*, Englewood Cliffs, NJ: Prentice-Hall.
162. Silvey, S. D. (1969), "Multicollinearity and Imprecise Estimation," *Journal of the Royal Statistical Society (B)*, 31, 539–552.

163. Simonoff, J. S. (2003), *Analyzing Categorical Data*, New York: Springer-Verlag.
164. Snedecor, G. W. and Cochran, W. G. (1980), *Statistical Methods*, 7th ed., Ames, IA: Iowa State University Press.
165. Staudte, R. G. and Sheather, S. J. (1990), *Robust Estimation and Testing*, New York: John Wiley & Sons.
166. Strang, G. (1988), *Linear Algebra and Its Applications*, 3rd ed., San Diego, CA: Harcourt Brace Jovanovich.
167. Thomson, A. and Randall-Maciver, R. (1905), *Ancient Races of the Thebaid*, Oxford: Oxford University Press.
168. Velleman, P. F. (1999), *Data Desk*, Ithaca, NY: Data Description.
169. Velleman, P. F. and Welsch, R. E. (1981), "Efficient Computing of Regression Diagnostics," *The American Statistician*, 35, 234–243.
170. Verzani, J. (2004), *Using R for Introductory Statistics*, Boca Raton, FL: Chapman & Hall/CRC.
171. Vinod, H. D. and Ullah, A. (1981), *Recent Advances in Regression Methods*, New York: Marcel Dekker.
172. Wahba, G., Golub, G. H., and Health, C. G. (1979), "Generalized Cross-Validation as a Method for Choosing a Good Ridge Parameter," *Technometrics*, 21, 215–223.
173. Welsch, R. E. and Kuh, E. (1977), "Linear Regression Diagnostics," *Technical Report* 923-77, Sloan School of Management, Cambridge, MA.
174. Wickham, H. (2014) *Advanced R*, 2nd ed., New York: Chapman & Hall/CRC.
175. Wickham, H. (2016), *ggplot2: Elegant Graphics for Data Analysis (Use R!)*, 2nd ed., New York: Springer-Verlag.
176. Wickham, H. and Bryon, J. (2015), *R Packages: Organize, Test, Document, and Share Your Code*, Sebastopol, CA: O'Reilly Media, Inc.
177. Wickham, H. and Gorlemund, G. (2017), *R for Data Science*, Sebastopol, CA: O'Reilly Media. https://r4ds.had.co.nz/.
178. Wildt, A. R. and Ahtola, O. (1978), *Analysis of Covariance*, Beverly Hills, CA: Sage Publications.
179. Wood, F. S. (1973), "The Use of Individual Effects and Residuals in Fitting Equations to Data," *Technometrics*, 15, 677–695.
180. Yan, X. and Su, X. G. (2009), *Linear Regression Analysis: Theory and Computing*, Hackensack, NJ: World Scientific Publishing Co.
181. Zumel, N. and Mount, J. (2019) *Practical Data Science with R*, Shelter Island, NY: Manning Publication Co.

INDEX

A
Added-variable plot, 163, 180, 181, 399
Adjusted multiple correlation coefficient, 376
Akaike Information Criterion (AIC), 377, 382, 383, 388, 399, 413, 419, 436
   bias corrected, 378
Analysis of covariance, 11, 14
Analysis of variance, 11, 14, 129, 213
ANOVA, 213
   multiple regression, 129, 130
   simple regression, 134
Anscombe quartet, 75, 157, 166
Arithmetic
   Operators, 48
Assumption
   constant variance, 153
   homogeneity, 153
   independent-errors, 153
   linearity, 84, 152
   normality, 90, 152
Autocorrelation, 18, 153, 258, 277, 281, 399

B
Backward elimination, 381
Bankruptcy, 413
Bayes Information Criterion (BIC), 378, 382, 383, 388, 399, 413, 419
Best linear unbiased estimator, 121, 147
Beta coefficients, 121
Biased estimation, 342
Binary
   response data, 275
   response variable, 410, 415
   variable, 11, 410, 415
BLUE, 121, 147
Book's Website, xvii, 3, 36, 38, 49, 54

C
Centering, 117
Classification, 428
Cochrane–Orcutt procedure, 286
Coefficient of determination, 97, 122
Coefficients Table, 86
Collinearity, 18, 131, 154, 302
Collinearity-influential observations, 332

Regression Analysis By Example Using R, Sixth Edition. Ali S. Hadi and Samprit Chatterjee
© 2024 John Wiley & Sons, Inc. Published 2024 by John Wiley & Sons, Inc.
Companion website: www.wiley.com/go/hadi/regression_analysis_6e

Command level prompt, 23
Computer repair data, 77, 82, 83, 89, 97
Concordance Index, 421
Condition indices, 315, 318, 319
Condition number, 319, 362, 379
Confidence
 interval, 92
 interval for β_j, 124
 region, 91
Constant variance assumption, 153
Cook's distance, 173
Correlated errors, 214
Correlation coefficient, 71, 74, 75, 119
 between Y and $\hat{Y}$, 94, 122
 population, 90
 sample, 90, 94
Correlation matrix, 188, 318, 332, 333
cov($\hat{\beta}_i$, $\hat{\beta}_j$), 148
Covariance, 71, 73
Cross-sectional data, 202, 296, 297
Cumulative distribution function, 411

D

DASL Website, 3, 7
Data
 collection, 10
 information about, 35
 reexpression, 12
 transformation, 12
data.frame, 34, 38
Dataset
 advertising, 313, 331
 air pollution, 392
 Animals, 248
 Anscombe quartet, 76, 166
 bacteria deaths, 230
 bankruptcy, 413
 cigarette consumption, 1, 144, 145, 404
 college expense, 260
 computer repair, 45, 77, 78
 consumer expenditure, 264, 278, 279, 299
 corn yields, 222
 cost of education, 261
 cost of health care, 8, 9
 Diabetes, 423

domestic immigration, 5
Dow Jones Industrial Average, 299
education expenditure, 215, 216, 264
Egyptian skulls, 6
equal educational opportunity, 303
examination, 140
exponential growth, 179
field goal kicking, 431
gasoline consumption, 324
Hald, 349
Hamilton, 162
hard disk prices, 255
heights of husbands and wives, 103, 142
homicide, 388, 389
housing starts, 289
import, 308, 321, 342
Industrial Production, 7
injury incidents, 237
labor force, 103
Longley, 362
milk production, 4
mtcars, 40
new drugs, 436
New York Rivers, 7, 167, 168, 170, 175
newspapers, 104, 105
nonlinear, 75
oil production, 254, 298
PCB, 440
preemployment, 204, 205
presidential election, 222, 253, 324, 325, 403
quantitative sensory testing, 13
real estate, 401
right-to-work laws, 5
salary survey, 194
Scottish hills races, 182
Ski sales, 214, 292
Ski sales expanded, 294
Space Shuttle, 8, 431
students information, 224
supervised workers, 239, 240, 259
supervisor performance, 110–112, 125, 384
wind chill factor, 252

dataset
  supervisor performance, 383
  gasoline consumption, 324
  in R packages, 40
Degrees of freedom, 113, 156
dev.copy2eps, 47
dev.copy2pdf, 47
DFITS, 174
Discriminant analysis, 428
Distance
  Cook's, 173
  orthogonal, 81
  perpendicular, 81
  vertical, 81
Distribution
  binomial, 434
  chi-square, 415
  gamma, 434
  inverse Gaussian, 434
  normal, 235
  Poisson, 236, 239, 434
Dose–response curve, 258, 273
download.file, 303
Draftsman's matrix, 161
Dummy variable, 193
Durbin–Watson statistic, 281, 288, 290–292, 295

E

Eigenvalues, 319, 320
Eigenvector, 320
Externally studentized residual, 157

F

F-Test, 127
Factors, 2
File format, 45
Fitted regression equation, 91
Fitted value, 14, 82, 113
Fitting models to data, 13
Forecast interval, 92
Forward selection, 381
Full model, 126
Function, 21, 61
  all.equal, 48
  anyNA, 30
  apply, 59

arguments, 61
as.character, 28, 29
as.data.frame, 47
as.na, 30
as.numeric, 30, 195
attach, 36, 37, 77
birthday, 31
cat, 30
cbind, 50, 51, 116, 178, 195, 215
ceiling, 48
choose.dir, 42
class, 35, 37
colMeans, 51, 77
colMedians, 51
colnames, 35
concatenate, 25, 28
cor, 51, 77
cov, 77
cumulative distribution, 411
data.matrix, 35
date, 30
detach, 77
dim, 37, 51
download.file, 45
DurbinWatsonTest, 282, 283, 285
elapsed.time, 30
exam, 64
excel_sheets, 44
extracting information, 51
filter, 40
floor, 48
ftable, 51
geom_point, 46, 47
getwd, 41
ggplot, 46
glm, 21
head, 35, 41, 43, 44
if, 56
ifelse, 56
install.packages, 39, 40
intersect, 48
intrinsically nonlinear, 12
IQR, 51
is.na, 38
is.nan, 38

Function (*continued*)
  lag, 40
  lapply, 59, 61
  length, 35, 37, 51
  levels, 33, 35
  library, 39
  linear, 12
  linearizable, 12
  link, 435
  lm, 21, 82
  logistic, 274
  ls, 42, 61, 63
  max, 51
  mean, 30, 51, 56, 57
  median, 30, 51, 57
  min, 51
  mult.table, 62, 63
  names, 35, 37
  namespaces, 40
  nchar, 28
  ncol, 35
  near, 38, 48
  nlevels, 33, 35
  nonlinear, 12, 274
  nrow, 35
  order, 51, 52
  pchisq, 419
  pf, 128
  princomp, 329
  pt, 87, 124
  q, 28
  qchisq, 418
  quantile, 51, 56, 57
  rank, 51, 54
  rbind, 27, 51
  read_csv2, 45
  read_tsv, 45
  read.csv, 44, 45
  read.table, 43–45
  rep, 51, 215
  replicate, 51
  require, 39
  rm, 42
  rnorm, 30, 66
  round, 48, 50, 178
  rownames, 35
  sample, 51
  sapply, 59, 61
  sd, 51
  search, 40
  seq, 26
  setdiff, 48
  setequal, 48
  sim.die, 65
  sink, 46
  sink.number, 46
  sort, 51, 54
  sqrt, 25
  str, 35
  strptime, 31
  sum, 51, 56, 57
  summary, 51
  sweep, 59, 60
  syntax, 61
  Sys.time, 30
  t, 28
  table, 51
  tail, 35, 41, 43, 44
  testing object type, 38
  trunc, 48
  union, 48
  unique, 48, 51, 56
  updateR, 40
  vapply, 59, 61
  var, 51
  View, 41, 43, 44
  weekdays, 31
  weighted.mean, 51
  which.max, 51
  write.xlsx, 47
Function name, 61
Functional, 59

G

Generalized linear models, 433, 434
Goodness-of-fit index, 97
Graph
  save, 46

H

Hadi's influence measure, 174
Hat matrix, 147, 155
Heteroscedasticity, 153, 235, 258

INDEX **459**

High leverage, 169, 351
Homogeneity (Homoscedasticity), 153, 234

I

Independent-errors assumption, 153
Indicator variable, 190, 193, 294
Influence measures, 172
    Cook's distance, 173, 176, 188
    Hadi's, 174, 176, 188
    Welsch and Kuh (DFITS), 174–176
Influential observations, 159, 167
install.packages, 23, 40
Integrated development environment, 22
Interaction effects, 199
Intercept, 113
Internally studentized residual, 157
Interpretation of regression coefficients, 114
Iterative process, 15

L

L-R plot, 177
Ladder of transformation, 248
Lagged variables, 297
Least absolute value, 106
Least squares, 13
    estimates, 81
    line, 81
    method, 80, 81, 112
    properties, 121
    weighted, 18, 243, 258
Leverage-residual plot, 177
Leverage values, 155
Linear Models, 82
Linearity assumption, 84, 152
Link function, 435
Logistic
    distribution, 411
    function, 254, 274
    models, 258, 275
    regression, 11, 14, 18, 214, 409, 411
    regression diagnostics, 415
    response function, 410
Logistic regression, 434
    binary, 18
    multinomial, 18

ordered response variable, 18
    polytomous, 422
Logit, 412
    link function, 435
    model, 275, 435
    transformation, 412

M

Masking problem, 170
Matrix
    calculations, 51
    correlation, 318, 332, 333
    draftsman's, 161
    hat, 155
    plot, 161, 188
    projection, 155
    variance-covariance, 121, 365
Maximum likelihood method, 14, 412, 423, 426, 434
Model
    fitting, 13
    full, 126
    no-intercept, 262, 266
    probit, 275, 411
    random walk, 300
    reduced, 126
    through the origin, 97
Multicollinearity, 302
Multinomial logit model, 422
Multiple correlation coefficient, 122, 129, 133, 153, 376
Multiple regression, 14, 110, 122
    ANOVA table, 130
    assumptions, 121
Multiplication Table, 61
Multiplicative effect, 199
Multivariate regression, 13, 14

N

NA, 30
Nested models, 126, 419
No-intercept model, 98, 262, 266
Nominal model, 426
Nonlinear regression, 14
Nonparametric statistics, 281
Normal
    distribution, 235

460 INDEX

Normal (*continued*)
  equations, 112, 147
  scores, 165
Normality assumption, 86, 152
Normalizing transformations, 227

O

Odds ratio, 412
One-sample *t*-Test, 100
Operator, 48
  !, 48
  %/%, 48
  %%, 48
  %in%, 48
  &, 48
  |, 48
Ordered response category, 426
Ordered response variable, 18
Ordinal logistic regression, 426
Ordinal model, 426
Orthogonal predictors, 301
Orthogonal regression, 81, 106
Outliers, 169, 351

P

P-R plot, 177
P-value for
  F-Test, 128
  t-Test, 87, 124
Package
  base, 39–41, 48, 178
  car, 282, 283
  curl, 45
  datasets, 40, 76
  DescTools, 21, 282
  dplyr, 38, 40, 48, 265
  ggplot2, 46
  install, 40
  installing, 39
  lubridate, 30
  MASS, 248, 251
  nlme, 21, 39
  nnet, 39
  olsrr, 21, 130, 167, 178
  RCurl, 45
  readr, 45
  readxl, 44

require, 39
robustbase, 39, 41, 51, 70
robustX, 39
stats, 21, 40, 178
tidyverse, 39
updateR, 40
utils, 45
vcdExtra, 69
writexl, 47
Packages
  install, 23
Parameter estimation, 13
Parsimony, 132
Partial
  regression coefficients, 110, 115
  regression plot, 181
  residual plot, 181
pchisq, 419
pf, 128
Plot
  added-variable, 163, 180, 181, 399
  L-R, 177
  matrix, 161, 188
  P-R, 177
  partial regression, 181
  partial residual, 181
  potential-residual, 177, 188
  residual plus component, 163, 180, 181, 399
  sequence, 280
Poisson regression, 19, 435
Potential-residual plot, 177, 188
Predicted value, 15
Prediction errors, 299
Prediction interval, 92
Principal components, 14, 328, 363, 364
Probit model, 275, 411
Projection matrix, 147, 155
Property Valuation, 401
Proportional odds model, 422, 426, 427
pt, 87, 124
Pure error, 263

Q

qf, 127, 131, 140
qt, 87, 88, 123, 140